Techniques
in Fractal Geometry

Techniques in
Fractal Geometry

Kenneth Falconer

University of St Andrews

JOHN WILEY & SONS

Chichester · New York · Weinheim · Brisbane · Singapore · Toronto

Other Wiley Editorial Offices

John Wiley & Sons, Inc., 605 Third Avenue,
New York, NY 10158-0012, USA

VCH Verlagsgesellschaft mbH, Pappelallee 3,
D-69469 Weinheim, Germany

Jacaranda Wiley Ltd, 33 Park Road, Milton,
Queensland 4064, Australia

John Wiley & Sons (Asia) Pte Ltd, 2 Clementi Loop #02-01,
Jin Xing Distripark, Singapore 129809

John Wiley & Sons (Canada) Ltd, 22 Worcester Road,
Rexdale, Ontario M9W 1L1, Canada

British Library Cataloguing in Publication Data

A catalogue record for this book is available from the British Library

ISBN 0 471 95724 0

Typeset in 10/12pt Times by Thomson Press (India) Ltd., New Delhi

This book is printed on acid-free paper responsibly manufactured from sustainable forestation, for which at least two trees are planted for each one used for paper production.

Printed and bound by Antony Rowe Ltd, Eastbourne

Contents

Preface

This book describes a variety of techniques in current use for studying the mathematics of fractals. It is an instructional and reference work for those researching in fractal geometry and for those who encounter fractals in other areas of mathematics or science, and it contains material suitable for advanced courses. The book is a sequel to 'Fractal Geometry — Mathematical Foundations and Applications' which was published in 1990, and which contains central material on the mathematics of fractals. 'Fractal Geometry' was originally aimed at a postgraduate audience, but with the explosion of interest in the subject it has also been used as the basis of undergraduate courses.

This book presupposes a reasonable competence in mathematical analysis, and in several places some knowledge of probability theory will be helpful. Familiarity with the basic material in 'Fractal Geometry' is assumed, particularly that on dimensions and iterated function systems; the main ideas and notation are reviewed here in Chapters 1 and 2. Specific references to 'Fractal Geometry' are often made and these are denoted by FG.

Much of the material presented in this book has come to the fore in the last few years. This includes a variety of methods for studying dimensions and other parameters of fractal sets and measures, as well as more sophisticated techniques, such as the thermodynamic formalism and tangent measures, which are now used routinely in fractal geometry and have many applications. The book also includes several 'big theorems' from probabilistic analysis, such as the ergodic theorem and renewal theorem, which have been applied effectively to fractals. As well as general theory, many examples and applications are described, in areas such as differential equations and harmonic analysis. Some results appear for the first time, and proofs have often been simplified.

The style of 'Techniques in Fractal Geometry' is similar to that of 'Fractal Geometry'. The book is mathematically precise, but aims to give an intuitive feel for the subject without getting unnecessarily involved in formal detail. The underlying concepts are presented as simply as possible and much of the theory is developed in detail in fairly specific cases with more general analogues summarised afterwards. For example, the thermodynamic formalism is presented for a simple non-linear generalisation of the Cantor set. As in 'Fractal Geometry', technicalities of measure theory are played down, with the existence of 'intuitively obvious' properties of measures taken for granted. An asterisk * indicates parts that can be omitted on first reading without losing the intuitive development.

No attempt has been made to include the most general results known. The author believes strongly that it is more important to communicate ideas and concepts than technical detail. Too often in mathematical writing, simple but elegant ideas are concealed by excessive generality. Often if the underlying ideas are understood then it is clear how they can be developed or combined to give more general results. It is hoped that readers will be able to extrapolate from the cases discussed here to more general situations.

Each chapter ends with brief notes on the history and current state of the subject. Given the scope of the topics covered, a comprehensive bibliography would be enormous, so we merely reference recent and key works for those interested in pursuing any topics further. Exercises are included to reinforce the text and to indicate further theory and examples.

With the wide range of topics included it is impossible to be entirely consistent as regards notation. In places a compromise has been made between standard notation and self-consistency within the book. There are some differences in notation from that in 'Fractal Geometry'.

Inevitably errors will have crept into the text during writing and rewriting. I regret this, and express the hope that such errors are obvious rather than misleading! I struggled to cope with correcting and revising an electronic version of the book. From experience with both approaches I can assure potential authors that the traditional method of correcting a double-spaced typescript by hand, whilst curled up in an armchair, is far less effort and less stressful and probably more accurate than working at a computer screen!

I am most grateful to all those who have assisted with the preparation of this book. In particular, John Howroyd, Maarit Järvenpää, Pertti Mattila, Lars Olsen and Toby O'Neil made very useful comments on early drafts of the book. Ben Soares produced some of the diagrams and, with Toby O'Neil, designed and produced the cover picture. I am greatly indebted to Gill Gardner for converting my almost illegible handwriting into an electronic form, and to the staff of John Wiley and Sons, in particular Stuart Gale, David Ireland and Helen Ramsey, for overseeing the production of the book.

Finally, I thank my family for their considerable patience and understanding whilst I was writing the book.

Kenneth J. Falconer
St Andrews, April 1996

Notes

References to the author's earlier book 'Fractal Geometry—Mathematical Foundations and Applications' are indicated by FG.

Parts of the book which may be omitted on a first reading are indicated by an asterisk * .

Introduction

The name 'fractal', from the latin 'fractus' meaning broken, was given to highly irregular sets by Benoit Mandelbrot in his foundational essay in 1975. Since then, fractal geometry has attracted widespread, and sometimes controversial, attention. The subject has grown on two fronts: on the one hand many 'real fractals' of science and nature have been identified. On the other hand, the mathematics that is available for studying fractal sets, much of which has its roots in geometric measure theory, has developed enormously with new tools emerging for fractal analysis. This book is concerned with the mathematics of fractals.

Various attempts have been made to give a mathematical definition of a fractal, but such definitions have not proved satisfactory in a general context. Here we avoid giving a precise definition, preferring to consider a set E in Euclidean space to be a fractal if it has all or most of the following features:

(i) E has a fine structure, that is irregular detail at arbitrarily small scales.
(ii) E is too irregular to be described by calculus or traditional geometrical language, either locally or globally.
(iii) Often E has some sort of self-similarity or self-affinity, perhaps in a statistical or approximate sense.
(iv) Usually the 'fractal dimension' of E (defined in some way) is strictly greater than its topological dimension.
(v) In many cases of interest E has a very simple, perhaps recursive, definition.
(vi) Often E has a 'natural' appearance.

Examples of fractals abound, but certain classes have attracted particular attention. Fractals that are invariant under simple families of transformations include self-similar, self-affine, approximately self-similar and statistically self-similar fractals, examples of which are shown in Figure 0.1. Certain self-similar fractals are especially well known: the middle-third Cantor set, the von Koch curve, the Sierpinski triangle (or gasket) and the Sierpinski carpet, see Figure 0.2. Fractals that occur as attractors or repellers of dynamical systems, for example the Julia sets resulting from iteration of complex functions, have also received wide coverage.

Fractal geometry is the study of sets with properties such as (i)–(vi). Many of the questions that are of interest about fractals are parallel to those that have been asked over the centuries about classical geometrical objects. These include:

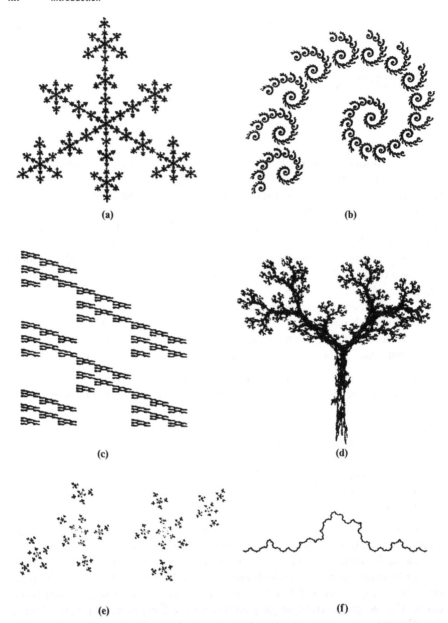

Figure 0.1 Fractals that are invariant under families of transformations. (a) and (b) are self-similar, (c) and (d) are self-affine, (e) is self-conformal, and (f) is statistically self-similar

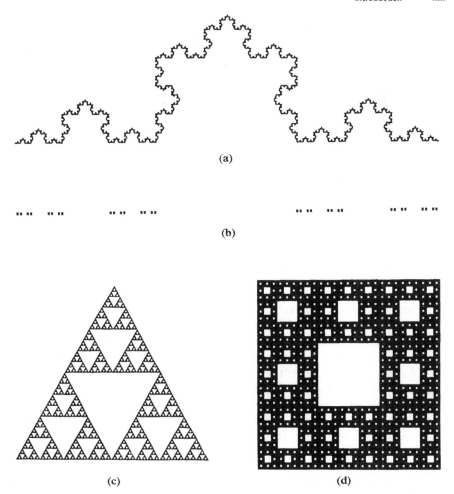

Figure 0.2 Well-known self-similar sets. (a) The von Koch curve (dimension log 4/log 3 = 1.262), (b) the middle-third Cantor set (dimension log 2/log 3 = 0.631), (c) the Sierpinski triangle or gasket (dimension log 3/log 2 = 1.585), (d) the Sierpinski carpet (dimension log 8/log 3 = 1.893)

(a) *Specification.* We seek efficient ways of defining fractals. For example, iterated functions systems provide one way of specifying fractals of certain classes.

(b) *Local description.* Locally a smooth curve looks like a line segment. Whilst fractals do not have such simple local structure, notions such as densities and tangent measures provide some local information.

(c) *Measurement of fractals.* The usual way of 'measuring' a fractal is by some form of dimension. Nevertheless dimension provides only limited information and other ways of quantifying aspects of fractality are being introduced. For example 'lacunarity' and 'porosity' are used to describe the small-scale preponderance of 'holes' in a set. For such quantities to have more than just descriptive use, their definitions and properties need a sound mathematical foundation.

It may be argued that there is too much emphasis on dimension in fractal analysis. Certainly, dimension (with its various definitions) tends to be mathematically tractable and can often be estimated experimentally. Moreover, the dimension of an object is often related to other features, for example the rate of heat flow through the boundary of a domain depends on the dimension of the boundary, and the dimension of the attractor of a dynamical system is related to other dynamical parameters such as the Liapunov exponents. However, many fractal aspects of an object are not reflected by dimension alone and other suitable measures of fractality are much needed.

(d) *Classification*. We seek ways of classifying fractals according to significant geometrical properties. One approach is to regard two sets as 'equivalent' if there is a bi-Lipschitz mapping between them (just as in topology two sets are considered equivalent if they are homeomorphic) and to seek 'invariants' for equivalent sets. For example two sets that are bi-Lipschitz equivalent have the same dimension, but dimension is far from a 'complete invariant' in that, except for certain rather specific classes of sets, there can be many non-equivalent sets of the same dimension.

(e) *Geometrical properties*. Properties of orthogonal projections, intersections, products, etc., are often of interest.

(f) *Occurrence in other areas of mathematics*. Fractals arise naturally in many areas of mathematics, for example dynamical systems or hyperbolic geometry. The general theory of fractals ought to relate easily to these areas.

(g) *Use of fractals to model physical phenomena*. There are many 'approximate fractals' in physics and nature, and these can often be modelled by 'mathematical' fractals. Ideally the mathematical theory should then tell us more about the physical situations.

In some areas the mathematics and physics tie together nicely, for example, Wiener's model of Brownian motion gives a reasonable probabilistic description of the irregular path described by a particle moving under molecular bombardment. However in other areas there is often a gulf between the fractals that are encountered in science or nature and the mathematics that is available. In many instances questions such as 'Why does an object have a fractal structure?' or 'If certain fractal features are present, what can we deduce?' have not been entirely satisfactorily answered. Nevertheless, progress is being made. Increasingly fractals are being studied in a 'dynamic' context, for example phenomena such as the diffusion of heat through fractal domains are being modelled mathematically.

Fractal features are often exhibited by measures rather than just by sets. 'Multifractal analysis' reveals a (sometimes very rich) fractal structure of measures, and a single measure may lead to a whole spectrum of fractal sets. Many of (a)–(g) above apply to measures just as to sets and multifractal measures are being studied in ways parallel to those for fractal sets.

This book presents some of the techniques that have been developed for studying aspects of fractals and multifractals. We briefly outline the material covered.

Chapter 1 brings together some general definitions and notation which will be needed throughout the book. Some inequalities involving submultiplicative sequences and convex functions are discussed. Basic ideas from measure theory are presented, and some results on convergence of measures are derived for later reference.

Chapter 2 reviews some standard aspects of fractal geometry which are discussed in much more detail in the earlier volume, FG. The basic definitions of dimension (Hausdorff, packing and box dimensions) and methods for their calculation are reviewed, and there is a discussion on representing fractals by iterated function systems.

In Chapter 3 we introduce two useful techniques for studying dimension. Firstly, implicit methods enable properties of certain fractals to be investigated without the need for a handle on the actual value of their dimension. In particular, sets that are 'approximately self-similar' in a weak sense must display considerable regularity from the point of view of dimension. Secondly, we address the relationship between the box dimension of sets of real numbers and the lengths of the complementary intervals of the set. In a certain sense, the box dimension describes the complement of a set whereas the Hausdorff dimension describes the set itself.

The next two chapters take the notion of approximate self-similarity further, leading to the 'thermodynamic formalism'. This powerful technique (which has roots in statistical mechanics) extends the 'linear' theory of strictly self-similar sets to the 'non-linear' setting of 'approximately self-similar' sets. We develop the thermodynamic formalism in the special case of 'cookie-cutter' sets, which may be thought of as 'non-linear Cantor sets'. After deriving the 'bounded distortion' principle for such sets, we obtain a formula for their dimension in terms of the 'pressure' of a certain function.

Chapters 6–8 present three corner-stone results of probabilistic analysis: the ergodic theorem, the renewal theorem and the martingale convergence theorem. These results are proved and applied to topics such as average densities of fractals, box-counting numbers of self-similar sets, and the classification of fractals under bi-Lipschitz mappings.

Tangent measures, described in Chapter 9, are essentially limits of a sequence of enlargements of a measure about a point. Tangent measures are not unlike derivatives, in that they contain information about the local structure of a set or measure, but they have more regular behaviour than the original measures. We give sample applications to densities of sets; in particular we give a tangent measure proof that sets of non-integral dimension fail to have densities almost everywhere. We indicate how tangent measures can be applied to problems in harmonic analysis.

Often it is natural to study fractal properties of measures rather than sets, indeed many fractal sets, such as attractors of dynamical systems, are in essence already measures. Chapters 10 and 11 discuss fractal properties of measures. In particular we consider sets such as E_α, the set of x at which a given measure μ has local dimension α, that is where the measure of a small ball centred at x is (roughly) equal to the radius of the ball to the power α. For certain μ the sets E_α may be 'large' for a range of α, and the 'size' of E_α may be measured by either μ or by dimension. In Chapter 10 we consider $\mu(E_\alpha)$, leading to the 'dimension decomposition' of μ, and in Chapter 11 we look at the dimension of E_α, leading to the 'multifractal spectrum' of μ. The thermodynamic formalism is used to extend the theory to non-linear cases.

Chapter 12 describes several ways in which fractal geometry interacts with differential equation theory. This is an area where a number of important methods have been developed and where some of the techniques from earlier in the book may be applied. We describe a general approach for bounding the dimension of attractors of dynamical systems and of differential equations. Then the effect of a fractal boundary of a region on the solutions of partial differential equations is discussed, in particular the way in which fractality affects the asymptotic form of the solutions and the asymptotic distribution of eigenvalues. The final section is concerned with setting up differential equations on a region that is itself fractal. This chapter is selective and far ranging, and full proofs are not included.

Fractal geometry may be studied from many viewpoints, and inevitably the approach adopted in this book reflects the author's own background and experience. The topics included have been selected according to the author's interests and whim, but there are many other worthy techniques in use in fractal analysis, such as wavelet methods and the variants of iterated function systems used in image compression. Nevertheless, the methods described here are widely applicable, and, hopefully, will find further applications in the future.

Notes and references

Since the pioneering essays of Mandelbrot (1975, 1982), a wide variety of books have been written on fractals. The books by Edgar (1990), Falconer (1990), Mehaute (1991) and Peitgen, *et al.* (1992) provide basic mathematical treatments. Federer (1969), Falconer (1985) and Mattila (1995) concentrate on geometric measure theory, Rogers (1970) addresses the general theory of Hausdorff measures, and Wicks (1991) approaches the subject from the standpoint of non-standard analysis. Books with a computational emphasis include Peitgen and Saupe (1988) and Devaney and Keen (1989). Several books, including those by Barnsley (1988) and Peruggia (1993), are particularly concerned with iterated function systems, those by Barnsley and Hurd (1993) and Fisher (1995) concentrating on applications to image compression. Massopust (1994) discusses fractal functions and surfaces, and Tricot (1995) considers fractal curves. The books by Kahane (1985) and Stoyan and Stoyan (1994) include material on random fractals. The anthology of 'classic papers' on fractals by Edgar (1993) helps put the subject in historical perspective.

Much of interest may be found in the proceedings of conferences on fractal mathematics, including the volumes edited by Cherbit (1991), Bélair and Dubuc (1991), Bedford, *et al.* (1991), Bandt, *et al.* (1992) and Bandt, *et al.* (1995).

A great deal has been written on physical applications of fractals, for a sample see Pietronero and Tosatti (1986), Feder (1988), Fleischmann, *et al.* (1990), Smith (1991), Vicsek (1992) and Hastings (1993).

Chapter 1   Mathematical background

In this chapter we collect together several topics of a general mathematical nature for future reference. The first section sets out basic terminology and notation. We then discuss some inequalities that will be especially useful: the subadditive inequality and some properties of convex functions. The last two sections are concerned with measure theoretic ideas which play a fundamental rôle in fractal geometry. We sketch the rudiments of measure theory, and then go into a little more detail on weak convergence, perhaps a less familiar topic.

1.1 Sets and functions

We remind the reader of some standard definitions and notation that will frequently be encountered.

We use the usual notation for the real numbers $\mathbb{R}$, the integers $\mathbb{Z}$, and the rational numbers $\mathbb{Q}$, with $\mathbb{R}^+$, $\mathbb{Z}^+$ and $\mathbb{Q}^+$ for their positive subsets.

We normally work in n-dimensional Euclidean space, $\mathbb{R}^n$, where $\mathbb{R} = \mathbb{R}^1$ is just the real line and $\mathbb{R}^2$ is the Euclidean plane. Points in $\mathbb{R}^n$ are denoted by lower case letters, x, y, etc. We write $x + y$ for the (vectorial) sum of x and y and λx for x multiplied by the real scalar λ. We work with the usual Euclidean distance or metric on $\mathbb{R}^n$; thus the distance between points $x, y \in \mathbb{R}^n$ is $|x - y| = (\sum_{i=1}^n |x_i - y_i|^2)^{1/2}$, where, in coordinate form, $x = (x_1, \ldots, x_n)$ and $y = (y_1, \ldots, y_n)$.

We generally use capitals, A, E, X, Y etc. to denote subsets of $\mathbb{R}^n$. The *diameter* of a non-empty set X is given by $|X| = \sup\{|x - y| : x, y \in X\}$ with the convention that $|\emptyset| = 0$. We write $\text{dist}(X, Y) = \inf\{|x - y| : x \in X, y \in Y\}$ for the distance between the non-empty sets X and Y. For $r > 0$ the *r-neighbourhood* or *r-parallel body* of a set X is given by

$$X_r = \left\{ y : \inf_{x \in X} |x - y| \le r \right\}.$$

We define the *closed* and *open balls* with centre $x \in \mathbb{R}^n$ and radius $r > 0$ as

$$B(x, r) = \{y \in \mathbb{R}^n : |y - x| \le r\}$$

and

$$B^0(x, r) = \{y \in \mathbb{R}^n : |y - x| < r\}$$

respectively. Of course balls in $\mathbb{R}^1$ are just *intervals*, and in $\mathbb{R}^2$ are *discs*. A set $X \subset \mathbb{R}^n$ is *bounded* if $X \subset B(x, r)$ for some x and r; thus a non-empty set X is bounded if and only if $|X| < \infty$.

Open and closed sets are defined in the usual way. A set $A \subset \mathbb{R}^n$ is *open* if for all $x \in A$ there is some $r > 0$ with $B(x, r) \subset A$. A set $A \subset \mathbb{R}^n$ is *closed* if it contains all its limit points, that is if whenever $(x_k)_{k=1}^{\infty}$ is a sequence of points of A converging to $x \in \mathbb{R}^n$ then $x \in A$. A set is open if and only if its complement is closed. The *interior* of a set A, written int A, is the union of all open subsets of A, and the *closure* of A, written $\bar{A}$, is the intersection of all closed sets that contain A. The *boundary* of A is defined as $\partial A = \bar{A} \backslash \text{int } A$.

Formally a set A is defined to be *compact* if every collection of open sets which cover A has a finite subcollection which covers A. A subset A of $\mathbb{R}^n$ is compact if and only if it is closed and bounded, and this may be taken as the definition of compactness for subsets of $\mathbb{R}^n$.

The idea of constructing sets as unions or intersections of open or closed sets leads to the concept of Borel sets. Formally, the family of *Borel* subsets of $\mathbb{R}^n$ is the smallest family of sets such that

(a) every open set is a Borel set and every closed set is a Borel set,
(b) if $A_1, A_2, \ldots$ is any countable collection of Borel sets then $\cup_{i=1}^{\infty} A_i$, $\cap_{i=1}^{\infty} A_i$ and $A_1 \backslash A_2$ are Borel sets.

Any set that can be constructed starting with open or closed sets and taking countable unions or intersections a finite number of times will be a Borel set. Virtually all subsets of $\mathbb{R}^n$ that will be encountered in this book will be Borel sets.

Occasionally we use the symbol $\#$ to denote the number of points in a (usually finite) set.

As usual, $f : X \to Y$ denotes a *function* or *mapping* f with *domain* X and *range* or *codomain* Y. A function $f : X \to Y$ is an *injection* or is *one-one* (1–1) if $f(x_1) \neq f(x_2)$ whenever $x_1 \neq x_2$, and is a *surjection* or *onto* if $f(X) = Y$. It is a *bijection* or a 1–1 *correspondence* if it is both an injection and a surjection. If $f : X \to Y$ and $g : Z \to W$ where $Y \subset Z$ we define the *composition* $g \circ f : X \to W$ by $(g \circ f)(x) = g(f(x))$. For $f : X \to X$ we define $f^k : X \to X$, the k-th *iterate* of f, by $f^0(x) = x$, and $f^k(x) = f(f^{k-1}(x))$ for $k = 1, 2, 3, \ldots$; thus f^k is the k-fold composition of f with itself. For a bijection $f : X \to Y$, the *inverse* of f is the function $f^{-1} : Y \to X$ such that $f^{-1}(f(x)) = x$ for all $x \in X$ and $f(f^{-1}(y)) = y$ for all $y \in Y$.

For $A \subset X$, the function $1_A : X \to \{0, 1\}$ given by $1_A(x) = 0$ if $x \notin A$ and $1_A(x) = 1$ if $x \in A$ is called the *indicator function* or *characteristic function* of A; its value 'indicates' whether or not the point x is in the set A.

Certain classes of function are of particular interest. We write $C(X)$ for the vector space of continuous functions $f : X \to \mathbb{R}$, and $C_0(X)$ for the subspace of functions with bounded support (the *support* of $f : X \to \mathbb{R}$ is the smallest closed subset of X outside which $f(x) = 0$). For a suitable domain $X \subset \mathbb{R}^n$ we write

$C^1(X)$ for the space of functions $f: X \to \mathbb{R}$ with continuous derivatives and $C^2(X)$ for those with continuous second derivatives. Of particular interest in connection with fractals are the Lipschitz functions. We call $f: X \to \mathbb{R}^m$ a *Lipschitz function* if there exists a number c such that

$$|f(x) - f(y)| \leq c|x - y| \quad \text{for all} \quad x, y \in X. \tag{1.1}$$

The infimum value of c for which such an inequality holds is called the *Lipschitz constant* of f, written $\text{Lip} f$. We also write $\text{Lip} X$ to denote the space of Lipschitz functions from X to $\mathbb{R}^m$ for appropriate m.

Statements such as $\lim_{k\to\infty} a_k = a$ or $\lim_{x\to 0} f(x) = a$ will always imply that the limit exists as well as taking the stated value.

There are some useful conventions for describing the limiting behaviour of functions. For $f: \mathbb{R}^+ \to \mathbb{R}^+$ we write $f(x) = o(g(x))$ to mean that $f(x)/g(x) \to 0$ as $x \to \infty$, and $f(x) = O(g(x))$ to mean that $f(x)/g(x)$ remains bounded as $x \to \infty$. Similarly, we write $f(x) \sim g(x)$ if $f(x)/g(x) \to 1$, and $f(x) \asymp g(x)$ if there exists numbers c_1, c_2 such that $0 < c_1 \leq f(x)/g(x) \leq c_2 < \infty$ for all $x \in \mathbb{R}^+$. We occasionally write $f(x) \simeq g(x)$; this is used in a loose fashion to indicate that $f(x)$ is 'roughly comparable' to $g(x)$ for large x. We adapt this notation in the obvious way for functions on other domains and for x approaching other limiting values.

1.2 Some useful inequalities

We now discuss some simple but very useful inequalities.

Subadditive sequences occur surprisingly often, in analysis in general, and in fractal geometry and dynamical systems in particular. A sequence of real numbers $(a_k)_{k=1}^\infty$ is *subadditive* if it satisfies the inequality

$$a_{k+m} \leq a_k + a_m \tag{1.2}$$

for all $k, m \in \mathbb{Z}^+$. The fundamental property of such a sequence is that $(a_k/k)_{k=1}^\infty$ converges.

Proposition 1.1

Let $(a_k)_{k=1}^\infty$ be a subadditive sequence. Then $\lim_{k\to\infty} a_k/k$ exists and equals $\inf_{k \geq 1} a_k/k$ (which may be a real number or $-\infty$).

Proof Given a positive integer m we may write any integer k in the form $k = qm + r$ where $q \in \mathbb{Z}$ and $0 \leq r \leq m - 1$. Using (1.2) q times gives, for $k \geq m$,

$$\frac{a_k}{k} = \frac{a_{qm+r}}{qm+r} \leq \frac{qa_m + a_r}{qm} = \frac{a_m}{m} + \frac{a_r}{qm}.$$

As $k \to \infty$, so $q \to \infty$, giving

$$\limsup_{k \to \infty} a_k/k \le a_m/m.$$

This is true for all $m \in \mathbb{Z}^+$, so $\limsup_{k \to \infty} a_k/k \le \inf_k a_k/k$. We conclude that the limit exists and equality holds. □

Corollary 1.2

Let b be a real number such that $(a_k)_{k=1}^{\infty}$ satisfies

$$a_{k+m} \le a_k + a_m + b$$

for all $k, m = 1, 2, \ldots$. Then $a \equiv \lim_{k \to \infty} a_k/k$ exists and $a_k \ge ka - b$ for all k.

Proof We have $(a_{k+m} + b) \le (a_k + b) + (a_m + b)$, so applying Proposition 1.1 to the sequence $(a_k + b)_{k=1}^{\infty}$ gives that $\lim_{k \to \infty} a_k/k = \lim_{k \to \infty} (a_k + b)/k = \inf_{k \ge 1}(a_k + b)/k$. Writing a for this limit, $a \le (a_k + b)/k$ for all k. □

In the same way, we say that a sequence $(b_k)_{k=1}^{\infty}$ of positive real numbers is *submultiplicative* if $b_{k+m} \le b_k b_m$ for all $k, m \in \mathbb{Z}^+$.

Corollary 1.3

Let $(b_k)_{k=1}^{\infty}$ be a submultiplicative sequence. Then $\lim_{k \to \infty}(b_k)^{1/k}$ exists and equals $\inf_{k \ge 1}(b_k)^{1/k}$.

Proof The sequence $a_k = \log b_k$ is subadditive, so $\log b_k^{1/k} = a_k/k$ is convergent by Proposition 1.1, so $b_k^{1/k}$ is convergent. □

Next we consider some inequalities associated with convex functions. Let $X \subset \mathbb{R}$ be an interval. A function $\psi : X \to \mathbb{R}$ is *convex* if for all $x_1, x_2 \in X$ and all numbers $\alpha_1, \alpha_2 > 0$ with $\alpha_1 + \alpha_2 = 1$,

$$\psi(\alpha_1 x_1 + \alpha_2 x_2) \le \alpha_1 \psi(x_1) + \alpha_2 \psi(x_2); \tag{1.3}$$

geometrically this means that every chord of the graph of ψ lies above the graph (Figure 1.1). If ψ has a continuous second derivative then ψ is convex if and only if $\psi''(x) \ge 0$ for all $x \in X$. The function ψ is *strictly convex* if the inequality (1.3) is strict for all $x_1 \neq x_2$; this will happen if $\psi''(x) > 0$ for all $x \in X$. A function $\psi : X \to \mathbb{R}$ is called *concave* if $-\psi$ is convex.

The convexity condition (1.3) implies a similar inequality for more terms; this extension is known as Jensen's inequality.

Proposition 1.4

Let $\psi : X \to \mathbb{R}$ be convex, let $x_1, \ldots, x_m \in X$ and let $\alpha_1, \ldots, \alpha_m > 0$ satisfy $\sum_{i=1}^{m} \alpha_i = 1$. Then

$$\psi\left(\sum_{i=1}^{m} \alpha_i x_i \right) \leq \sum_{i=1}^{m} \alpha_i \psi(x_i). \tag{1.4}$$

If ψ is strictly convex then equality holds if and only if $x_1 = x_2 = \ldots = x_m$.

Proof For $m \geq 3$, we use the inductive step

$$\psi\left(\sum_{i=1}^{m} \alpha_i x_i \right) = \psi\left((1 - \alpha_m) \sum_{i=1}^{m-1} \alpha_i (1 - \alpha_m)^{-1} x_i + \alpha_m x_m \right)$$

$$\leq (1 - \alpha_m) \psi\left(\sum_{i=1}^{m-1} \alpha_i (1 - \alpha_m)^{-1} x_i \right) + \alpha_m \psi(x_m) \tag{1.5}$$

by (1.3), so (1.4) follows from the inequality for $m - 1$, since $\sum_{i=1}^{m-1} \alpha_i (1 - \alpha_m)^{-1} = 1$.

If ψ is strictly convex, then equality in (1.5) implies that $x_m = \sum_{i=1}^{m-1} \alpha_i (1 - \alpha_m)^{-1} x_i$, that is $x_m = \sum_{i=1}^{m} \alpha_i x_i$. By renumbering, we could work with any of the x_i as 'x_m'; thus equality in (1.4) implies $x_k = \sum_{i=1}^{m} \alpha_i x_i$ for all k. $\square$

Note that if $\psi : X \to \mathbb{R}$ is a concave function then the opposite inequality holds in (1.4).

Suitable choice of ψ in (1.4) yields the well-known arithmetic-geometric mean inequality, see Exercise 1.2.

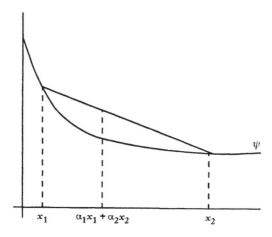

Figure 1.1 Graph of a convex function ψ

The following application will be especially important in Chapter 5 in connection with entropy. We make the convention that $0 \log 0 = 0$.

Corollary 1.5

Let $p_1, \ldots, p_m$ be 'probabilities' with $p_i \geq 0$ for all i and $\sum_{i=1}^m p_i = 1$, and let $q_1, \ldots, q_m$ be real numbers. Then

$$\sum_{i=1}^m p_i(-\log p_i + q_i) \leq \log \sum_{i=1}^m e^{q_i} \tag{1.6}$$

with equality if and only if $p_i = e^{q_i} / \sum_{j=1}^m e^{q_j}$ for all i.

Proof Defining $\psi(x) = x \log x$ $(x > 0), \psi(0) = 0$ gives that $\psi : [0, \infty) \to \mathbb{R}$ is a continuous strictly convex function, since $\psi''(x) > 0$ for $x > 0$. For convenience, write $s = (\sum_{j=1}^m e^{q_j})^{-1}$. Applying (1.4) with $\alpha_i = s e^{q_i}$ and $x_i = p_i / e^{q_i}$ we have

$$\psi(s) = \psi\left(\sum_{i=1}^m (s e^{q_i})(p_i/e^{q_i})\right) \leq \sum_{i=1}^m s e^{q_i} \psi(p_i/e^{q_i}),$$

that is

$$s \log s \leq \sum_{i=1}^m s e^{q_i} p_i e^{-q_i} \log\left(p_i/e^{q_i}\right)$$

$$= s \sum_{i=1}^m p_i (\log p_i - q_i),$$

which is (1.6). Since ψ is strictly convex, equality requires that $p_i/e^{q_i} = c$ where c is independent of i, and $1 = \sum_{i=1}^m p_i = c \sum_{i=1}^m e^{q_i}$. $\square$

1.3 Measures

Measures or 'mass distributions' have a central place in fractal geometry. They are a major tool in the mathematics of fractals, but also, measures may exhibit fractal features which may be studied in their own right. Basically, a measure is a way of ascribing a numerical size to sets so that the priniciple 'the whole is the sum of the parts' applies. Thus if a set is decomposed into a finite or countable number of pieces in a reasonable way then the measure of the whole set is the sum of the measures of the pieces. A measure is often thought of as a 'mass distribution' or a 'charge distribution', an interpretation that may be helpful to those less familiar with formal measure theory.

In general we try to play down the more technical aspects of measure theory. Since we shall just work with measures defined on subsets of $\mathbb{R}^n$ many of the awkward features of measures that can occur in a more general topological setting may be avoided. We give a formal definition of a measure to ensure precision, but it is perhaps more important that the reader develops an intuitive feel for the basic properties of measures.

Let $X \subset \mathbb{R}^n$. We call μ a *measure* on X if μ assigns a non-negative number, possibly ∞, to each subset of X such that

(a) $\mu(\emptyset) = 0$,   (1.7)
(b) if $A \subset B$ then $\mu(A) \leq \mu(B)$, and   (1.8)
(c) if $A_1, A_2, \ldots$ is a countable sequence of sets then

$$\mu\left(\bigcup_{i=1}^{\infty} A_i\right) \leq \sum_{i=1}^{\infty} \mu(A_i). \tag{1.9}$$

Thus (a) requires the empty set to have zero measure, and (b) states that 'the bigger the set the larger the measure'. Property (c) ensures that the measure of any set is no more than the sum of the measures of the pieces in any countable decomposition. For a measure to be useful we require more than this, namely that equality holds in (1.9) for 'nice' disjoint sets A_i. This leads to the idea of measurability.

Given a measure μ there is a family of subsets of X on which μ behaves in a nice additive way: a set $A \subset X$ is called μ-*measurable* (or just *measurable* if the measure in use is clear) if

$$\mu(E) = \mu(E \cap A) + \mu(E \backslash A) \quad \text{for all} \quad E \subset X. \tag{1.10}$$

We write $\mathcal{M}$ for the family of measurable sets which always form a σ-*field*, that is $\emptyset \in \mathcal{M}$, $X \in \mathcal{M}$, and if $A_1, A_2, \ldots \in \mathcal{M}$ then $\cup_{i=1}^{\infty} A_i \in \mathcal{M}$, $\cap_{i=1}^{\infty} A_i \in \mathcal{M}$ and $A_1 \backslash A_2 \in \mathcal{M}$. For reasonably defined measures, $\mathcal{M}$ will be a very large family of sets, and in particular will contain the σ-field of Borel sets.

Proposition 1.6

Let μ be a measure on X and let $\mathcal{M}$ be the family of all μ-measurable subsets of X.
(a) If $A_1, A_2, \ldots \in \mathcal{M}$ are disjoint then

$$\mu\left(\bigcup_{i=1}^{\infty} A_i\right) = \sum_{i=1}^{\infty} \mu(A_i). \tag{1.11}$$

(b) If $A_1 \subset A_2 \subset \ldots$ is an increasing sequence of sets in $\mathcal{M}$ then

$$\mu\left(\bigcup_{i=1}^{\infty} A_i\right) = \lim_{i \to \infty} \mu(A_i). \tag{1.12}$$

(c) If $A_1 \supset A_2 \supset \ldots$ is a decreasing sequence of sets in $\mathcal{M}$ and $\mu(A_1) < \infty$ then

$$\mu\left(\bigcap_{i=1}^{\infty} A_i\right) = \lim_{i \to \infty} \mu(A_i). \tag{1.13}$$

The continuity properties (*b*) and (*c*) follow easily from (*a*). Property (*a*) is the crucial property of a measure: that μ is additive on disjoint sets of some large class $\mathcal{M}$. For all the measures that we encounter $\mathcal{M}$ includes the Borel sets. However, in general $\mathcal{M}$ does not consist of *all* subsets of X, and (*a*) does not hold for arbitrary disjoint sets $A_1, A_2, \ldots$.

(Technical note: What is termed a 'measure' here is often referred to as an 'outer measure' in general texts on measure theory. Such texts define a measure μ only on the sets of some σ-field $\mathcal{M}$, with (1.7)–(1.9) holding for sets of $\mathcal{M}$, with equality in (1.9) if the A_i are disjoint sets in $\mathcal{M}$. However, μ can then be extended to *all* $A \subset X$ by setting $\mu(A) = \inf\{\sum_i \mu(A_i) : A \subset \cup_i A_i$ and $A_i \in \mathcal{M}\}$. In work relating to Hausdorff measures, etc., it is convenient to assume that measures are defined on all sets in the first place.)

In this book we will be concerned with measures on $\mathbb{R}^n$, or on a subset of $\mathbb{R}^n$, that behave nicely on the Borel sets. We term a measure μ a *Borel measure* on $X \subset \mathbb{R}^n$ if the Borel subsets of X are μ-measurable. It may be shown that μ is a Borel measure if and only if

$$\mu(A \cup B) = \mu(A) + \mu(B) \quad \text{whenever} \quad A, B \subset X \text{ and } \text{dist}(A, B) > 0. \quad (1.14)$$

A Borel measure μ is termed *Borel regular* if every subset of X is contained in a Borel set of the same measure; for such measures we can, for all practical purposes, work entirely with Borel sets.

Virtually all the measures that we will encounter (including Hausdorff and packing measures) will be Borel regular on $\mathbb{R}^n$ or on the pertinent subset thereof. Therefore, to avoid tedious repetition, we make the convention throughout this book that the term 'measure' means 'Borel regular measure'. Thus, for our purposes, a measure is a set-function that behaves nicely with respect to Borel sets. To avoid trivial cases we also assume that $\mu(X) > 0$ for all measures μ.

A measure μ on X with $\mu(X) < \infty$ is called *finite*; if $\mu(A) < \infty$ for every bounded set A it is *locally finite*. We call μ a *probability measure* if $\mu(X) = 1$ (this standard terminology does not necessarily mean that μ has probabilistic associations).

If μ is a locally finite (Borel regular) measure, we can approximate the measure of sets by compact sets and open sets, in the sense that

$$\mu(U) = \sup\{\mu(A) : A \subset U \text{ with } A \text{ compact}\} \quad (1.15)$$

for every non-empty open set U, and

$$\mu(E) = \inf\{\mu(U) : E \subset U \text{ with } U \text{ open}\} \quad (1.16)$$

for every set E, see Exercise 1.5.

The *support* of μ, written spt μ, is the smallest closed set with complement of measure 0, that is

$$\text{spt} \, \mu = X \setminus \cup \{U : U \text{ is open and } \mu(U) = 0\}.$$

We list below some basic examples of measures.

(1) For each $A \subset \mathbb{R}^n$ let $\mu(A)$ be the number of points in A (which may be ∞); this is the *counting measure* on $\mathbb{R}^n$.

(2) For given $a \in \mathbb{R}^n$ let $\mu(A) = 0$ if $a \notin A$ and $\mu(A) = 1$ if $a \in A$. Then μ is a measure with support $\{a\}$ that we think of as a unit *point mass* concentrated at a.

(3) Lebesgue measure on $\mathbb{R}^n$ is the natural extension to a large class of sets of 'n-dimensional volume' ('length' if $n = 1$, 'area' if $n = 2$ and 'volume' if $n = 3$). We define the *n-dimensional volume* of the 'coordinate parallelepiped' $A = \{(x_1, \ldots, x_n) \in \mathbb{R}^n : a_i \le x_i \le b_i\}$ by

$$\mathrm{vol}^n(A) = (b_1 - a_1)(b_2 - a_2) \ldots (b_n - a_n).$$

Then *n-dimensional Lebesgue measure* $\mathcal{L}^n$ is defined by

$$\mathcal{L}^n(A) = \inf\left\{ \sum_{i=1}^{\infty} \mathrm{vol}^n(A_i) : A \subset \bigcup_{i=1}^{\infty} A_i \right\},$$

where the infimum is over all coverings of A by countable collections of parallelepipeds. With some effort it may be shown that $\mathcal{L}^n$ is indeed a (Borel regular) measure on $\mathbb{R}^n$ such that $\mathcal{L}^n(A)$ equals the n-dimensional volume of A if A is a parallelpiped or any other set for which the volume can be calculated using the usual rules of mensuration.

(4) Let μ be a measure on X and let $E \subset X$. The *restriction of μ to E*, denoted by $\mu|_E$, is defined by

$$\mu|_E(A) = \mu(A \cap E) \tag{1.17}$$

for all $A \subset X$. It is easy to check that every μ-measurable set is $\mu|_E$-measurable and, provided E is measurable and $\mu(E) < \infty$, then μ is a (Borel regular) measure.

(5) A very useful method of defining a measure is by *repeated subdivision*, see Figure 1.2. For $m \ge 2$ we take a hierarchy of subsets of $\mathbb{R}^n$ indexed by sequences $\{(i_1, \ldots, i_k) : k \ge 0 \text{ and } 1 \le i_j \le m \text{ for each } j\}$. For every $(i_1, \ldots, i_k)$ let $X_{i_1, \ldots, i_k}$ be a bounded non-empty closed subset of $\mathbb{R}^n$, and write $\mathcal{E}$ for the family of all such sets. We assume that these sets are nested so that

$$X_{i_1, \ldots, i_k} \supset \bigcup_{i=1}^{m} X_{i_1, \ldots, i_k, i} \tag{1.18}$$

(frequently this union is disjoint, though it need not be). Suppose that $\mu(X_{i_1, \ldots, i_k}) < \infty$ is defined for $X_{i_1, \ldots, i_k} \in \mathcal{E}$ in such a way that

$$\mu(X_{i_1, \ldots, i_k}) = \sum_{i=1}^{m} \mu(X_{i_1, \ldots, i_k, i}) \tag{1.19}$$

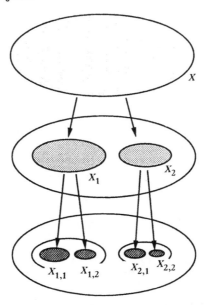

Figure 1.2   Construction of a measure by repeated subdivision. The measure on each set of $\mathscr{E}$ is divided between its subsets in the hierarchical construction

for each $(i_1, \ldots, i_k)$, that is the 'mass' $\mu(X_{i_1,\ldots,i_k})$ is subdivided between the subsets $X_{i_1,\ldots,i_k,i}, (1 \le i \le m)$. We assume that for every sequence $(i_1, i_2, \ldots)$ both the diameters of the sets $|X_{i_1,\ldots,i_k}|$ and their measures $\mu(X_{i_1,\ldots,i_k})$ tend to 0 as $k \to \infty$. We write $E_k = \cup_{i_1,\ldots,i_k} X_{i_1,\ldots,i_k}$ for each k, and $E = \cap_{k=0}^{\infty} E_k$, so that E is the intersection of a decreasing sequence of closed non-empty sets, and is therefore closed and non-empty. For $A \subset \mathbb{R}^n$ we define

$$\mu(A) = \inf\left\{ \sum_i \mu(V_i) : A \cap E \subset \cup_i V_i \text{ and } V_i \in \mathscr{E} \right\}. \qquad (1.20)$$

It is not hard to show that μ is a measure with support contained in E, such that $\mu(X_{i_1,\ldots,i_k})$ is the preassigned value for all $(i_1, \ldots, i_k)$. Thus if μ is defined by this 'repeated subdivision' procedure it may be extended to a measure on E.

For a simple instance of this procedure, let $m = 2$ and for each k let $X_{i_1,\ldots,i_k}$ comprise the set of 2^k closed binary subintervals of $[0, 1]$ of length 2^{-k}, nested in the obvious way. Taking $\mu(X_{i_1,\ldots,i_k}) = 2^{-k}$ for each such interval, (1.19) is readily verified and (1.20) then defines the restriction of Lebesgue measure to $[0, 1]$.

We say that a property holds for *almost all x* or *almost everywhere* (with respect to a measure μ) if the set for which the property fails has μ-measure 0.

For example, with respect to Lebesgue measure, almost all real numbers are irrational.

Occasionally we need the following density result, to the effect that almost all points of a set E are, from the point of view of measure, 'well inside' E. A point x at which (1.21) holds is called a *density point* of E.

Proposition 1.7

Let μ be a locally finite Borel measure on $\mathbb{R}^n$. Then, for every μ-measurable set E, we have that

$$\lim_{r \to 0} \mu(E \cap B(x,r))/\mu(B(x,r)) \tag{1.21}$$

exists and equals 1 for μ-almost all $x \in E$ and equals 0 for μ-almost all $x \notin E$.

Proof Since this is a local result, we may assume that μ is a finite measure. Take $c < 1$, and define

$$A = \{x \in E : \mu(E \cap B(x,r)) < c\mu(B(x, r)) \text{ for arbitrarily small } r\};$$

we will show that $\mu(A) = 0$. Given $\epsilon > 0$ there exists an open set $U \supset A$ such that $\mu(U) < \mu(A) + \epsilon$. Define a class $\mathcal{V}$ of balls by

$$\mathcal{V} = \{B : B \text{ has centre in } A, \text{ with } B \subset U \text{ and } \mu(E \cap B) < c\mu(B)\}.$$

Then $\mathcal{V}$ is a *Vitali cover* for A, which means that for all $x \in A$ and $\delta > 0$ there is a ball in $\mathcal{V}$ with centre x and radius less than δ. The Vitali covering theorem asserts that there is a sequence of *disjoint* balls $B_1, B_2, \ldots$ in $\mathcal{V}$ such that $\mu(A \setminus \cup_i B_i) = 0$. Then

$$\mu(A) = \mu(A \cap \cup_i B_i) + \mu(A \setminus \cup_i B_i)$$

$$= \sum_i \mu(A \cap B_i) + 0 \le c \sum_i \mu(B_i)$$

$$= c\mu(\cup_i B_i) \le c\mu(U) < c(\mu(A) + \epsilon).$$

This is true for all $\epsilon > 0$ so $\mu(A) \le c\mu(A)$, implying that $\mu(A) = 0$. We conclude that for all $c < 1$, for μ-almost all $x \in E$ we have $c\mu(B(x,r)) \le \mu(E \cap B(x,r)) \le \mu(B(x, r))$ for all sufficiently small r. Thus for μ-almost all $x \in E$ the limit (1.21) exists and equals 1.

Applying this with E replaced by $\mathbb{R}^n \setminus E$ now gives that the limit (1.21) equals 0 for μ-almost all $x \notin E$. $\square$

Sometimes we will work with several measures on the same set. We say that the measures μ and ν on X are *equivalent* if there exist numbers $c_1, c_2 > 0$ such that

$$c_1\mu(A) \le \nu(A) \le c_2\mu(A) \tag{1.22}$$

for all $A \subset X$.

Integration with respect to a measure μ on X is defined using the usual steps. A *simple function* $f: X \to \mathbb{R}$ is a function of the form

$$f(x) = \sum_{i=1}^{k} a_i 1_{A_i}(x)$$

where $a_1, \ldots, a_k \in \mathbb{R}$ and $A_1, \ldots, A_k$ are μ-measurable sets, with 1_{A_i} the indicator function of A_i. We define the *integral* of the simple function f with respect to μ as

$$\int f \mathrm{d}\mu = \sum_{i=1}^{k} a_i \mu(A_i).$$

Integration of more general functions is defined using approximation by simple functions. We term $f: X \to \mathbb{R}$ a *measurable function* if for all $c \in \mathbb{R}$ the set $\{x \in X : f(x) < c\}$ is a measurable set (in particular for a Borel measure μ all continuous functions are measurable). We define the *integral* of a measurable $f: X \to \mathbb{R}^+ \cup \{0\}$ by

$$\int f \mathrm{d}\mu = \sup \left\{ \int g \mathrm{d}\mu : g \text{ is simple, } 0 \leq g \leq f \right\}$$

(this value may be infinite). Finally, for a measurable function $f: X \to \mathbb{R}$ we write $f_+(x) = \max\{f(x), 0\}$ and $f_-(x) = \max\{-f(x), 0\}$, and define

$$\int f \mathrm{d}\mu = \int f_+ \mathrm{d}\mu - \int f_- \mathrm{d}\mu,$$

provided both $\int f_+ \, \mathrm{d}\mu$ and $\int f_- \, \mathrm{d}\mu$ are finite. This happens if $\int |f| \mathrm{d}\mu = \int f_+ \, \mathrm{d}\mu + \int f_- \, \mathrm{d}\mu < \infty$; such functions are called μ-*integrable*. All the usual properties of the integral hold, for example $\int (f+g) \mathrm{d}\mu = \int f \mathrm{d}\mu + \int g \, \mathrm{d}\mu$ and $\int (\lambda f) \mathrm{d}\mu = \lambda \int f \mathrm{d}\mu$ for real λ.

For a measurable set A we define the *integral of f over A* by $\int_A f \mathrm{d}\mu = \int f 1_A \mathrm{d}\mu$.

Some basic convergence theorems hold, that is conditions on a sequence of functions $f_k: X \to \mathbb{R}$ with $\lim_{k \to \infty} f_k(x) = f(x)$ for almost all x that guarantee that

$$\lim_{k \to \infty} \int f_k \, \mathrm{d}\mu = \int f \mathrm{d}\mu. \tag{1.23}$$

This is the case if (f_k) is a monotonic sequence of non-negative functions (the *monotone* convergence theorem), or if $\mu(X) < \infty$ and for some c we have $|f_k(x)| \leq c$ for all k and $x \in X$ (the *bounded* convergence theorem). The limit (1.23) is also valid if there is a function $g : X \to \mathbb{R}^+ \cup \{0\}$ with $\int g \, \mathrm{d}\mu < \infty$ and $|f_k(x)| \leq g(x)$ for all k and x (the *dominated* convergence theorem).

Closely related is Fatou's lemma, that for any sequence (f_k) of measurable functions

$$\int \liminf_{k \to \infty} f_k \mathrm{d}\mu \le \liminf_{k \to \infty} \int f_k \mathrm{d}\mu.$$

We will often wish to interchange the order of integration in a double integral, and this is generally permitted by versions of Fubini's theorem. If μ and ν are locally finite measures on subsets of Euclidean space then

$$\int \left(\int f(x,y) \mathrm{d}\mu(x) \right) \mathrm{d}\nu(y) = \int \left(\int f(x,y) \mathrm{d}\nu(y) \right) \mathrm{d}\mu(x)$$

for continuous $f : X \times Y \to \mathbb{R}^+ \cup \{0\}$. (This also holds if f is a Borel function, that is if $\{(x,y) : f(x,y) < c\}$ is a Borel subset of $X \times Y$ for all real numbers c.)

As usual, integration is denoted in a variety of ways, such as $\int f \mathrm{d}\mu$, $\int f$ or $\int f(x) \mathrm{d}\mu(x)$ depending on the emphasis required. When μ is n-dimensional Lebesgue measure $\mathcal{L}^n$, we usually write $\int f \mathrm{d}x$ or $\int f(x) \mathrm{d}x$ in place of $\int f \mathrm{d}\mathcal{L}^n(x)$.

We write $L^1(\mu)$ for the vector space of μ-integrable functions, that is functions $f : X \to \mathbb{R}$ with $\int |f| \mathrm{d}\mu < \infty$, and $L^1(\mathbb{R})$ for the *Lebesgue integrable* functions, that is $f : \mathbb{R} \to \mathbb{R}$ with $\int |f| \mathrm{d}\mathcal{L} < \infty$.

*1.4 Weak convergence of measures

We collect together here some properties of weak convergence of measures that will be needed mainly in Chapter 9. This section may be deferred until the properties are needed. Alternatively, the proofs may be omitted on a first reading with little loss of feeling for the subject.

Let $\mu, \mu_1, \mu_2, \ldots$ be locally finite measures on $\mathbb{R}^n$. We say that the sequence $(\mu_k)_{k=1}^{\infty}$ *converges weakly* to μ if

$$\lim_{k \to \infty} \int f \mathrm{d}\mu_k = \int f \mathrm{d}\mu \tag{1.24}$$

for every $f \in C_0(\mathbb{R}^n)$ (i.e. for every continuous f of compact support), and we denote this by $\mu_k \to \mu$ or $\lim_{k \to \infty} \mu_k = \mu$.

For a simple example on $\mathbb{R}$, if $\mu_k(A) = \frac{1}{k} \{\#i \in \mathbb{Z} : i/k \in A\}$, so that μ_k is an aggregate of point masses of $1/k$, then $\mu_k \to \mathcal{L}^1$.

Although weak convergence does not imply that $\mu_k(A) \to \mu(A)$ for every set A, some useful inequalities hold for open or compact sets.

Lemma 1.8

Let $(\mu_k)_{k=1}^{\infty}$ be a sequence of locally finite measures on $\mathbb{R}^n$ with $\mu_k \to \mu$. Then if A is compact

$$\mu(A) \ge \limsup_{k \to \infty} \mu_k(A) \tag{1.25}$$

and if U is open

$$\mu(U) \leq \liminf_{k \to \infty} \mu_k(U). \tag{1.26}$$

Proof Writing $A_\delta^0 = \{x : \text{dist}(x, A) < \delta\}$ for the *open* δ-neighbourhood of a compact set A, we have that $A_\delta^0 \searrow A$ as $\delta \searrow 0$ so $\mu(A_\delta^0) \to \mu(A)$ by (1.13). Thus, given $\epsilon > 0$ we may take $\delta > 0$ such that $\mu(A_\delta^0) \leq \mu(A) + \epsilon$. Let $f \in C_0(\mathbb{R}^n)$ be any function satisfying $0 \leq f(x) \leq 1$ with $f(x) = 1$ for $x \in A$ and $f(x) = 0$ for $x \notin A_\delta^0$ ($f(x) = \max\{0, 1 - \delta^{-1} \text{dist}(x, A)\}$ will do). Then

$$\mu(A) + \epsilon \geq \mu(A_\delta^0) \geq \int f d\mu = \lim_{k \to \infty} \int f d\mu_k \geq \limsup_{k \to \infty} \mu_k(A);$$

since this is true for arbitrarily small ϵ, (1.25) follows. Inequality (1.26) is similar using (1.12). $\square$

The importance of weak convergence lies in the following compactness property which allows us to extract weakly convergent subsequences from general sequences of measures.

Proposition 1.9

Let $\mu_1, \mu_2, \ldots$ be locally finite measures on $\mathbb{R}^n$ with $\sup_k \mu_k(A) < \infty$ for all bounded sets A. Then $(\mu_k)_{k=1}^\infty$ has a weakly convergent subsequence.

*Proof We note that $C_0(\mathbb{R}^n)$ has a countable dense subset of functions $(f_k)_{k=1}^\infty$ under the norm $\|f\|_\infty = \max\{|f(x)| : x \in \mathbb{R}^n\}$. (For example, setting $g_m(x) = \max\{0, m - |x|\}$, the set of functions $\{pg_m : p$ is a polynomial with rational coefficients and $m \in \mathbb{Z}^+\}$ is easily seen to be countable and dense using the Weierstrass approximation theorem.) A diagonal argument, using induction on k, gives sequences $(\mu_{k,i})_{i=1}^\infty$ with $\mu_{0,i} = \mu_i$ and with $(\mu_{k,i})_{i=1}^\infty$ a subsequence of $(\mu_{k-1,i})_{i=1}^\infty$ for $k = 1, 2, \ldots$, such that $\int f_k d\mu_{k,i} \to a_k$ as $i \to \infty$ for some $a_k \in \mathbb{R}$. Thus, $\int f_k d\mu_{i,i} \to a_k$ as $i \to \infty$ for all k. Since the (f_k) are dense it follows that for all $f \in C_0(\mathbb{R}^n)$

$$\int f d\mu_{i,i} \to a(f) \tag{1.27}$$

for some $a(f) \in \mathbb{R}$. Moreover, a is linear, that is $a(f + g) = a(f) + a(g)$ and $a(\lambda f) = \lambda a(f)$, and bounded, that is $|a(f)| \leq (\sup_k \mu_k(A))\|f\|_\infty$ if spt $f \subset A$. The Riesz representation theorem states that under these conditions there is a locally finite measure μ such that $a(f) = \int f d\mu$ for all $f \in C_0(\mathbb{R}^n)$; thus by (1.27) $\mu_{i,i} \to \mu$, with $(\mu_{i,i})_{i=1}^\infty$ a subsequence of $(\mu_k)_{k=1}^\infty$. $\square$

It is sometimes convenient to express weak convergence of measures in terms of convergence with respect to metrics. For $R > 0$ we define d_R on the set of locally finite measures on $\mathbb{R}^n$ by

$$d_R(\mu, \nu) = \sup\left\{ \left| \int f \, d\mu - \int f \, d\nu \right| : f \in \text{Lip}_R \right\} \tag{1.28}$$

where here Lip_R denotes the set of Lipschitz functions $f : \mathbb{R}^n \to [0, \infty)$ with spt $f \subset B(0, R)$ and with $\text{Lip } f \leq 1$ (that is with $|f(x) - f(y)| \leq |x - y|$ for $x, y \in \mathbb{R}^n$). Then for each R we have that d_R is a pseudo-metric (that is, it is non-negative, symmetric and satisfies the triangle inequality). However, $d_R(\mu, \nu) = 0$ need not imply $\mu = \nu$. Nevertheless, d_R is a metric on the set of locally finite measures with support contained in the open ball $B^0(0, R)$, see Exercise 1.10. Clearly, if $R_1 \leq R_2$ then $d_{R_1}(\mu, \nu) \leq d_{R_2}(\mu, \nu)$.

Lemma 1.10

Let $\mu_1, \mu_2, \ldots$ and μ be measures on $\mathbb{R}^n$. Then $\mu_k \to \mu$ weakly if and only if $d_R(\mu_k, \mu) \to 0$ for all $R > 0$.

Proof Suppose first that $d_R(\mu_k, \mu) \not\to 0$ for some R. By passing to a subsequence and renumbering, there exists $\epsilon > 0$ and functions $f_k \in \text{Lip}_R$ such that $|\int f_k \, d\mu_k - \int f_k \, d\mu| \geq \epsilon$ for all k. By the Arzelà–Ascoli theorem there is a subsequence of (f_k), which by renumbering we may again take to be the whole sequence, such that $f_k \to f$ uniformly for some f which must be in Lip_R, using that the f_k vanish outside $B^0(0, R)$. Then

$$\int f \, d\mu_k - \int f \, d\mu = \left(\int f \, d\mu_k - \int f_k \, d\mu_k \right) + \left(\int f_k \, d\mu_k - \int f_k \, d\mu \right)$$
$$+ \left(\int f_k \, d\mu - \int f \, d\mu \right).$$

As $k \to \infty$, the first term of this sum tends to 0 (since $f_k \to f$ uniformly and $(\mu_k(B(0, R)))_{k=1}^{\infty}$ is bounded by (1.25)), the third term tends to 0 (as $f_k \to f$ uniformly), but the absolute value of the middle term is bounded below by ϵ, so $\mu_k \not\to \mu$ weakly.

For the converse suppose that $d_R(\mu_k, \mu) \to 0$ for all $R > 0$. Let $f : \mathbb{R}^n \to \mathbb{R}$ be continuous, with spt $f \subset B(0, R)$. Then given $\epsilon > 0$, there exists a Lipschitz $g : \mathbb{R}^n \to \mathbb{R}$ with spt $g \subset B(0, R)$ that approximates f in the sense that $|f(x) - g(x)| < \epsilon$ for all $x \in \mathbb{R}^n$. (Using the mean value theorem, g can be any sufficiently close approximation to f that is continuously differentiable.) Then using the triangle inequality and that g has Lipschitz positive and negative parts

$$\left| \int f \, d\mu_k - \int f \, d\mu \right| \leq \int |f - g| \, d\mu_k + \left| \int g \, d\mu_k - \int g \, d\mu \right| + \int |g - f| \, d\mu$$
$$\leq \epsilon \mu_k(B(0, R)) + 2(\text{Lip } g) d_R(\mu_k, \mu) + \epsilon \mu(B(0, R))$$
$$\leq 3\epsilon \mu(B(0, R)) + 2(\text{Lip } g)\epsilon$$

if k is sufficiently large, using (1.25). Hence $\mu_k \to \mu$ weakly. $\quad\square$

The other property we need is that the d_R are separable, that is there exists a countable set of measures that is dense in these pseudo-metrics.

Lemma 1.11

There is a countable set of locally finite measures $\mu_1, \mu_2, \ldots$ on $\mathbb{R}^n$ such that, for every locally finite measure μ on $\mathbb{R}^n$ and all $R, \eta > 0$, there exists k such that $d_R(\mu, \mu_k) < \eta$.

*Proof It is enough to show that such a set of measures exists for each $R = 1, 2, \ldots$ and use that a countable union of countable sets is countable. For $j = 1, 2, \ldots$ let $C_{j,1}, \ldots C_{j,m_j}$ be the (half-open) binary cubes of side-lengths 2^{-j} which meet $B(0, R)$, and let $\delta_{j,i}$ be a unit point mass at the centre of $C_{j,i}$. Let $(\mu_k)_{k=1}^\infty$ be an enumeration of the set of all measures of the form $\sum_{i=1}^{m_j} q_{j,i,p} \delta_{j,i}$, where for each j and i the sequence $(q_{j,i,1}, q_{j,i,2}, \ldots)$ is an enumeration of the non-negative rational numbers.

Given μ, let $\nu_j = \sum_{i=1}^{m_j} \mu(C_{j,i}) \delta_{j,i}$; by choosing j large enough we may ensure that $d_R(\mu, \nu_j) < \frac{1}{2}\eta$. Now choose $\mu_k = \sum_{i=1}^{m_j} q_{j,i,p} \delta_{j,i}$ so that $d_R(\nu_j, \mu_k) < \frac{1}{2}\eta$, which may be achieved by taking $q_{j,i,p}$ sufficiently close to $\mu(C_{j,i})$ for each i. Then $d_R(\mu, \mu_k) < \eta$, as required. $\square$

1.5 Notes and references

Most of the material in this chapter may be found in much more detail in any basic text on measure theory, for example Doob (1994) or Kingman and Taylor (1966). The treatments in Falconer (1985) and Mattila (1995a) are specifically directed towards fractal geometry, and also include more details of Vitali covering results and density properties.

Exercises

1.1 Let $f : [0, 1] \to [0, 1]$ be a differentiable function. Show that the limit $\lim_{k \to \infty} b_k^{1/k}$ exists, where $b_k = \sup_{0 \le x \le 1} |\frac{d}{dx} f^k(x)|$ and f^k is the kth iterate of f. (Hint: use the chain rule to show that (b_k) is submultiplicative.)

1.2 Use (1.4) to prove the arithmetic-geometric mean inequality: that $(\prod_{i=1}^m x_i)^{1/m} \le \frac{1}{m} \sum_{i=1}^m x_i$ where $x_1, \ldots, x_m > 0$. (Hint: $-\log x$ is convex.)

1.3 Let $\mathbb{Q} = (q_1, q_2, \ldots)$ be an enumeration of the rational numbers. For $A \subset \mathbb{R}$ define $\mu(A) = \sum_{q_i \in A} 2^{-i}$. Verify that μ is a measure with all subsets of $\mathbb{R}$ measurable. Show that $\mathrm{spt}\mu = \mathbb{R}$, even though $\mu(\mathbb{R} \backslash \mathbb{Q}) = 0$, that is μ is 'concentrated' on $\mathbb{Q}$.

1.4 Let μ be a measure on $\mathbb{R}^n$ such that for all $x \in \mathbb{R}^n$ there is a ball $B(x, r)$ with $\mu(B(x, r)) < \infty$. Show that μ is locally finite.

1.5 Verify (1.15) and (1.16) from the definition of a locally finite (Borel regular) measure and Proposition 1.6.

1.6   For each k, let $\{X_{i_1,\dots,i_k} : i_j = 1 \text{ or } 2\}$ be the set of 2^k k-th level intervals of length 3^{-k} that occur in the usual construction of the middle-third Cantor set E, nested in the usual way. Verify that setting $\mu(X_{i_1,\dots,i_k}) = 2^{-k}$ and using (1.20) leads to a measure on E. Show that the same is true setting $\mu(X_{i_1,\dots,i_k}) = (1/3)^{n_1}(2/3)^{n_2}$ where n_1 and n_2 are the number of occurrences of the digits 1 and 2 respectively in $(i_1,\dots,i_k)$.

1.7   Let $f : [0,1] \to \mathbb{R}^+$ be continuous, and define μ on $[0,1]$ by $\mu(A) = \int_A f(x)\mathrm{d}x$. Show that μ is equivalent to $\mathcal{L}$, where $\mathcal{L}$ is Lebesgue measure. (Note that $0 < c_1 \leq f(x) \leq c_2$ for all $x \in [0,1]$ for some c_1, c_2.)

1.8   Let μ_k be the measure on $\mathbb{R}$ that assigns unit mass to the point $1 + 1/k$. Find the weak limit μ of the sequence of measures (μ_k). Does $\mu_k([0,1]) \to \mu([0,1])$?

1.9   Show that if $\mu_k \to \mu$ and if A is a bounded set with $\mu(\partial A) = 0$, where ∂A is the boundary of A, then $\mu_k(A) \to \mu(A)$.

1.10 Verify that d_R defined by (1.28) is a pseudo-metric on the locally finite measures and a metric on the measures with support in $B^0(0,R)$.

Chapter 2 Review of fractal geometry

In this chapter we review some of the basic ideas of fractal geometry that will crop up frequently throughout this book. We first discuss fractal dimensions, and in particular the definitions and properties of Hausdorff, packing and box-counting dimensions. We then review iterated function systems which provide a convenient way of representing many fractals and fractal measures.

These basic definitions, properties and notation are collected together here for convenient reference; almost all of this material is discussed in much more detail and with full proofs in FG.

2.1 Review of dimensions

The notion of 'fractal dimension' of a set is central to nearly all fractal mathematics. We shall usually be interested in the dimensions of subsets of $\mathbb{R}^n$, though essentially the same definitions hold in general metric spaces.

Most definitions of dimension depend on a 'measurement at scale r' of a set E, which quantifies the irregularity of the set when viewed at that scale. The dimension is then usually defined in terms of the power law behaviour of these measurements as $r \searrow 0$.

We shall mainly be concerned with the Hausdorff, packing and box-counting dimensions of sets; these are by far the most common definitions of dimension in use, although a variety of other definitions have been proposed. It should be emphasised that the value of the dimension of a set may vary according to the definition used, although the usual definitions often give the same values for 'reasonably regular' sets. Thus it is important to be clear about the definition of dimension in use in any particular context.

Box-counting dimension

Box-counting dimension (also variously termed entropy dimension, capacity dimension, logarithmic density, etc.) is conceptually the simplest dimension in use, see FG, Section 3.1. For E a non-empty bounded subset of $\mathbb{R}^n$ let $N_r(E)$ be the smallest number of sets of diameter r that can cover E. (Recall that the *diameter* of a set U is defined as $|U| = \sup\{|x - y| : x, y \in U\}$, that is the greatest distance apart of any pair of points in U.) The *lower* and *upper box-counting* (or *box*) *dimensions* of E are defined as

$$\underline{\dim}_B E = \liminf_{r \to 0} \frac{\log N_r(E)}{-\log r} \tag{2.1}$$

and

$$\overline{\dim}_B E = \limsup_{r \to 0} \frac{\log N_r(E)}{-\log r} \tag{2.2}$$

respectively. If these are equal we refer to the common value as the *box-counting dimension* or *box dimension* of E

$$\dim_B E = \lim_{r \to 0} \frac{\log N_r(E)}{-\log r}. \tag{2.3}$$

Thus the least number of sets of diameter r which can cover E is roughly of order r^{-s} where $s = \dim_B E$.

There are a number of equivalent forms of these definitions which are often useful. The value of the limits (2.1)–(2.3) remain unaltered if $N_r(E)$ is taken to be any of the following:

   (i) the smallest number of sets of diameter r that cover E,
   (ii) the smallest number of closed balls of radius r that can cover E,
   (iii) the smallest number of cubes of side r that cover E,
   (iv) the largest number of disjoint balls of radius r with centres in E,
   (v) the number of r-mesh cubes that intersect E, hence the name 'box-counting'.

(An *r-mesh cube* is a cube of the form $[m_1 r, (m_1 + 1)r) \times \cdots \times [m_n r, (m_n + 1))$ where $m_1, \ldots, m_n$ are integers).

The equivalence of these forms of the definition follows on comparing the values of $N_r(E)$ in each case, see FG, Equivalent definitions 3.1.

An equivalent definition of box dimension of a rather different nature involves the n-dimensional volume of the *r-neighbourhood* or *r-parallel body* E_r of E, given by

$$E_r = \{x \in \mathbb{R}^n : |x - y| \le r \text{ for some } y \in E\}.$$

Then for $E \subset \mathbb{R}^n$

$$\underline{\dim}_B E = n - \limsup_{r \to 0} \frac{\log \mathcal{L}^n(E_r)}{\log r} \tag{2.4}$$

$$\overline{\dim}_B E = n - \liminf_{r \to 0} \frac{\log \mathcal{L}^n(E_r)}{\log r} \tag{2.5}$$

and

$$\dim_B E = n - \lim_{r \to 0} \frac{\log \mathcal{L}^n(E_r)}{\log r} \tag{2.6}$$

if this limit exists, where $\mathcal{L}^n$ is n-dimensional volume or n-dimensional Lebesgue measure. In the context of this definition, box dimension is sometimes referred to as *Minkowski dimension*.

Hausdorff and packing dimensions

Hausdorff and packing dimensions are more sophisticated than box dimensions, being defined in terms of measures. A finite or countable collection of subsets $\{U_i\}$ of $\mathbb{R}^n$ is called a δ-*cover* of a set $E \subset \mathbb{R}^n$ if $|U_i| \leq \delta$ for all i and $E \subset \cup_{i=1}^{\infty} U_i$. Let E be a subset of $\mathbb{R}^n$ and $s \geq 0$. For all $\delta > 0$ we define

$$\mathcal{H}_{\delta}^s(E) = \inf\left\{\sum_{i=1}^{\infty} |U_i|^s : \{U_i\} \text{ is a } \delta\text{-cover of } E\right\}. \tag{2.7}$$

As δ increases, the class of δ-covers of E is reduced, so this infimum increases and approaches a limit as $\delta \searrow 0$. Thus we define

$$\mathcal{H}^s(E) = \lim_{\delta \to 0} \mathcal{H}_{\delta}^s(E). \tag{2.8}$$

This limit exists, perhaps as 0 or ∞, for all $E \subset \mathbb{R}^n$. We term $\mathcal{H}^s(E)$ the *s-dimensional Hausdorff measure* of E.

It may be shown that $\mathcal{H}^s(E)$ is a Borel regular measure on $\mathbb{R}^n$ (see (1.14)), so in particular

$$\mathcal{H}^s\left(\bigcup_{i=1}^{\infty} E_i\right) \leq \sum_{i=1}^{\infty} \mathcal{H}^s(E_i) \tag{2.9}$$

for all sets $E_1, E_2, \ldots$, with equality if the E_i are disjoint Borel sets.

Hausdorff measures generalise Lebesgue measures, so that $\mathcal{H}^1(E)$ gives the 'length' of a set or curve E, and $\mathcal{H}^2(E)$ gives the (normalised) 'area' of a region or surface, etc. In general, $\mathcal{L}^n = 2^{-n} v_n \mathcal{H}^n$, where v_n is the volume of the n-dimensional unit ball.

We often wish to consider the Hausdorff measure of the image of a set under a Lipschitz mapping. For a Lipschitz $f : E \to \mathbb{R}^n$ such that

$$|f(x) - f(y)| \leq c|x - y| \quad \text{for all} \quad x, y \in E, \tag{2.10}$$

we have

$$\mathcal{H}^s(f(E)) \leq c^s \mathcal{H}^s(E). \tag{2.11}$$

Similarily, if $f : E \to \mathbb{R}^m$ is *bi-Lipschitz*, so that for some $c_1, c_2 > 0$

$$c_1|x - y| \leq |f(x) - f(y)| \leq c_2|x - y| \text{ for all } x, y \in E,$$

then

$$c_1^s \mathcal{H}^s(E) \leq \mathcal{H}^s(f(E)) \leq c_2^s \mathcal{H}^s(E). \tag{2.12}$$

A special case of this is when f is a *similarity transformation* of ratio r, so $|f(x) - f(y)| = r|x - y|$ for all $x, y \in E$, in which case

$$\mathcal{H}^s(f(E)) = r^s \mathcal{H}^s(E). \tag{2.13}$$

This is the *scaling property* of Hausdorff measures, which generalises the familiar scaling properties of length, area, volume, etc.

It is easy to show from (2.7) and (2.8) that for all sets $E \subset \mathbb{R}^n$ there is a number $\dim_H E$, called the *Hausdorff dimension* of E, such that $\mathcal{H}^s(E) = \infty$ if $s < \dim_H E$ and $\mathcal{H}^s(E) = 0$ if $s > \dim_H E$. Thus

$$\dim_H E = \inf\{s : \mathcal{H}^s(E) = 0\} = \sup\{s : \mathcal{H}^s(E) = \infty\},$$

so that the Hausdorff dimension of a set E may be thought of as the number s at which $\mathcal{H}^s(E)$ 'jumps' from ∞ to 0, see Figure 2.1. When $s = \dim_H E$ the measure $\mathcal{H}^s(E)$ can be zero or infinite, but in the nicest situation (which occurs in many familiar examples) $0 < \mathcal{H}^s(E) < \infty$, in which case E is sometimes termed an *s-set*.

The definition of packing dimension parallels that of Hausdorff dimension. Here we define a δ-*packing* of $E \subset \mathbb{R}^n$ to be a finite or countable collection of *disjoint* balls $\{B_i\}$ of radii at most δ and with centres in E. For $\delta > 0$ we define

$$\mathcal{P}_\delta^s(E) = \sup\left\{ \sum_{i=1}^{\infty} |B_i|^s : \{B_i\} \text{ is a } \delta\text{-packing of } E \right\}.$$

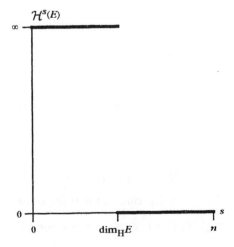

Figure 2.1   The Hausdorff dimension of E is the number s at which $\mathcal{H}^s(E)$ jumps from ∞ to 0

Then $\mathcal{P}_\delta^s(E)$ decreases as δ increases, so we may take the limit

$$\mathcal{P}_0^s(E) = \lim_{\delta \to 0} \mathcal{P}_\delta^s(E).$$

Unfortunately $\mathcal{P}_0^s$ is not a measure (it need not be countable subadditive); to overcome this difficulty we define

$$\mathcal{P}^s(E) = \inf\left\{ \sum_{i=1}^{\infty} \mathcal{P}_0^s(E_i) : E \subset \bigcup_{i=1}^{\infty} E_i \right\}$$

which *is* a Borel measure on $\mathbb{R}^n$, called the *s-dimensional packing measure* of E. Again $\mathcal{P}^1, \mathcal{P}^2$, give the length, area, etc. of smooth sets, but for fractals $\mathcal{H}^s$ and $\mathcal{P}^s$ can be very different measures.

Packing measures behave in the same way as Hausdorff measures in respect of Lipschitz mappings, and $(2.11)-(2.13)$ remain valid with $\mathcal{H}^s$ replaced by $\mathcal{P}^s$. It may be shown that $\mathcal{H}^s(A) \leq \mathcal{P}^s(A)$ for all sets A.

As for Hausdorff dimension, there is a number $\dim_P E$, called the *packing dimension* of E, such that $\mathcal{P}^s(E) = \infty$ for $s < \dim_P E$ and $\mathcal{P}^s(E) = 0$ for $s > \dim_P E$. Thus

$$\dim_P E = \inf\{s : \mathcal{P}^s(E) = 0\} = \sup\{s : \mathcal{P}^s(E) = \infty\}.$$

It is sometimes convenient to express packing dimension in terms of upper box dimension. For $E \subset \mathbb{R}^n$ it is the case that

$$\dim_P E = \inf\left\{ \sup_i \overline{\dim}_B E_i : E \subset \bigcup_{i=1}^{\infty} E_i \right\}$$

(the infimum is over all countable covers $\{E_i\}$ of E), see FG, Proposition 3.6.

Basic properties of dimensions

We will have frequent recourse to certain basic properties of dimensions. The following properties hold with 'dim' denoting any of Hausdorff, packing, lower box or upper box dimension.

Monotonicity. If $E_1 \subset E_2$ then $\dim E_1 \leq \dim E_2$.
Finite sets. If E is finite then $\dim E = 0$.
Open sets. If E is a (non-empty) open subset of $\mathbb{R}^n$ then $\dim E = n$.
Smooth manifolds. If E is a smooth m-dimensional manifold in $\mathbb{R}^n$ then $\dim E = m$.
Lipschitz mappings. If $f : E \to \mathbb{R}^m$ is Lipschitz then $\dim f(E) \leq \dim E$. (For Hausdorff and packing dimensions this follows from (2.11) and its packing measure analogue, for box dimensions it may be deduced from the definitions.) Note in particular that this is true if $E \subset X$ for an open set X

on which $f: X \to \mathbb{R}^n$ is differentiable with bounded derivative, using the mean value theorem.

Bi-Lipschitz invariance. If $f: E \to f(E)$ is bi-Lipschitz then $\dim f(E) = \dim E$.

Geometric invariance. If f is a similarity or affine transformation then $\dim f(E) = \dim E$ (this is a special case of bi-Lipschitz invariance).

Hausdorff, packing and upper box dimensions are *finitely stable*, that is $\dim \cup_{i=1}^{k} E_i = \max_{1 \le i \le k} \dim E_i$ for any finite collection of sets $\{E_1, \ldots, E_k\}$. However lower box dimension is not finitely stable.

Hausdorff and packing dimensions are *countably stable*, that is $\dim \cup_{i=1}^{\infty} E_i = \sup_{1 \le i < \infty} \dim E_i$. (This may be deduced from the countable subadditivity of Hausdorff and packing measures). Countable stability is one of the main advantages of these dimensions over box dimensions; in particular it implies that *countable sets* have Hausdorff and packing dimensions zero.

We also recall that $\underline{\dim}_B E = \underline{\dim}_B \bar{E}$ and $\overline{\dim}_B E = \overline{\dim}_B \bar{E}$ where $\bar{E}$ is the closure of E. In fact this is a disadvantage of box dimensions, since we often wish to study a fractal E that is dense in an open region of $\mathbb{R}^n$ and which therefore has full box dimension n.

There are some basic inequalities between these dimensions. For any non-empty set E

$$\dim_H E \le \dim_P E \le \overline{\dim}_B E \quad \text{and} \quad \dim_H E \le \underline{\dim}_B E \le \overline{\dim}_B E \qquad (2.14)$$

(for the inequalities involving box dimensions we assume E is non-empty and bounded). In practice, most definitions of dimension take values between the Hausdorff and upper box dimensions, so if it can be shown that $\dim_H E = \overline{\dim}_B E$ then all the normal definitions of dimension take this common value.

Calculating dimensions

We frequently wish to estimate the dimensions of sets; usually it is harder to get lower estimates than upper estimates. There are various approaches to finding the dimension of a set, but most methods involve studying a suitable measure supported by the set (other methods can usually be reduced to this process). One basic but very useful technique is termed the 'mass distribution principle'.

Proposition 2.1   (mass distribution principle)

Let $E \subset \mathbb{R}^n$ and let μ be a finite measure with $\mu(E) > 0$. Suppose that there are numbers $s \ge 0$, $c > 0$ and $\delta_0 > 0$ such that

$$\mu(U) \le c|U|^s$$

for all sets U with $|U| \leq \delta_0$. Then $\mathcal{H}^s(E) \geq \mu(E)/c$ and

$$s \leq \dim_H E \leq \underline{\dim}_B E \leq \overline{\dim}_B E.$$

Proof We recall the very simple proof from FG, Principle 4.2. If $\{U_i\}$ is any cover of E by sets of diameter at most δ_0 then

$$\mu(E) \leq \mu(\cup_i U_i) \leq \sum_i \mu(U_i) \leq c \sum_i |U_i|^s,$$

so that $\mu(E) \leq c\, \mathcal{H}^s_\delta(E)$ for $\delta \leq \delta_0$. The result follows on letting $\delta \to 0$. $\square$

Developing this idea we can estimate the Hausdorff and packing measures and dimensions of a set E if we can find a measure μ satisfying certain 'local density' conditions on E. Observe the (near) symmetry between Hausdorff and packing measures and lower and upper estimates in the following propositions.

Proposition 2.2

Let $E \subset \mathbb{R}^n$ be a Borel set, let μ be a finite Borel measure on $\mathbb{R}^n$ and $0 < c < \infty$.

(a) If $\limsup_{r \to 0} \mu(B(x,r))/r^s \leq c$ for all $x \in E$ then $\mathcal{H}^s(E) \geq \mu(E)/c$.
(b) If $\limsup_{r \to 0} \mu(B(x,r))/r^s \geq c$ for all $x \in E$ then $\mathcal{H}^s(E) \leq 2^s\mu(E)/c$.
(c) If $\liminf_{r \to 0} \mu(B(x,r))/r^s \leq c$ for all $x \in E$ then $\mathcal{P}^s(E) \geq 2^s\mu(E)/c$.
(d) If $\liminf_{r \to 0} \mu(B(x,r))/r^s \geq c$ for all $x \in E$ then $\mathcal{P}^s(E) \leq 2^s\mu(E)/c$.

Proof Parts (a) and (b) are proved in FG, Proposition 4.9; part (a) requires little more than the definition of Hausdorff measure, whilst (b) requires the Vitali covering lemma. The packing measure analogues are proved in a very similar way, with (d) following easily from the definition of packing measure and with (c) requiring a covering lemma. $\square$

We shall more often be interested in the dimension rather than the measure of sets, so we give a version of Proposition 2.2 for dimensions. This may be conveniently expressed in terms of local dimensions of measures. We define the *lower* and *upper local dimension s* of μ at $x \in \mathbb{R}^n$ (also called the *pointwise dimension* or *Hölder exponent*) by

$$\underline{\dim}_{\,\mathrm{loc}}\,\mu(x) = \liminf_{r \to 0} \frac{\log \mu(B(x,r))}{\log r} \tag{2.15}$$

$$\overline{\dim}_{\,\mathrm{loc}}\,\mu(x) = \limsup_{r \to 0} \frac{\log \mu(B(x,r))}{\log r}. \tag{2.16}$$

These local dimensions express the power law behaviour of $\mu(B(x,r))$ for small r. Note that $\underline{\dim}_{\,\mathrm{loc}}\,\mu(x) = \overline{\dim}_{\,\mathrm{loc}}\,\mu(x) = \infty$ if $\mu(B(x,r)) = 0$ for some $r > 0$.

Proposition 2.3

Let $E \subset \mathbb{R}^n$ be a Borel set and let μ be a finite measure.

(a) If $\underline{\dim}_{\mathrm{loc}}\mu(x) \geq s$ for all $x \in E$ and $\mu(E) > 0$ then $\dim_H E \geq s$.
(b) If $\underline{\dim}_{\mathrm{loc}}\mu(x) \leq s$ for all $x \in E$ then $\dim_H E \leq s$.
(c) If $\overline{\dim}_{\mathrm{loc}}\mu(x) \geq s$ for all $x \in E$ and $\mu(E) > 0$ then $\dim_P E \geq s$.
(d) If $\overline{\dim}_{\mathrm{loc}}\mu(x) \leq s$ for all $x \in E$ then $\dim_P E \leq s$.

Proof This follows from Proposition 2.2 noting that the hypothesis in (a) implies that $\limsup_{r \to 0}\mu(B(x,r))/r^{s-\epsilon} = 0$ for all $\epsilon > 0$, and similarly for (b), (c) and (d). $\square$

We remark that in (a) and (c) of Proposition 2.3 it is enough for the hypothesis to hold for x in a subset of E of positive μ-measure.
We record the following partial converses to these results which stipulate that a set of given dimension carries a measure with corresponding local dimensions.

Proposition 2.4

Let $E \subset \mathbb{R}^n$ be a non-empty Borel set.

(a) If $\dim_H E > s$ there exists μ with $0 < \mu(E) < \infty$ and $\underline{\dim}_{\mathrm{loc}}\mu(x) \geq s$ for all $x \in E$.

(b) If $\dim_H E < s$ there exists μ with $0 < \mu(\bar{E}) < \infty$ and $\underline{\dim}_{\mathrm{loc}}\mu(x) \leq s$ for all $x \in E$.

(c) If $\dim_P E > s$ there exists μ with $0 < \mu(E) < \infty$ and $\overline{\dim}_{\mathrm{loc}}\mu(x) \geq s$ for μ-almost all x.

(d) If $\dim_P E < s$ there exists μ with $0 < \mu(\bar{E}) < \infty$ and $\overline{\dim}_{\mathrm{loc}}\mu(x) \leq s$ for all $x \in E$.

Proof Part (a) is FG, Corollary 4.12 ('Frostman's Lemma') expressed in local dimension form. Parts (b), (c) and (d) require similar delicate arguments, and the technical details may be found in the literature. $\square$

Note that, in Propositions 2.3 and 2.4, it is the lower local dimensions that relate to the Hausdorff dimension of a set, and the upper local dimensions to the packing dimension of a set.
Propositions 2.3(a) and 2.4(a) may be 'integrated' to get potential theoretic criteria which are often useful when calculating Hausdorff dimensions and measures. For $s \geq 0$ we define the *s-energy* of a measure μ on $\mathbb{R}^n$ by

$$I_s(\mu) = \int \int |x - y|^{-s} \mathrm{d}\mu(x)\mathrm{d}\mu(y).$$

Proposition 2.5

Let $E \subset \mathbb{R}^n$.

(a) If there is a finite measure μ on E with $I_s(\mu) < \infty$ then $\mathcal{H}^s(E) = \infty$ and $\dim_H E \geq s$.

(b) If E is a Borel set with $\mathcal{H}^s(E) > 0$ then there exists a finite measure μ on E with $I_t(\mu) < \infty$ for all $t < s$.

Proof   See FG, Theroem 4.13.   □

Densities and rectifiability

Densities have played a major rôle in the development of geometric measure theory. Although it is possible to define the densities of any finite measure, we restrict attention here to densities of an *s*-set, that is a Borel set $E \subset \mathbb{R}^n$ with $0 < \mathcal{H}^s(E) < \infty$, where $s = \dim_H E$. The *lower* and *upper* (*s-dimensional*) *densities* of E at x are given by

$$\underline{D}^s(x) = \underline{D}^s(E, x) = \liminf_{r \to 0} \mathcal{H}^s(E \cap B(x, r))/(2r)^s \qquad (2.17)$$

and

$$\overline{D}^s(x) = \overline{D}^s(E, x) = \limsup_{r \to 0} \mathcal{H}^s(E \cap B(x, r))/(2r)^s \qquad (2.18)$$

(we write $\underline{D}^s(x)$ rather than $\underline{D}^s(E, x)$ when the set E under consideration is clear). When $\underline{D}^s(E, x) = \overline{D}^s(E, x)$ we say that the *density* $D^s(E, x)$ of E at x exists and equals this common value. (It is possible to define the densities of a more general measure μ by replacing $\mathcal{H}^s$ by μ in (2.17) and (2.18).)

We remark that if X is an open subset of $\mathbb{R}$ and $f : X \to \mathbb{R}$ is a C^1 mapping, then for $E \subset \mathbb{R}$ we have

$$\underline{D}^s(E, x) = \underline{D}^s(f(E), f(x)) \quad \text{and} \quad \overline{D}^s(E, x) = \overline{D}^s(f(E), f(x)) \qquad (2.19)$$

for all x where $f'(x) \neq 0$. Intuitivily this is because f may be regarded as a local similarity of ratio $|f'(x)|$ near x, scaling the diameter of $B(x, r)$ by a factor $|f'(x)|$ and the $\mathcal{H}^s$-measure of $B(x, r)$ by a factor $|f'(x)|^s$, using the scaling property (2.13), see Exercise 2.6. In fact (2.19) also holds for a differentiable conformal mapping $f : X \to \mathbb{R}^n$ where $X \subset \mathbb{R}^n$. (The mapping f is *conformal* if $f'(x)$, regarded as a linear transformation on $\mathbb{R}^n$, is a similarity transformation for all $x \in \mathbb{R}^n$.)

The classical result on densities is the Lebesgue density theorem. This states that if E is a Lebesgue measurable subset of $\mathbb{R}^n$ then for $\mathcal{L}^n$-almost all x

$$\lim_{r \to 0} \mathcal{L}^n(E \cap B(x, r))/\mathcal{L}^n(B(x, r)) = \begin{cases} 1 & \text{if} \quad x \in E \\ 0 & \text{if} \quad x \notin E \end{cases} ; \qquad (2.20)$$

this is a special case of Proposition 1.7 taking μ as $\mathcal{L}^n$. Expressed in terms of densities this is

$$D^n(E,x) = \lim_{r \to 0} \mathcal{H}^n(E \cap B(x,r))/(2r)^n = \begin{cases} 1 & \text{if} \quad x \in E \\ 0 & \text{if} \quad x \notin E \end{cases} \tag{2.21}$$

for $\mathcal{H}^n$-almost all x, since $\mathcal{L}^n = 2^{-n} v_n \mathcal{H}^n$, where v_n is the volume of the n-dimensional unit ball. It is natural to ask to what extent analogues of (2.21) hold for general values of s.

For $E \subset \mathbb{R}^n$ an s-set, it is easy to show that $D^s(E,x) = 0$ for $\mathcal{H}^s$-almost all $x \notin E$. Moreover, $2^{-s} \leq \overline{D}^s(E,x) \leq 1$ for almost all $x \in E$, see FG, Proposition 5.1. A rather deeper property is that unless s is an integer $\underline{D}^s(E,x) < \overline{D}^s(E,x)$ for $\mathcal{H}^s$-almost all $x \in E$. We will prove this in Corollary 9.8 using tangent measures; a special case is given in FG, Proposition 5.3. In some ways, this non-existence of densities is a reflection of fractality.

The situation when s is an integer is more involved. In this case an s-set $E \subset \mathbb{R}^n$ may be decomposed as $E = E_R \cup E_I$, where E_R is *regular*, that is, with $\underline{D}^s(E_R,x) = \overline{D}^s(E_R,x) = 1$ for $\mathcal{H}^s$-almost all $x \in E_R$, and E_I is *irregular*, with $\underline{D}^s(E_I,x) \leq c\overline{D}^s(E_I,x)$ for $\mathcal{H}^s$-almost all $x \in E_I$, where $c < 1$ depends only on n and s, see Figure 2.2.

Regular and irregular sets have geometrical characterisations in terms of rectifiability. If E is regular then E is *rectifiable*, which means that almost all of E may be covered by a countable collection of Lipschitz pieces $E_1, E_2, \ldots$ such that $\mathcal{H}^s(E \setminus \bigcup_{i=1}^{\infty} E_i) = 0$. (A set E_0 is a *Lipschitz piece* if $E_0 = f(A)$ where

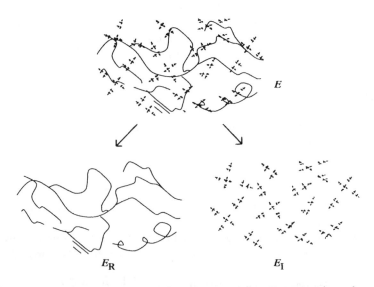

E

E_R E_I

Figure 2.2   Decomposition of a 1-set E into a regular part E_R and an irregular part E_I

$A \subset \mathbb{R}^s$ and $f : A \to \mathbb{R}^n$ is a Lipschitz mapping.) Thus a rectifiable set is built up from a countable number of Lipschitz pieces which look like subsets of curves, surfaces or classical s-dimensional sets.

On the other hand, if E is irregular then E is *totally unrectifiable*. This means that $\mathcal{H}^s(E \cap E_0) = 0$ for every Lipschitz piece E_0, so that E has negligible intersection with rectifiable s-dimensional sets. Thus, irregular s-sets might be considered to be fractal and the regular, or rectifiable, sets non-fractal, with non-existence of the density being a characteristic fractal property for sets of integer dimension. The relationship between the metric or density properties of s-sets and the geometric or rectifiability structure of sets is difficult, and has been central in the development of geometric measure theory; see FG, Chapter 5 for more details.

2.2 Review of iterated function systems

Iterated function systems (or schemes) provide a very convenient way of representing and reconstructing many fractals that in some way are made up of small images of themselves. We work in a non-empty closed subset X of $\mathbb{R}^n$, often $X = \mathbb{R}^n$. An *iterated function system* (IFS) consists of a family of contractions $\{F_1, \ldots, F_m\}$ on X, where $m \geq 2$. Thus for $i = 1, \ldots, m$ we have $F_i : X \to X$ and

$$|F_i(x) - F_i(y)| \leq r_i|x - y| \quad \text{for all } x, y \in X, \tag{2.22}$$

where $r_i < 1$. We write

$$r_{\max} = \max_{1 \leq i \leq m} r_i \tag{2.23}$$

so $r_{\max} < 1$.

The fundamental property of an IFS is that it determines a unique non-empty compact set E satisfying $E = \cup_{i=1}^m F_i(E)$; these sets are frequently fractals. For example, with $F_1, F_2 : \mathbb{R} \to \mathbb{R}$ given by

$$F_1(x) = \tfrac{1}{3}x \quad \text{and} \quad F_2(x) = \tfrac{1}{3}x + \tfrac{2}{3}, \tag{2.24}$$

the set satisfying $E = F_1(E) \cup F_2(E)$ is the middle-third Cantor set, see Figure 2.3.

To establish this fundamental property, we work with the class $\mathcal{S}$ of non-empty compact subsets of X. We may define a metric or distance d on $\mathcal{S}$ by

$$d(A, B) = \inf\{\delta : A \subset B_\delta \quad \text{and} \quad B \subset A_\delta\} \tag{2.25}$$

where A_δ is the δ-neighbourhood of A. Thus d satisfies the three requirements for a metric ((i) $d(A, B) \geq 0$ with equality if and only if $A = B$, (ii) $d(A, B) = d(B, A)$, (iii) $d(A, B) \leq d(A, C) + d(C, B)$ for all A, B, C), and is termed the *Hausdorff metric* on $\mathcal{S}$.

It may be shown that d is a *complete* metric on $\mathcal{S}$, that is every Cauchy sequence of sets in $\mathcal{S}$ is convergent to a set in $\mathcal{S}$. We use this fact to give the slick

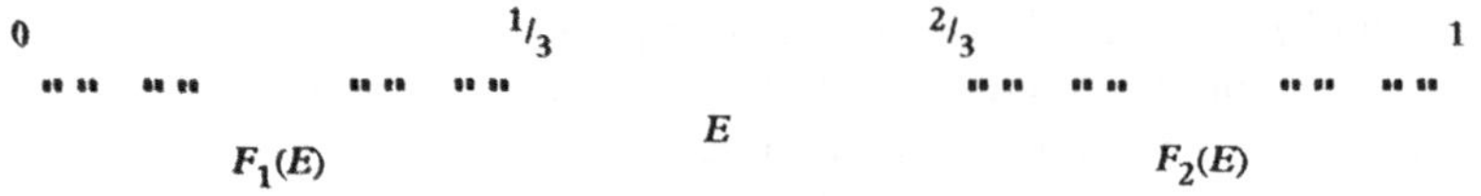

Figure 2.3 The middle-third Cantor set is made up of two scale $\frac{1}{3}$ copies of itself; thus $E = F_1(E) \cup F_2(E)$ where F_1 and F_2 are given by (2.24)

'contraction mapping theorem' proof of the fundamental property of IFSs; an alternative proof is given in FG, Theorem 9.1.

Theorem 2.6

Let $\{F_1, \ldots, F_m\}$ be an IFS on $X \subset \mathbb{R}^n$. Then there exists a unique, non-empty compact set $E \subset X$ that satisfies

$$E = \bigcup_{i=1}^{m} F_i(E).$$ (2.26)

Moreover, defining a transformation $F : S \to S$ by

$$F(A) = \bigcup_{i=1}^{m} F_i(A)$$ (2.27)

for $A \in S$, we have that for all $A \in S$

$$F^k(A) \to E$$

in the metric d as $k \to \infty$, where F^k is the k-th iterate of F. Furthermore, if $A \in S$ is such that $F_i(A) \subset A$ for all i, then

$$E = \bigcap_{k=0}^{\infty} F^k(A).$$ (2.28)

Proof   If $A, B \in S$ then

$$d(F(A), F(B)) = d\left(\bigcup_{i=1}^{m} F_i(A), \bigcup_{i=1}^{m} F_i(B) \right)$$

$$\leq \max_{1 \leq i \leq m} d(F_i(A), F_i(B)),$$

using the definition of the metric d and noting that if the δ-neighbourhood $(F_i(A))_\delta$ contains $F_i(B)$ for all i then $\left(\cup_{i=1}^{m} F_i(A) \right)_\delta$ contains $\cup_{i=1}^{m} F_i(B)$ and vice versa. By (2.22)

$$d(F(A), F(B)) \leq \left(\max_{1 \leq i \leq m} r_i \right) d(A, B).$$ (2.29)

Since $\max_{1\leq i\leq m} r_i < 1$, the mapping F is a contraction on the complete metric space (S, d). By Banach's contraction mapping theorem F has a unique fixed point, that is to say, there is a unique set $E \in S$ such that $F(E) = E$, which is (2.26), and moreover $F^k(A) \to E$ as $k \to \infty$. In particular, if $F_i(A) \subset A$ for all i then $F(A) \subset A$, so that $F^k(A)$ is a decreasing sequence of non-empty compact sets containing E with intersection $\cap_{k=0}^{\infty} F^k(A)$ which must equal E. $\square$

The unique non-empty compact set E satisfying (2.26) is called the *attractor* or *invariant set* of the IFS $\{F_1, \ldots, F_m\}$. The IFS may be thought of as defining or representing the set E.

There are two main problems that arise in connection with IFSs. The first is, given a fractal E, to find an IFS with attractor E or, at least, a close approximation to E. In many cases, including many familiar self-similar sets, a suitable IFS with a small number of contractions can be written down by inspection, for example (2.24) for the middle-third Cantor set. In such instances the IFS provides a very efficient way of representing or 'coding' the set. This has led to the more general problem of fractal image compression: how to find a relatively small family of contractions that represent *any* given set or picture, see FG, Section 9.5.

The inverse problem is to reconstruct the attractor E of a given IFS. Computationally this is very easy, since it follows by iterating (2.29) that

$$d(F^k(A), E) \leq \left(\max_{1\leq i\leq m} r_i\right)^k d(A, E) \tag{2.30}$$

for every $A \in S$. Thus $F^k(A)$ converges to E at a geometric rate, so plotting $F^k(A) = \cup_{I_k} F_{i_1} \circ F_{i_2} \circ \cdots \circ F_{i_k}(A)$ for a suitable k gives an approximation to E (the union is over the set I_k of all k-term sequences $(i_1, i_2, \ldots i_k)$ with $i_j \in \{1, 2, \ldots, m\}$). These sets $F^k(A)$ are sometimes called *pre-fractals* for E, see Figures 2.4 and 2.5 for examples. An alternative, but often effective, way of reconstructing E is to take any initial point x_0, and select a sequence $F_{i_1}, F_{i_2}, \ldots$ independently at random from $\{F_1, \ldots, F_m\}$, say with equal probability. Then the points defined by

$$x_k = F_{i_k}(x_{k-1}) \quad \text{for} \quad k = 1, 2, \ldots \tag{2.31}$$

are indistinguishably close to E for large enough k, and also appear randomly distributed across E. A plot of the sequence (x_k) starting, say, with $k = 100$, may give a good impression of E. In some instances better results will be obtained by weighting the probabilities of choosing the F_i. The reason for the name 'iterated function system' should be obvious from these reconstruction procedures.

An IFS provides a natural way of coding the attractor E and the components of the pre-fractals for E, analogous to the way in which the points of the middle-third Cantor set may be identified with the numbers whose base 3 expansions contain only the digits 0 and 2. Let $\{F_1, \ldots, F_m\}$ be an IFS

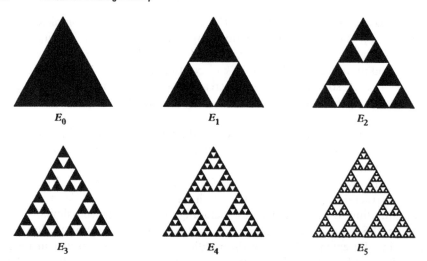

Figure 2.4   The usual sequence of pre-fractals in the construction of the Sierpinski triangle, with $E_k = F^k(A)$ where A is an equilateral triangle

with attractor E. For $k = 0, 1, 2, \ldots$ we define I_k to be the set of all k-term sequences of integers selected from $\{1, 2, \ldots, m\}$, that is

$$I_k = \{(i_1, i_2, \ldots, i_k) : 1 \leq i_j \leq m\}; \tag{2.32}$$

we regard I_0 as just containing the empty sequence. We often abbreviate a sequence of I_k by

$$\boldsymbol{i} = (i_1, i_2, \ldots, i_k). \tag{2.33}$$

We write

$$I = \bigcup_{k=0}^{\infty} I_k \tag{2.34}$$

for the set of all such finite sequences, and I_∞ for the corresponding set of infinite sequences, so

$$I_\infty = \{(i_1, i_2, \ldots) : 1 \leq i_j \leq m\}. \tag{2.35}$$

It is convenient to write $\boldsymbol{i}, \boldsymbol{j}$ for the sequence defined by juxtaposition of $\boldsymbol{i}$ and $\boldsymbol{j}$. In particular, with $\boldsymbol{i} = (i_1, \ldots, i_k)$, we have $\boldsymbol{i}, i = (i_1, \ldots, i_k, i)$.

Let $A \in \mathcal{S}$ be such that $F_i(A) \subset A$ for all i, so that $F(A) \subset A$. From (2.28) E is the intersection of the decreasing sequence of sets

$$F^k(A) = \bigcup_{I_k} F_{i_1} \circ \cdots \circ F_{i_k}(A), \tag{2.36}$$

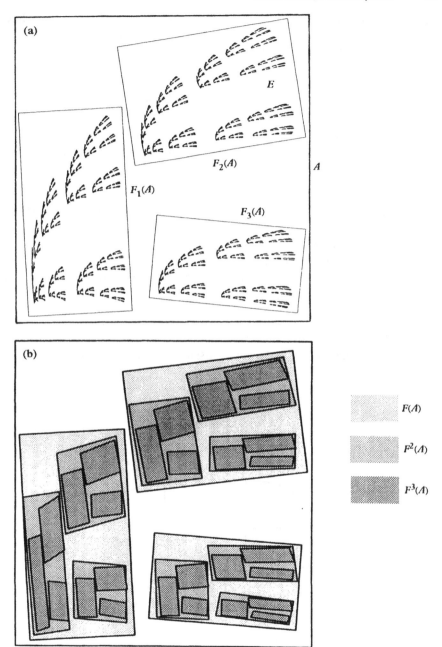

Figure 2.5 An IFS consisting of three affine transformations $\{F_1, F_2, F_3\}$ which map the square A onto the rectangles in (a) in the obvious way. The convergence of the pre-fractals $F^k(A)$ to E is apparent in (b)

with the union over $(i_1, \ldots, i_k) \in I_k$. Moreover, for all $(i_1, \ldots, i_k)$ we have $F_{i_1} \circ \cdots \circ F_{i_k}(A) \subset F_{i_1} \circ \cdots \circ F_{i_{k-1}}(A)$ and also that $|F_{i_1} \circ \cdots \circ F_{i_k}(A)| \leq r_{\max} |F_{i_1} \circ \cdots \circ F_{i_{k-1}}(A)|$. Thus for all $(i_1, i_2, \ldots) \in I_\infty$ we have $|F_{i_1} \circ \cdots \circ F_{i_k}(A)| \to 0$ as $k \to \infty$ with

$$x_{i_1, i_2, \ldots} \equiv \bigcap_{k=0}^{\infty} F_{i_1} \circ \cdots \circ F_{i_k}(A), \tag{2.37}$$

the single point of intersection of this decreasing sequence of sets. Since every point of E is in such an intersection for at least one sequence $(i_1, i_2, \ldots) \in I_\infty$,

$$E = \bigcup_{I_\infty} \{x_{i_1, i_2, \ldots}\}. \tag{2.38}$$

We conclude that E may be constructed using the hierarchy of sets $F_{i_1} \circ \cdots \circ F_{i_k}(A)$ for $(i_1, \ldots, i_k) \in I$, see Figure 2.6; this is analogous to the usual construction of the middle-third Cantor set.

This coding of the components $F_{i_1} \circ \cdots \circ F_{i_k}(A)$ and of points $x_{i_1, i_2, \ldots}$ is very useful indeed in analysis of attractors of IFSs. For convenience we shall write

$$A_i = A_{i_1, \ldots, i_k} = F_{i_1} \circ \ldots \circ F_{i_k}(A) \tag{2.39}$$

for $i = (i_1, \ldots, i_k) \in I$ and $A \subset X$.

In the simplest situation the sets $F_1(A), \ldots, F_m(A)$ may be mutually disjoint, in which case the unions (2.36) will be disjoint and each point of E will have a

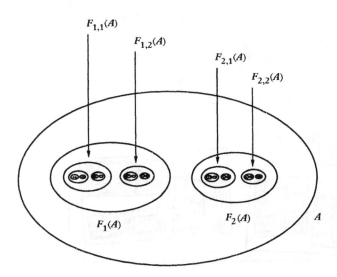

Figure 2.6 Notation for iterated function systems. The contractions F_1 and F_2 map the large ellipse A onto the regions $F_1(A)$ and $F_2(A)$ respectively. The sets $F^k(A) = \cup F_{i_1} \circ \cdots \circ F_{i_k}(A)$ decrease to the IFS attractor E

unique representation as $x_{i_1,i_2,...}$. In fact, if this happens then $F_1(E),...,F_m(E)$ are mutually disjoint, since $E \subset A$ for any A with $F(A) \subset A$. When the union $E = \cup_{i=1}^m F_i(E)$ is disjoint we say that the IFS $\{F_1,...F_m\}$ satisfies the *strong separation condition*. This is the case with the middle-third Cantor set IFS (2.24) and with other IFSs with a totally disconnected attractor. However, strong separation is rather too strong for many purposes, and we often work with a weaker separation condition. An IFS $\{F_1,...,F_m\}$ satisfies the *open set condition* (OSC) if there exists a non-empty bounded open set $U \subset X$ such that

$$\bigcup_{i=1}^k F_i(U) \subset U \tag{2.40}$$

with this union disjoint. The OSC is easily verified for sets such as the von Koch curve, see Figure 2.7, and the Sierpinski carpet.

In the particularly nice situation when $F_1(X),...,F_m(X)$ are mutually disjoint, we may define $f : \cup_{i=1}^m F_i(X) \to X$ by

$$f(x) = F_i^{-1}(x) \quad \text{if} \quad x \in F_i(X). \tag{2.41}$$

(If the $F_i(X)$ are not disjoint, but the strong separation condition holds, then this situation will pertain on replacing X by an appropriate subset, perhaps even E itself.) For certain purposes it is more convenient to work with this single mapping f rather than the m mappings F_i; we make considerable use of this approach in Chapters 3 and 4. In particular the attractor E is invariant for f in the sense that $E = f(E) = f^{-1}(E)$.

There are many classes of IFS of special interest. If the $\{F_i,...,F_m\}$ are similarities, the attractor E is called *self-similar*, if they are affine transformations E is called *self-affine*, and if they are conformal transformations (so that the derivatives $F_i'(x)$ are similarities for all i and all $x \in X$) then E is called *self-conformal*: see Figures 0.1 and 0.2 for some examples. In Chapter 4 we shall see

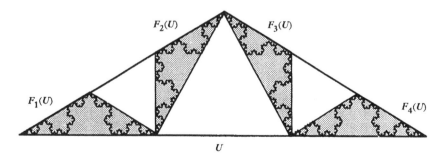

Figure 2.7 The open set condition for the von Koch curve. The open set U is the interior of the bounding triangle with $F_1,...,F_4$ the obvious similarities

how 'cookie-cutter' sets associated with certain dynamical systems may be identified with attractors of IFSs.

A great deal of effort has been devoted to calculating the dimensions of IFS attractors. A variety of estimates are given in FG, Chapter 9, and others will be obtained in this book. We recall perhaps the most important result, the dimension formula for self-similar sets.

Theorem 2.7

Let E be the attractor of a family of similarity transformations $\{F_1, \ldots, F_m\}$ where F_i has similarity ratio r_i. If the open set condition (2.40) is satisfied then $\dim_H E = \dim_P E = \underline{\dim}_B E = \overline{\dim}_B E = s$ and moreover $0 < \mathcal{H}^s(E), \mathcal{P}^s(E) < \infty$, where s is the unique positive solution of

$$\sum_{i=1}^{m} r_i^s = 1. \tag{2.42}$$

Proof The standard proof of this is given in FG, Theorem 9.3. For an alternative approach see Example 3.3 later in this book. □

Theorem 2.7 immediately gives, for example, that the von Koch curve has (box and Hausdorff) dimensions $\log 4/\log 3$ and that the Sierpinski carpet has dimensions $\log 8/\log 3$, these sets satisfying the OSC. Of course, Theorem 2.7 holds in particular for systems for which the strong separation condition applies, giving that the middle-third Cantor set has dimensions $\log 2/\log 3$.

The notion of an iterated function system may be extended to define natural invariant measures supported by the attractor of the system. Let $\{F_1, \ldots, F_m\}$ be an IFS on $X \subset \mathbb{R}^n$ and let $p_1, \ldots, p_m$ be probabilities, with $0 \le p_i \le 1$ for all i and $\sum_{i=1}^m p_i = 1$; such a system is called a *probabilistic iterated function system*. At least in the strong separation case, it is not hard to see how this leads to a measure on E. Here the sets $F_1(E), \ldots, F_m(E)$ are disjoint, so that $E_{i,1}, \ldots, E_{i,m}$ are disjoint subsets of E_i for all $i \in I$. We may define a measure μ on this hierarchy of sets by repeated subdivision of the measure in the ratio $p_1 : p_2 : \ldots : p_m$, so that

$$\mu(E_{i_1,i_2,\ldots,i_k}) = p_{i_1} p_{i_2} \cdots p_{i_k}, \tag{2.43}$$

and this extends to a Borel measure supported by E in the usual way, see Section 1.3. In the general case the existence of such measures is assured by the following result.

Theorem 2.8

Let $\{F_1, \ldots, F_m\}$ be an iterated function system on $X \subset \mathbb{R}^n$ with associated probabilities $\{p_1, \ldots, p_m\}$. Then there exists a unique Borel probability measure

μ (*that is with* $\mu(X) = 1$) *such that*

$$\mu(A) = \sum_{i=1}^{m} p_i \mu(F_i^{-1}(A)) \tag{2.44}$$

for all Borel sets A, and

$$\int g(x) d\mu(x) = \sum_{i=1}^{m} p_i \int g(F_i(x)) d\mu(x) \tag{2.45}$$

for all continuous $g : X \to \mathbb{R}$. *Moreover,* spt $\mu = E$, *where E is the attractor of the IFS* $\{F_i : 1 \le i \le m$ *and* $p_i \ne 0\}$. *If the strong separation condition is satisfied then* (2.43) *holds.*

Proof   The easiest way to prove this is using the contraction mapping theorem. Let $\mathcal{M}$ be the class of Borel probability measures on X with bounded support. Endow $\mathcal{M}$ with the metric

$$d(\nu_1, \nu_2) = \sup\left\{ \left| \int g \, d\nu_1 - \int g \, d\nu_2 \right| : \text{Lip} \, g \le 1 \right\} \tag{2.46}$$

where Lip g is the Lipschitz constant of g, see (1.1). It is easily verified that d is a metric on $\mathcal{M}$, and with a little bit of effort it may be shown that d is a complete metric. We define a mapping $\psi : \mathcal{M} \to \mathcal{M}$ by

$$\psi(\nu)(A) = \sum_{i=1}^{m} p_i \nu(F_i^{-1}(A)) \tag{2.47}$$

for all Borel sets A. This implies that

$$\int g \, d\psi(\nu) = \sum_{i=1}^{m} p_i \int (g \circ F_i) \, d\nu \tag{2.48}$$

for every measurable function $g : X \to \mathbb{R}$. To see that ψ is a contraction on $\mathcal{M}$ we note that

$$d(\psi(\nu_1), \psi(\nu_2)) = \sup\left\{ \left| \int g \, d\psi(\nu_1) - \int g \, d\psi(\nu_2) \right| : \text{Lip} \, g \le 1 \right\}$$

$$= \sup\left\{ \left| \sum_{i=1}^{m} p_i \left(\int (g \circ F_i) d\nu_1 - \int (g \circ F_i) d\nu_2 \right) \right| : \text{Lip} \, g \le 1 \right\}$$

$$\le \sum_{i=1}^{m} p_i \sup\left\{ \left| \int (g \circ F_i) d\nu_1 - \int (g \circ F_i) d\nu_2 \right| : \text{Lip} \, g \le 1 \right\}$$

$$\le \sum_{i=1}^{m} p_i \sup\left\{ r_i \left| \int r_i^{-1}(g \circ F_i) d\nu_1 - \int r_i^{-1}(g \circ F_i) d\nu_2 \right| : \text{Lip} \, g \le 1 \right\}$$

$$\le \sum_{i=1}^{m} p_i r_i \sup\left\{ \left| \int g \, d\nu_1 - \int g \, d\nu_2 \right| : \text{Lip} \, g \le 1 \right\}$$

$$\le r_{\max} d(\nu_1, \nu_2)$$

with r_i and $r_{\max}$ as in (2.22) and (2.23), since $\text{Lip}(r_i^{-1}(g \circ F_i)) \leq 1$ for all i. Thus ψ is a contraction on $\mathcal{M}$, so by the contraction mapping theorem there is a unique $\mu \in \mathcal{M}$ satisfying $\psi(\mu) = \mu$, that is, satisfying (2.44) and (2.45).

From (2.44) we have that $\text{spt}\mu = \cup F_i(\text{spt}\mu)$ where the union is over those i with $p_i \neq 0$, so that $\text{spt}\mu$ is the unique non-empty compact attractor of the IFS $\{F_i : 1 \leq i \leq m \text{ and } p_i \neq 0\}$. Finally, if the strong separation condition holds, then taking A as $E_{i_1,\ldots,i_k}$ in (2.44) gives $\mu(E_{i_1,\ldots,i_k}) = p_{i_1}\mu(E_{i_2,\ldots,i_k})$, so iterating gives (2.43). $\square$

The probability measure μ satisfying (2.44) is called the *invariant measure* for the probabilistic IFS. We remark that in the strongly separated case, μ is invariant under the mapping f given by (2.41), in the sense that $\mu(f^{-1}(A)) = \mu(A)$ for all Borel sets A.

There is a random algorithm for constructing the invariant measure μ. Let $(i_1, i_2, \ldots)$ be a random sequence such that $i_j = i$ with probability p_i, independently for each j. Fixing $x \in \text{spt}\mu$, we define for each Borel set A

$$\mu_x(A) = \lim_{k \to \infty} \frac{1}{k} \#\{k' \leq k \text{ such that } F_{i_{k'}} \circ \cdots \circ F_{i_1}(x) \in A\}. \tag{2.49}$$

Then for μ-almost all x we have $\mu_x(A) = \mu(A)$, see Exercise 2.10. Thus on

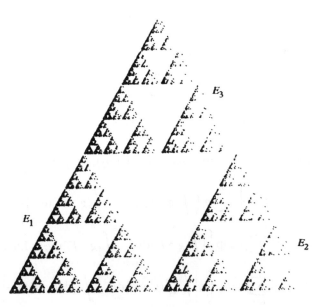

Figure 2.8 A self-similar measure supported on the Sierpinski triangle (the density of the measure is indicated by the concentration of dots). Here $p_1 = 0.8, p_2 = 0.05$, $p_3 = 0.15$

iterating x under a random sequence of mappings with F_i chosen with probability p_i, the proportion of iterates lying in a set A approximates $\mu(A)$. This property is very useful for computer study of invariant measures.

Certain classes of invariant measures are of particular interest. A measure μ resulting from a family $\{F_1, \ldots, F_m\}$ of similarity transformations is called a *self-similar measure*. For example, taking F_1 and F_2 as in (2.24) and $p_1 = p_2 = \frac{1}{2}$ gives the 'Cantor measure', evenly distributed over the middle-third Cantor set. For an example based on the Sierpinski triangle, see Figure 2.8. In the same way a measure resulting from a family of affine transformations is called a *self-affine measure*.

2.3 Notes and references

The material on dimensions, densities and rectifiability may be found in FG, Chapters 2–5; for a fuller discussion see the books by Falconer (1985) and Mattila (1995a). A detailed treatment of the local dimension characterisations of dimensions, Propositions 2.2–2.4, is given by Cutler (1986, 1995). Interestingly, Hausdorff (1919) introduced the measures bearing his name many years before Tricot (1982) proposed packing measures.

The idea of representing sets by IFSs is due to Hutchinson (1981), although much of the theory was given by Moran (1946). For more detailed discussions of IFSs see, for example, FG, Chapter 9, Barnsley (1988), Barnsley and Demko (1985) and Edgar (1990). There is a very considerable literature on the dimensions of sets represented by certain IFSs. Many results on self-similar sets may be found in these references; for self-affine sets see, for example, Bedford and Urbanski (1990), Falconer (1988, 1992) and Hueter and Lalley (1995).

Exercises

2.1 Verify the dimension inequalities (2.14).

2.2 Let $\{F_1, \ldots, F_m\}$ be an IFS consisting of similarity transformations satisfying the open set condition, with attractor E of dimension s. Show that $\mathcal{H}^s(F_i(E) \cap F_j(E)) = 0$ if $i \neq j$.

2.3 Find the Hausdorff and box dimensions of the set $\{(1/p, 1/q) : p, q \in \mathbb{Z}^+\} \subset \mathbb{R}^2$.

2.4 Let E be the middle-third Cantor set. Use Proposition 2.2 to find estimates for $\mathcal{H}^s(E)$ and $\mathcal{P}^s(E)$ where $s = \log 2 / \log 3$.

2.5 Show that the lower density $\underline{D}^s(E, x)$ is a Borel function of x, that is for all $c \in \mathbb{R}$ we have that $\{x : \underline{D}^s(E, x) \leq c\}$ is a Borel set. Show that the same is true for the upper density.

2.6 Verify the mapping properties for densities (2.19).

2.7 Fix $0 < \lambda < \frac{1}{2}$ and let $F_1, F_2 : \mathbb{R} \to \mathbb{R}$ be given by $F_1(x) = \lambda x$, $F_2(x) = \frac{1}{2}x + \frac{1}{2}$. Describe the attractor of $\{F_1, F_2\}$ and find an expression for its Hausdorff and packing dimensions.

2.8   Find IFSs for the fractals depicted in Figure 0.2 and hence find expressions for their Hausdorff and box dimensions.

2.9   Verify that (2.25) and (2.46) define metrics.

2.10 Prove that for μ-almost all x the random measure μ_x given by (2.49) equals the invariant measure μ, in the case where $\mathrm{spt}\mu$ satisifies the strong separation condition. (Hint: Define $f : \mathrm{spt}\mu \to \mathrm{spt}\mu$ by (2.41), so that $f(x_{i_1,i_2,\ldots}) = x_{i_2,i_3,\ldots}$.)

2.11 Verify that if E is the attractor of an IFS satisfying the strong separation condition, then there exist numbers $0 < c_1, c_2 < \infty$ such that $c_1 r^s \le \mathcal{H}^s(E \cap B(x,r)) \le c_2 r^s$ for all $x \in E$ and $0 < r \le 1$. Extend this to the case of an IFS satisfying the open set condition. (Hint: look for similar copies of E contained in, and containing, $E \cap B(x,r)$.)

Chapter 3 Some techniques for studying dimension

As we have remarked, there are many possible ways of defining 'fractal dimension'. Being based on a measure, Hausdorff dimension is particularly suited for developing general mathematical theory. On the other hand, box-counting dimension is often rather easier to calculate or estimate in practice, and a variety of other definitions of dimension only apply to sets of specific types, for example to curves. Many familiar examples of fractals, including self-similar sets such as the von Koch curve, have equal Hausdorff, lower and upper box-counting dimensions. However, other fractals, such as self-affine sets, may have Hausdorff dimension strictly less than their box dimensions.

Since almost all definitions of dimension that have been proposed give values between the Hausdorff dimension and upper box dimension, these dimensions are perhaps of special interest. It can be particularly useful to know that a set has equal Hausdorff and upper box dimensions, with the intermediate definitions also taking this value. In Section 3.1 we give some 'approximate self-similarity' conditions on a set that ensure that these dimensions are equal and that often allow a straightforward evaluation of the dimension. In Section 3.2 we obtain expressions for the box dimensions of a set in terms of the geometry of its complement, leading to a criterion for equality of lower and upper box dimensions. These techniques complement the more direct approaches to dimension calculation discussed in FG.

3.1 Implicit methods

The usual method for finding the Hausdorff dimension of a set E is to calculate the s-dimensional Hausdorff measure $\mathcal{H}^s(E)$ for $s \geq 0$ and find the value of s at which this jumps from infinity to zero. Such calculations, and the corresponding calculations for box dimensions, can be quite intricate. In this section we discuss a different approach. We give geometrical conditions on a set E that guarantee that $0 < \mathcal{H}^s(E)$ or $\mathcal{H}^s(E) < \infty$ where $s = \dim_H E$, without the need of first calculating s. With this knowledge s can often be found easily. (For example, given that the middle-third Cantor set has positive finite Hausdorff measure at the dimensional value, it is very easy to show that its Hausdorff

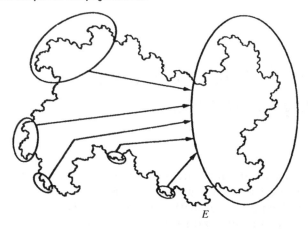

E

Figure 3.1   A set E satisfying the conditions of Theorem 3.1, with mappings from small neighbourhoods of E into E satisfying (3.1)

dimension is $\log 2/\log 3$ using the scaling property of Hausdorff measure.) We give similar conditions that guarantee that $\dim_H E = \underline{\dim}_B E = \overline{\dim}_B E$.

Theorems that enable us to draw conclusions about dimensions without their explicit calculation are called *implicit* theorems. We prove two such theorems here which apply to sets with small parts 'approximately similar' to large parts in a sense that can be made precise in terms of Lipschitz functions. Theorem 3.1 applies to a set E if every small neighbourhood of E may be mapped into a large part of E under a Lipschitz mapping with the Lipschitz constant controlled by the size of the neighbourhood, see Figure 3.1.

Theorem 3.1

Let E be a non-empty compact subset of $\mathbb{R}^n$ and let $a > 0$ and $r_0 > 0$. Suppose that for every set U that intersects E with $|U| < r_0$ there is a mapping $g : E \cap U \to E$ satisfying

$$a|U|^{-1}|x - y| \le |g(x) - g(y)| \qquad (x, y \in E \cap U). \qquad (3.1)$$

Then, writing $s = \dim_H E$, we have $\mathcal{H}^s(E) \ge a^s > 0$ and $\underline{\dim}_B E = \overline{\dim}_B E = s$.

Proof   It is enough to show that for all $d > 0$, if $\mathcal{H}^d(E) < a^d$ then $\overline{\dim}_B E < d$, and so by (2.14) $\dim_H E < d$. By taking d arbitraily close to $\dim_H E$ this also implies that $\overline{\dim}_B E \le \dim_H E$, giving equality.

If $\mathcal{H}^d(E) < a^d$, there are sets $U_1, \ldots U_m$ which intersect E with $|U_i| < \min\{\frac{1}{2}a, r_0\}$ such that $E \subset \cup_{i=1}^m U_i$ and $\sum_{i=1}^m |U_i|^d < a^d$. (Since we may take the U_i to be open in estimating these sums, the compactness of E allows a finite collection of covering sets.) By taking t close to d, we may find $0 < t < d$

such that

$$a^{-t} \sum_{i=1}^{m} |U_i|^t < 1. \qquad (3.2)$$

By hypothesis there exist $g_i : E \cap U_i \to E$ $(i = 1, 2, \ldots, m)$ such that

$$|x - y| \le a^{-1} |U_i| |g_i(x) - g_i(y)| \quad (x, y \in E \cap U_i). \qquad (3.3)$$

We treat the inverses of these functions $\{g_1^{-1}, \ldots, g_m^{-1}\}$ taken on appropriate domains rather like an iterated function system. Let $I_k = \{(i_1, \ldots, i_k) : 1 \le i_j \le m\}$ be the set of k-term sequences formed using the integers $\{1, 2, \ldots, m\}$, and let $I = \cup_{k=0}^{\infty} I_k$. For each $\boldsymbol{i} = (i_1, \ldots, i_k) \in I_k$ define

$$U_{i_1, \ldots, i_k} = g_{i_1}^{-1}(g_{i_2}^{-1}(\ldots(g_{i_k}^{-1}(E))\ldots)).$$

Note that some of these sets may be empty, since we have $g_i^{-1}(A) = \emptyset$ if $A \cap g_i(E \cap U_i) = \emptyset$, but nevertheless $E \subset \cup_{i \in I_k} U_i$ for each k. For $x, y \in U_{i_1, \ldots, i_k}$ repeated application of (3.3) gives

$$|x - y| \le a^{-k} |U_{i_1}| \cdots |U_{i_k}| |g_{i_1} \circ \cdots \circ g_{i_k}(x) - g_{i_1} \circ \cdots \circ g_{i_k}(y)|.$$

In particular,

$$|U_{i_1, \ldots, i_k}| \le a^{-k} |U_{i_1}| \cdots |U_{i_k}| |E|.$$

Let $b = a^{-1} \min_{1 \le i \le m} |U_i|$. Given $r < |E|$, for all $x \in E$ there exists $\boldsymbol{i} = (i_1, \ldots, i_k) \in I$ such that $x \in U_i$ and $br \le a^{-k} |U_{i_1}| \cdots |U_{i_k}| |E| < r$. Hence, with $N(r)$ denoting the least number of sets of diameter at most r which can cover E, we have

$$N(r) \le \#\{i \in I : br \le a^{-k} |U_{i_1}| \cdots |U_{i_k}| |E|\}$$

$$\le \sum_{i \in I} (br)^{-t} (a^{-k} |U_{i_1}| \cdots |U_{i_k}| |E|)^t$$

$$\le |E|^t b^{-t} r^{-t} \sum_{k=0}^{\infty} a^{-kt} \sum_{i \in I_k} (|U_{i_1}| \cdots |U_{i_k}|)^t$$

$$= |E|^t b^{-t} r^{-t} \sum_{k=0}^{\infty} \left(a^{-t} \sum_{i=1}^{m} |U_i|^t \right)^k$$

$$\le c_1 r^{-t}$$

for some $c_1 < \infty$, using (3.2). It follows from (2.2) that $\overline{\dim}_B E \le t < d$, as required. $\quad \square$

The hypotheses of the preceding theorem involves mappings from small neighbourhoods into large parts of a set. The next theorem is complementary, in that we require that all small neighbourhoods contain 'not too contracted' images of the whole set, see Figure 3.2.

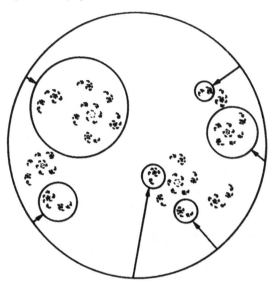

Figure 3.2 A set E satisfying the conditions of Theorem 3.2, with mappings from E into small balls satisfying (3.4)

Theorem 3.2

Let E be a non-empty compact subset of $\mathbb{R}^n$ and let $a > 0$ and $r_0 > 0$. Suppose that for every closed ball B with centre in E and radius $r < r_0$ there is a mapping $g : E \rightarrow E \cap B$ satisfying

$$ar|x - y| \leq |g(x) - g(y)| \qquad (x, y \in E). \tag{3.4}$$

Then, writing $s = \dim_H E$, we have that $\mathcal{H}^s(E) \leq 4^s a^{-s} < \infty$ and $\underline{\dim}_B E = \overline{\dim}_B E = s$.

Proof For the purposes of this proof we take $N(r)$ to be the maximum number of disjoint closed balls of radius r with centres in E. We suppose that for some $r < \min\{a^{-1}, r_0\}$ we have

$$N(r) > a^{-s} r^{-s} \tag{3.5}$$

and derive a contradiction. Given (3.5), there exists $t > s$ such that

$$m \equiv N(r) > a^{-t} r^{-t}; \tag{3.6}$$

thus there are disjoint balls $B_1, \ldots, B_m$ of radii r with centres in E.
By hypothesis there exist mappings $g_i : E \rightarrow E \cap B_i$ $(1 \leq i \leq m)$ such that

$$ar|x - y| \leq |g_i(x) - g_i(y)|. \tag{3.7}$$

Essentially, $\{g_1, \ldots, g_m\}$ is a (not necessarily contracting) IFS with an attractor that is a subset of E. We find a lower bound for the dimension of the attractor, and thus of E, by a routine method.

Let $d = \min_{i \neq j} \text{dist}(B_i, B_j) > 0$. Using (3.7) $(q - 1)$ times, we see that

$$\text{dist}(g_{i_1} \circ \cdots \circ g_{i_k}(E), g_{j_1} \circ \cdots \circ g_{j_k}(E)) \geq (ar)^{q-1} \text{dist}(B_{i_q}, B_{j_q})$$
$$\geq (ar)^q d \qquad (3.8)$$

where q is the least integer such that $i_q \neq j_q$, recalling that $r < a^{-1}$. Let μ be the measure on E defined by repeated subdivision (see (1.19)) such that $\mu(g_{i_1} \circ \cdots \circ g_{i_k}(E)) = m^{-k}$ for all $(i_1, \ldots, i_k)$. Let U be any subset of $\mathbb{R}^n$ that intersects E with $|U| < d$, and let k be the least integer such that

$$(ar)^{k+1} d \leq |U| < (ar)^k d. \qquad (3.9)$$

By (3.8) U intersects $g_{i_1} \circ \cdots \circ g_{i_k}(E)$ for at most one k-term sequence $(i_1, \ldots, i_k)$, so

$$\mu(U) \leq m^{-k} < (ar)^{kt} \leq (dar)^{-t} |U|^t$$

by (3.6) and (3.9). It follows by the mass distribution principle, Proposition 2.1, that $\dim_H E \geq t > s$.

We conclude that if $\dim_H E = s$, then $N(r) \leq a^{-s} r^{-s}$ for all sufficiently small r. This implies $\overline{\dim}_B E \leq s$, so equality of the dimensions follows from (2.14). Moreover, by using balls of double the radius, E can be covered by $N(r)$ balls of radius $2r$ (otherwise the $N(r)$ disjoint balls of radius r with centres in E would not be a maximal collection). Hence $\mathcal{H}^s_{4r}(E) \leq a^{-s} r^{-s} (4r)^s = 4^s a^{-s}$, giving $\mathcal{H}^s(E) \leq 4^s a^{-s}$. $\quad\square$

The hypotheses of Theorem 3.2 also imply that $\mathcal{P}^s(E) < \infty$, see Exercise 3.2.

The middle-third Cantor set E illustrates very simply how these theorems can be used. If U intersects E and $3^{-k-1} \leq |U| < 3^{-k}$ for $k \in \mathbb{Z}^+$ then there is an obvious similarity mapping of ratio 3^k from $U \cap E$ into E, so (3.1) is satisfied with $a = \frac{1}{3}$. Similarly, if B is an interval (one-dimensional ball) with centre in E and length $2r$ with $3^{-k} \leq r < 3^{-k+1}$ there is a similarity transformation of ratio 3^{-k} of E into $E \cap B$, giving (3.4) with $a = \frac{1}{3}$. We conclude from Theorems 3.1 and 3.2 that, with $s = \dim_H E$, we have $\underline{\dim}_B E = \overline{\dim}_B E = s$ and $0 < \mathcal{H}^s(E) < \infty$. Writing E_L and E_R for the left and right 'parts' of the Cantor set E, we obtain

$$\mathcal{H}^s(E) = \mathcal{H}^s(E_L) + \mathcal{H}^s(E_R) = 3^{-s}\mathcal{H}^s(E) + 3^{-s}\mathcal{H}^s(E)$$

by the scaling property (2.13) of Hausdorff measures, see Figure 2.3. Thus $1 = 2 \times 3^{-s}$, giving immediately that the dimension of the middle-third Cantor set is $s = \log 2 / \log 3$.

The next example extends this argument to more general self-similar sets, providing an alternative proof for Theorem 2.6 under the strong separation condition.

Corollary 3.3   (self-similar sets)

Let E be the self-similar set defined by the IFS consisting of similarities $\{F_1, \ldots, F_m\}$ where F_i has ratio r_i with $0 < r_i < 1$. If $\dim_H E = s$ then $\mathcal{H}^s(E) < \infty$ and $\underline{\dim}_B E = \overline{\dim}_B E = s$. Moreover if $\{F_i(E)\}_{i=1}^m$ are disjoint sets, then $0 < \mathcal{H}^s(E)$ and s satisfies $\sum_{i=1}^m r_i^s = 1$.

Proof   Write $r_{\min} = \min_{1 \leq i \leq m} r_i$. Let $x \in E$ and $r \leq |E|$. There is a (not necessarily unique) sequence $(i_1, i_2, \ldots)$ such that $x \in F_{i_1} \circ \cdots \circ F_{i_k}(E)$ for all k. Choose k so $r_{\min} r < r_{i_1} \cdots r_{i_k} |E| \leq r$. Then $F_{i_1} \circ \cdots \circ F_{i_k} : E \to E \cap B(x, r)$ is a similarity of ratio at least $r_{\min} |E|^{-1} r$ so Theorem 3.2 gives equality of the dimensions and that $\mathcal{H}^s(E) < \infty$.

Now suppose $\min_{i \neq j} \mathrm{dist}(F_i(E), F_j(E)) = d > 0$. Then we have that $\mathrm{dist}(F_{i_1} \circ \cdots \circ F_{i_k}(E), F_{j_1} \circ \cdots \circ F_{j_k}(E)) \geq r_{i_1} \cdots r_{i_{k-1}} d$ if $(j_1, \ldots, j_k)$ is distinct from $(i_1, \ldots, i_k)$. If U intersects E with $|U| < d$ and $x \in E \cap U$, we may find $(i_1, \ldots, i_k)$ such that $x \in F_{i_1} \circ \cdots \circ F_{i_k}(E)$ and $dr_{i_1} \cdots r_{i_k} \leq |U| < dr_{i_1} \cdots r_{i_{k-1}}$. Thus U is disjoint from $F_{j_1} \circ \cdots \circ F_{j_k}(E)$ for all $(j_1, \ldots, j_k) \neq (i_1, \ldots, i_k)$ and so $E \cap U \subset F_{i_1} \circ \cdots \circ F_{i_k}(E)$. Hence $(F_{i_1} \circ \cdots \circ F_{i_k})^{-1} : E \cap U \to E$ is a similarity of ratio $(r_{i_1} \cdots r_{i_k})^{-1} \geq d|U|^{-1}$, and Theorem 3.1 implies that $0 < \mathcal{H}^s(E)$ as well as equality of the dimensions again.

Finally, in the disjoint case, the scaling property (2.13) of Hausdorff measure gives $\mathcal{H}^s(E) = \sum_{i=1}^m \mathcal{H}^s(F_i(E)) = \sum_{i=1}^m r_i^s \mathcal{H}^s(E)$; since $0 < \mathcal{H}^s(E) < \infty$ we get $1 = \sum_{i=1}^m r_i^s$, where $s = \dim_H E$. $\square$

In fact the conclusion that $0 < \mathcal{H}^s(E)$ is true if $\{F_1, \ldots, F_m\}$ satisfies the open set condition; this may be deduced in a similar manner using a strengthened version of Theorem 3.1, see Exercise 3.3.

Note that for a self-similar set E we have $\dim_H E = \underline{\dim}_B E = \overline{\dim}_B E$ and $\mathcal{H}^s(E) < \infty$ without any separation condition on the IFS. In particular this is true even if the sets $\{F_i(E)\}$ overlap substantially enough for the dimension to be strictly less than the solution of $\sum_{i=1}^m r_i^s = 1$.

A trivial modification of the last proof allows the implicit theorems to be applied to many subsets of self-similar sets. The sets in the next corollary might be termed *super-self-similar* (3.10) and *sub-self-similar* (3.11), see Figure 3.3.

Corollary 3.4   (super-self-similar and sub-self-similar sets)

Let $\{F_1, \ldots, F_m\}$ be contacting similarities with attractor E.

(a)   If A is a non-empty compact subset of E satisfying

$$A \supset \bigcup_{i=1}^m F_i(A) \tag{3.10}$$

and $\dim_H A = s$ then $\underline{\dim}_B A = \overline{\dim}_B A = s$ and $\mathcal{H}^s(A) < \infty$.

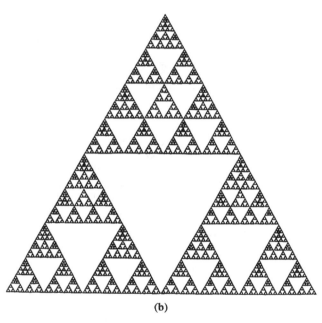

Figure 3.3 Examples of (a) a sub-self-similar set and (b) a super-self-similar set, based on the Sierpinski triangle

(b) *If A is a non-empty compact set satisfying*

$$A \subset \bigcup_{i=1}^{m} F_i(A) \tag{3.11}$$

and $\dim_H A = s$ *then, provided that the sets* $\{F_i(E)\}_{i=1}^{m}$ *are disjoint,* $\underline{\dim}_B A = \overline{\dim}_B A = s$ *and* $\mathcal{H}^s(A) > 0.$

Proof (a) If $x \in A \subset E$ and r is small enough, the restriction of the mapping $F_{i_1} \circ \cdots \circ F_{i_k} : A \to A \cap B(x, r)$ defined in the proof of Corollary 3.3 satisfies the conditions of Theorem 3.2, giving the result.

(b) Condition (3.11) implies that $A \subset E$ (see (2.28)). Thus if U is a set of small enough diameter that intersects A, the restricted mapping $(F_{i_1} \circ \cdots \circ F_{i_k})^{-1} :$ $A \cap U \to A$ from the proof of Corollary 3.3 satisfies the conditions of Theorem 3.1, and the conclusion follows. $\square$

Graph-directed sets, which generalise self-similar sets, provide our next example. Let $\mathcal{V}$ be a set of 'vertices' which we label $\{1, 2, \ldots, q\}$, and let $\mathcal{E}$ be a set of 'directed edges' with each edge starting and ending at a vertex so that $(\mathcal{V}, \mathcal{E})$ is a directed graph. A pair of vertices may be joined by several edges and we also allow edges starting and ending at the same vertex. We write $\mathcal{E}_{i,j}$ for the set of edges from vertex i to vertex j, and $\mathcal{E}_{i,j}^k$ for the set of sequences of k edges $(e_1, e_2, \ldots, e_k)$ which form a directed path from vertex i to vertex j. We assume a *transitivity condition*, that there is a positive integer p_0 such that for all i, j there is an integer p with $1 \leq p \leq p_0$ such that $\mathcal{E}_{i,j}^p$ is non-empty; this means that there are paths in the graph joining every pair of vertices.

For each edge $e \in \mathcal{E}$, let $F_e : \mathbb{R}^n \to \mathbb{R}^n$ be a contracting similarity of ratio r_e with $0 < r_e < 1$. Then there is a unique family of non-empty compact sets $E_1, \ldots, E_q$ such that

$$E_i = \bigcup_{j=1}^{q} \bigcup_{e \in \mathcal{E}_{i,j}} F_e(E_j). \tag{3.12}$$

(The proof of this is similar to that of Theorem 2.6 for a conventional IFS, see Exercise 3.6.) The set of contractions $\{F_e : e \in \mathcal{E}\}$ is called a *graph-directed iterated function system* and the sets $\{E_1, \ldots, E_q\}$ are called a family of *graph-directed sets*. By iterating (3.12) we see that

$$E_i = \bigcup_{j=1}^{q} \bigcup_{(e_1, \ldots, e_k) \in \mathcal{E}_{i,j}^k} F_{e_1} \circ \cdots \circ F_{e_k}(E_j). \tag{3.13}$$

We assume that the unions in (3.12) are disjoint for all i; this separation condition may be relaxed to an open set condition.

We shall find the dimension of graph-directed sets in terms of associated $q \times q$ matrices $A^{(s)}$ with (i, j)-th entry given by

$$A_{i,j}^{(s)} = \sum_{e \in \mathcal{E}_{i,j}} r_e^s. \tag{3.14}$$

We write $\rho(A^{(s)})$ to denote the largest eigenvalue of $A^{(s)}$ (in absolute value) which must be real; in fact $\rho(A^{(s)}) = \lim_{k \to \infty} \| (A^{(s)})^k \|^{1/k}$, which is the spectral

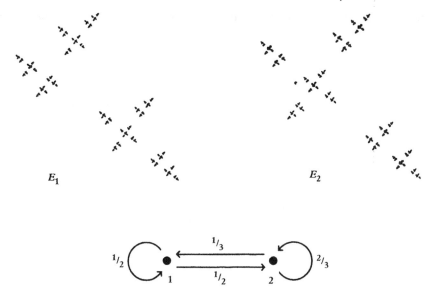

Figure 3.4 A pair of graph-directed sets, with its graph labelled by the similarity ratios. Thus E_1 comprises a scale $\frac{1}{2}$ copy of itself rotated by $90°$ together with a scale $\frac{1}{2}$ copy of E_2, and E_2 comprises a $\frac{2}{3}$ scale copy of itself rotated by $90°$ and a $\frac{1}{3}$ scale copy of E_1

radius of $A^{(s)}$. It may be shown (see Exercise 3.7) that $\rho(A^{(s)})$ is strictly decreasing in s, so that there is a unique positive s such that $\rho(A^{(s)}) = 1$. This value of s turns out to be the dimension of each E_i, a fact which follows easily once we establish (in Corollary 3.5) that $0 < \mathcal{H}^s(E_i) < \infty$ for all i.

A pair of graph-directed sets E_1, E_2 is displayed in Figure 3.4. In this case

$$A^{(s)} = \begin{pmatrix} \left(\frac{1}{2}\right)^s \left(\frac{1}{2}\right)^s \\ \left(\frac{1}{3}\right)^s \left(\frac{2}{3}\right)^s \end{pmatrix}$$

It is easy to check that $\rho(A^{(1)}) = 1$, so Corollary 3.5 will imply that $\dim_H E_i = \dim_B E_i = 1$ for $i = 1, 2$.

The following analysis of graph-directed sets parallels that of self-similar sets in Corollary 3.3.

Corollary 3.5 (graph-directed sets)

Let $E_1, \ldots, E_q$ be a family of graph-directed sets as above. Then there is a number s such that $\dim_H E_i = \underline{\dim}_B E_i = \overline{\dim}_B E_i = s$ and $0 < \mathcal{H}^s(E_i) < \infty$ for all $i = 1, \ldots, q$. Moreover, s is the unique positive number satisfying $\rho(A^{(s)}) = 1$, where $A^{(s)}$ is given by (3.14).

Proof A consequence of the transitivity condition is that for each pair $i, j \, (1 \leq i, j \leq m)$ there is a similarity

$$F_{e_1} \circ \cdots \circ F_{e_p} : E_j \to E_i \tag{3.15}$$

with $(e_1, \ldots, e_p) \in \mathcal{E}_{i,j}^p$ and $p \leq p_0$. In particular, this implies that $\dim_H E_i \geq \dim_H E_j$ for all i, j, so that there is a number s with $\dim_H E_i = s$ for all i.

Let $r_{\min} = \min_{e \in \mathcal{E}_{i,j}} r_e$. Given $x \in E_i$ and $r \leq |E_i|$, there is an integer j and a sequence of edges $(e_1, \ldots, e_k) \in \mathcal{E}_{i,j}^k$ such that $x \in F_{e_1} \circ \cdots \circ F_{e_k}(E_j)$, by (3.13). Choose k so that $rr_{\min}|E_i|^{-1} < r_{e_1} r_{e_2} \cdots r_{e_k} \leq r|E_i|^{-1}$. By (3.15) we may find $(e_{k+1}, \ldots, e_{k+p}) \in \mathcal{E}_{j,i}^p$ where $p \leq p_0$, so $x \in F_{e_1} \circ \cdots \circ F_{e_{k+p}}(E_i)$. Since $r|E_i|^{-1} \geq r_{e_1} \cdots r_{e_k} r_{e_{k+1}} \cdots r_{e_{k+p}} > rr_{\min}|E_i|^{-1} r_{\min}^{p_0}$, the similarity $F_{e_1} \circ \cdots \circ F_{e_{k+p}} : E_i \to E_i \cap B(x, r)$ is of ratio at least $rr_{\min}^{p_0+1}|E_i|^{-1}$. By Theorem 3.2, $s = \dim_H E_i = \underline{\dim}_B E_i = \overline{\dim}_B E_i$ and $\mathcal{H}^s(E_i) < \infty$.

Now fix i and write $d = \min \text{dist}(F_e(E_j), F_{e'}(E_{j'})) > 0$ where the minimum is over distinct e and e' such that $e \in \mathcal{E}_{i,j}$ and $e' \in \mathcal{E}_{i,j'}$. Then

$$\text{dist}(F_{e_1} \circ \cdots \circ F_{e_k}(E_j), F_{e'_1} \circ \cdots \circ F_{e'_k}(E_{j'})) \geq dr_{e_1} \cdots r_{e_{k-1}} \tag{3.16}$$

for distinct edge sequences $(e_1, \ldots, e_k) \in \mathcal{E}_{i,j}^k$ and $(e'_1, \ldots, e'_k) \in \mathcal{E}_{i,j'}^k$. If U intersects E_i and $|U| < d$ and $x \in E_i \cap U$, we may by (3.13) find j, k and $(e_1, \ldots, e_k) \in \mathcal{E}_{i,j}^k$ such that $x \in F_{e_1} \circ \cdots \circ F_{e_k}(E_j)$ and $dr_{e_1} \cdots r_{e_k} \leq |U| < dr_{e_1} \cdots r_{e_{k-1}}$. By (3.13) and (3.16) $E_i \cap U \subset F_{e_1} \circ \cdots \circ F_{e_k}(E_j)$, that is $(F_{e_1} \circ \cdots \circ F_{e_k})^{-1} : E_i \cap U \to E_j$. Using (3.15) we may find $F_{e_{k+1}} \circ \cdots \circ F_{e_{k+p}} : E_j \to E_i$ with $p \leq p_0$ so $(F_{e_{k+1}} \circ \cdots \circ F_{e_{k+p}})(F_{e_1} \circ \cdots \circ F_{e_k})^{-1} : E_i \cap U \to E_i$ is a similarity of ratio $(r_{e_{k+1}} \cdots r_{e_{k+p}})(r_{e_1} \cdots r_{e_k})^{-1} \geq r_{\min}^{p_0} d|U|^{-1}$. By Theorem 3.1, $\mathcal{H}^s(E_i) > 0$.

The unions in (3.12) are assumed to be disjoint, so for each i

$$\mathcal{H}^s(E_i) = \sum_{j=1}^{q} \sum_{e \in \mathcal{E}_{i,j}} \mathcal{H}^s(F_e(E_j))$$

$$= \sum_{j=1}^{q} \sum_{e \in \mathcal{E}_{i,j}} r_e^s \mathcal{H}^s(E_j).$$

In matrix form

$$\begin{pmatrix} \mathcal{H}^s(E_1) \\ \vdots \\ \mathcal{H}^s(E_q) \end{pmatrix} = A^{(s)} \begin{pmatrix} \mathcal{H}^s(E_1) \\ \vdots \\ \mathcal{H}^s(E_q) \end{pmatrix}$$

where $A^{(s)}$ is the matrix given by (3.14). By the Perron–Frobenius theorem any matrix which has non-negative entries has an eigenvector with non-negative components that is unique to within a scalar multiple and which corresponds to the unique eigenvalue of largest absolute value. In this case, taking $s = \dim_H E_i$

(for all i), we know that $(\mathcal{H}^s(E_1), \ldots, \mathcal{H}^s(E_q))^T$ is an eigenvector of $A^{(s)}$ with positive components and eigenvalue 1, which must therefore be the largest eigenvalue. Thus $\rho(A^{(s)}) = 1$. Since $\rho(A^{(s)})$ is strictly decreasing with s (see Exercise 3.7), s is completely specified by this condition. $\quad\square$

We give an application of these implicit methods to cookie-cutter sets in Corollary 4.6.

3.2 Box-counting dimensions of cut-out sets

In this section we investigate how the box-counting dimensions of certain fractals can be found. We will be particularly concerned with fractals that may be obtained by removing or 'cutting out' a sequence of disjoint regions from an initial set. All compact subsets of $\mathbb{R}$ can be obtained in this way. For example, the middle-third Cantor set may be obtained from $[0, 1]$ by removing the sequence of open intervals $(\frac{1}{3}, \frac{2}{3}), (\frac{1}{9}, \frac{2}{9}), (\frac{7}{9}, \frac{8}{9}), (\frac{1}{27}, \frac{2}{27}), \ldots$. Similarily, in the plane, the Sierpinski triangle is obtained by removing a sequence of equilateral triangles from an initial equilateral triangle. We treat subsets of $\mathbb{R}$ in some detail, and then briefly discuss higher-dimensional analogues. We shall see that the box dimensions of a subset of $\mathbb{R}$ depend only on the size of the complementary intervals and not on their arrangement. Thus, in a sense, box dimension describes the complement of a set rather than the set itself.

Let A be a bounded closed interval in $\mathbb{R}$, and let $A_1, A_2, \ldots$ be a sequence of disjoint open subintervals of A with $|A| = \sum_{i=1}^{\infty} |A_i|$. (Of course $|A_i|$ is just the length of A_i). Let $E = A \backslash \bigcup_{i=1}^{\infty} A_i$ so that E is a compact set of Lebesgue measure (or length) zero, with complementary intervals A_k. We call such a set a *cut-out set* when we wish to emphasise its construction by cutting out a sequence of intervals. We write $a_i = |A_i|$ for the length of A_i, and assume that these intervals are ordered by decreasing length, so that $a_1 \geq a_2 \geq a_3 \geq \ldots$, see Figure 3.5(a).

Whilst we could work with any of the equivalent definitions of lower and upper box dimension (see Section 2.1) it is convenient to use the Minkowski definition in terms of the size of the r-neighbourhood E_r of E, see (2.4)–(2.6). We write $V(r) = \mathcal{L}^1(E_r)$ for the 1-dimensional Lebesgue measure (or length) of E_r. It is easy to express $V(r)$ in terms of the lengths of the intervals A_i. Assuming $r \leq \frac{1}{2}a_1$, let k be an integer such that $a_{k+1} \leq 2r \leq a_k$. Then E_r consists of all the intervals A_i with $i \geq k+1$, together with two intervals of length r inside each A_i with $1 \leq i \leq k$, and an interval of length r at each end of the set E, see Figure 3.5(b). Thus

$$V(r) = 2(k+1)r + \sum_{k+1}^{\infty} a_i \quad \text{where} \quad a_{k+1} \leq 2r \leq a_k. \tag{3.17}$$

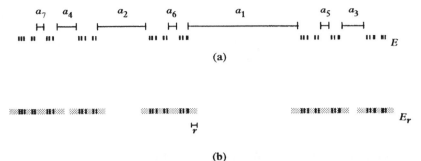

Figure 3.5   (a) The gap lengths of a cut-out set E. (b) The r-neighbourhood E_r of E

In principle, this formula in conjunction with (2.4)–(2.6) enables us to find the box dimensions of E from a knowledge of the lengths a_k. In particular, the dimensions depend only on these lengths and not on the arrangement of the corresponding disjoint intervals A_i within A. The next proposition bounds the box dimensions of E in terms of the limiting behaviour of the a_i.

We will need the following inequality. If $(a_k)_{k=1}^\infty$ is a decreasing sequence convergent to 0 and $0 < \alpha < 1$ then

$$\sum_{i=k}^\infty a_i^{-\alpha}(a_i - a_{i+1}) \le (1 - \alpha)^{-1} a_k^{1-\alpha}. \tag{3.18}$$

To see this, note that the left-hand side is a lower sum for the integral of $x^{-\alpha}$ over the interval $[0, a_k]$.

Proposition 3.6

Write

$$a = -\lim_{k\to\infty} \inf \frac{\log a_k}{\log k} \quad and \quad b = -\lim_{k\to\infty} \sup \frac{\log a_k}{\log k}.$$

Then $1 \le b \le a$ and

$$1/a \le \underline{\dim}_B E \le \overline{\dim}_B E \le 1/b. \tag{3.19}$$

Proof Since $ka_k \le |E|$, we have $1 \le b \le a$. Using approximation it is enough to deduce (3.19) from the assumption that

$$c_1 k^{-a} \le a_k \le c_2 k^{-b} \tag{3.20}$$

for all sufficiently large k where $0 < c_1, c_2 < \infty$. If r is small enough and

$$a_{k+1} \le 2r < a_k \tag{3.21}$$

then (3.17), (3.20) and (3.21) give

$$V(r) \geq 2(k+1)r \geq 2c_1^{1/a}a_{k+1}^{-1/a}r \geq 2c_1^{1/a}2^{-1/a}r^{1-1/a}.$$

Using (2.4) it follows that $1/a \leq \underline{\dim}_B E$.

For the upper bound we can assume $b > 1$ in (3.20). If r is small enough and k is chosen to satisfy (3.21), identity (3.17) gives

$$V(r) = 2(k+1)r + \sum_{k+1}^{\infty} a_i$$

$$= (k+1)(2r - a_{k+1}) + \sum_{k+1}^{\infty}(i+1)(a_i - a_{i+1})$$

$$\leq 4kr + 2\sum_{k+1}^{\infty} i(a_i - a_{i+1})$$

$$\leq 4c_2^{1/b}a_k^{-1/b}r + 2c_2^{1/b}\sum_{k+1}^{\infty} a_i^{-1/b}(a_i - a_{i+1})$$

$$\leq 4c_2^{1/b}2^{-1/b}r^{1-1/b} + 2c_2^{1/b}(1 - 1/b)^{-1}a_{k+1}^{1-1/b}$$

$$\leq c_3 r^{1-1/b}$$

for a constant c_3, using (3.20), (3.21) and (3.18). The estimate $\overline{\dim}_B E \leq 1/b$ follows using (2.5). □

Of course (3.19) remains true for $b = \infty$, where '$1/\infty = 0$'.

It is immediate from Proposition 3.6 that $\dim_B E = -1/\lim(\log a_k/\log k)$ if this limit exists, a formula which allows the box dimension of many sets to be found. For example, if E is the middle-third Cantor set, the lengths $(a_k)_{k=1}^{\infty}$ of the complementary intervals are given by $a_k = 3^{-m-1}$ where m is the integer such that $2^m \leq k \leq 2^{m+1} - 1$. Then we have that $\lim_{k\to\infty}\log a_k/\log k = \lim_{m\to\infty} -(m+1)\log 3/m\log 2 = -\log 3/\log 2$, so $\dim_B E = \log 2/\log 3$.

Similarly, for the 'convergent sequence sets' defined by $E^{(p)} = \{0, 1, 2^{-p}, 3^{-p}, 4^{-p}, \ldots\}$ for $p > 0$, we have $a_k = k^{-p} - (k+1)^{-p} \sim pk^{-p-1}$ (using the mean value theorem), so $\lim_{k\to\infty}\log a_k/\log k = \lim_{k\to\infty}\log pk^{-p-1}/\log k = -(p+1)$, giving $\dim_B E^{(p)} = 1/(p+1)$.

The following partial converse to Proposition 3.6 indicates what we may deduce about the complementary interval lengths $(a_k)_{k=1}^{\infty}$ given the box dimensions.

Proposition 3.7

Suppose

$$t = \underline{\dim}_B E \quad and \quad s = \overline{\dim}_B E \tag{3.22}$$

where $0 < t \leq s < 1$. *Then*

$$-(1-t)/(t(1-s)) \leq \liminf_{k \to \infty} \frac{\log a_k}{\log k} \leq \limsup_{k \to \infty} \frac{\log a_k}{\log k} \leq -1/s. \qquad (3.23)$$

Proof From the Minkowski definition of the box dimension (2.5)–(2.6) it is enough to derive (3.23) on the assumption that for all small enough r

$$c_1 r^{1-t} \leq V(r) \leq c_2 r^{1-s} \qquad (3.24)$$

for positive constants c_1, c_2 with $0 < t < s < 1$. Using (3.17) this gives

$$c_1 r^{1-t} \leq 2(k+1)r + \sum_{k+1}^{\infty} a_i \leq c_2 r^{1-s} \quad \text{where} \quad a_{k+1} \leq 2r \leq a_k. \qquad (3.25)$$

Taking $r = \frac{1}{2}a_k$, the right-hand inequality gives $a_k(k+1) \leq c_2 2^{s-1} a_k^{1-s}$, for the right-hand inequality of (3.23).

Now write $\gamma = (1-s)/(1-t)$ and choose $b \geq 1$ such that

$$c_1 b^{1-t} \geq 2c_2. \qquad (3.26)$$

Take $r = ba_k^\gamma$ and let q be the integer such that $a_{q+1} \leq 2r < a_q$ (so $q < k$). Provided that k is large enough, (3.17) gives

$$V(r) - V(\tfrac{1}{2}a_k) = 2(q+1)r + \sum_{q+1}^{\infty} a_i - (k+1)a_k - \sum_{k+1}^{\infty} a_i$$

$$= 2(q+1)r - (k+1)a_k + \sum_{q+1}^{k} a_i$$

$$= (q+1)(2r - a_{q+1}) + \sum_{q+1}^{k-1}(i+1)(a_i - a_{i+1})$$

$$\leq k(2r - a_{q+1}) + k(a_{q+1} - a_k)$$

$$= k(2r - a_k)$$

$$\leq 2kr.$$

Hence from (3.26) and (3.24)

$$c_2 a_k^{1-s} \leq c_1 b^{1-t} a_k^{1-s} - c_2 a_k^{1-s}$$

$$= c_1 r^{1-t} - c_2 a_k^{1-s}$$

$$\leq V(r) - V(a_k)$$

$$\leq V(r) - V(\tfrac{1}{2}a_k)$$

$$\leq 2kba_k^\gamma.$$

Thus $c_3 k^{-1} \leq a_k^{\gamma-1+s} = a_k^{t(1-s)/(1-t)}$, giving $a_k \geq c_4 k^{-(1-t)/t(1-s)}$ where c_3, c_4 are independent of k, as required for the left-hand inequality of (3.23). $\square$

It is easy to check that the right-hand inequality (3.23) holds if $s = 0$ or $s = 1$, and if $t = s = 1$ then $\lim \log a_k / \log k = -1$.

It can be shown that the bounds stated in (3.23) are the best that can be achieved. In particular, taking a_k to be constant for blocks of rapidly increasing length of consecutive k shows that the lower bound is best-possible.

A satisfying corollary of Propositions 3.6 and 3.7 is that the box dimension of E exists precisely when the limit of $\log a_k / \log k$ does.

Corollary 3.8

Let $E \subset \mathbb{R}$ be a cut-out set, as above. Then $\underline{\dim}_B E = \overline{\dim}_B E$ if and only if $\lim_{k \to \infty} \log a_k / \log k$ *exists, in which case*

$$\dim_B E = -1 / (\lim_{k \to \infty} \log a_k / \log k).$$

Proof From Proposition 3.6, if $1 \le a = b \le \infty$ then $1/a \le \underline{\dim}_B \le \overline{\dim}_B E \le 1/b = 1/a$.

Conversely, if $0 \le t = s \le 1$, Proposition 3.7 and the remark following give that $\lim \log a_k / \log k = -1/s$. $\square$

A similar analysis may be carried out for cut-out sets in higher-dimensional space, though the applicability is less general. We illustrate this for a set constructed by removing a sequence of discs from a plane region.

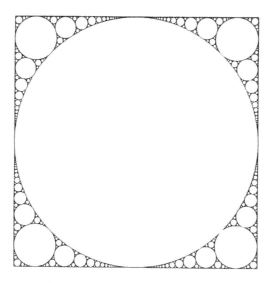

Figure 3.6 A cut-out set in the plane. Here, the largest possible disc is removed at each step. The family of discs removed is called the Apollonian packing of the square, and the cut-out set remaining is called the residual set, which has Hausdorff and box dimension about 1.31

For convenience let A be a plane compact convex region of perimeter length p, and let $A_1, A_2, \ldots$ be a sequence of disjoint open discs contained in A with total area equal to that of A. Let $E = A \setminus \bigcup_{i=1}^{\infty} A_i$, so that E is a set of area zero with a cellular appearance, see Figure 3.6. Let r_i be the radius of A_i, and assume that $r_1 \geq r_2 \geq r_3 \geq \ldots$.

Just as in the case of subsets of the line we can find the area $V(r)$ of the r-neighbourhood E_r of E. If $r_{k+1} \leq r \leq r_k$ then $E_r \setminus E$ consists of a band of points outside A but within distance r of A, an annulus with inner and outer radii $r_i - r$ and r_i inside each A_i for $1 \leq i \leq k$, together with the discs A_i for $i \geq k+1$. Thus

$$V(r) = (pr + \pi r^2) + \sum_{i=1}^{k} \pi(r_i^2 - (r_i - r)^2) + \sum_{i=k+1}^{\infty} \pi r_i^2$$

$$= pr + 2\pi r \sum_{i=1}^{k} r_i + \pi \sum_{i=k+1}^{\infty} r_i^2 + \pi r^2(1 - k).$$

It is possible to relate $V(r)$, and thus $\dim_B E$, to the r_i as in the one-dimensional case. For example, suppose $r_k \asymp k^{-a}$ where $\frac{1}{2} < a < 1$. (Recall that $a_k \asymp b_k$ means that for some $c_1, c_2 \geq 0$ we have $c_1 \leq a_k/b_k \leq c_2$ for all k.) Then for $r_{k+1} \leq r \leq r_k$ we have

$$V(r) \asymp r + r \sum_{1}^{k} i^{-a} + \sum_{k+1}^{\infty} i^{-2a} - r^2 k$$

$$\asymp r k^{1-a} + k^{1-2a}$$

$$\asymp r r_k^{(a-1)/a} + r_k^{(2a-1)/a}$$

$$\asymp r^{2-1/a}.$$

From (2.6) $\dim_B E = 1/a$.

Clearly a similar approach may be adopted to find the box dimensions of sets obtained by cutting out regions of other shapes.

3.3 Notes and references

A basic 'implicit theorem' was given by McLaughlin (1987) with further results and applications in Falconer (1989). For further details of graph-directed sets see Bedford (1986), Mauldin and Williams (1988) and Edgar (1990) and for sub-self-similar sets see Bandt (1989) and Falconer (1995a). The Perron–Frobenius theorem is discussed, for example, in Bellman (1960). Besicovitch and Taylor (1954) studied the relationship between the lengths of complementary intervals and the dimension of a subset of $\mathbb{R}$. Results on box dimension akin to Propositions 3.6 and 3.7 may be found in Lapidus and Pommerance (1993) and Falconer (1995b).

Exercises

3.1 Give examples to show that Theorems 3.1 and 3.2 fail without the stipulation that E is compact.

3.2 Show that the hypotheses of Theorem 3.2 imply that $\mathcal{P}^s(E) \leq \mathcal{P}_0^s(E) < \infty$ where $\mathcal{P}^s$ and $\mathcal{P}_0^s$ are the packing measure and pre-measure, see Section 2.1. (Modify the proof of Theorem 3.2 by replacing (3.5) with the assumption $\mathcal{P}_r^s(E) > a^{-s}$ to get disjoint balls $B_1, \ldots, B_m$ with $\sum |B_i|^t > a^{-t}$ for $t > s$. Then take $g_i : E \to E \cap B_i$ satisfying $a|B_i||x - y| \leq |g_i(x) - g_i(y)|$ and deduce in a similar way that $\dim_H E > s$.)

3.3 Show that Theorem 3.1 remains true with the second sentence replaced by: 'Suppose that there is a number q such that for every set U that intersects E with $|U| < r_0$, there exist, for $i = 1, \ldots, q$, sets U_i such that $U \subset \cup_{i=1}^q U_i$, and mappings $g_i : E \cap U_i \to E$ satisfying $a|U|^{-1}|x - y| \leq |g_i(x) - g_i(y)|$.'

3.4 Show using Exercise 3.3 that the final conclusion of Corollary 3.3 holds if we merely require that E satisfies the open set condition (2.40).

3.5 Let E be the subset of the middle-third Cantor set consisting of those numbers in $[0, 1]$ with base 3 expansion containing only the digits 0 and 2, and where two consecutive digits 2 are not allowed. Let $E_1 = E \cap [0, \frac{1}{3}]$ and $E_2 = E \cap [\frac{2}{3}, 0]$. Show that E_1, E_2 may be represented as a system of graph-directed sets, and show that $\dim_H E_1 = \dim_H E_2 = \log((1 + \sqrt{5})/2) /\log 3$.

3.6 Show that there is a unique family of non-empty compact sets E_i satisfying (3.12). (Hint: define a metric on q-tuples of sets so that the distance between two q-tuples is the maximum of the Hausdorff distances between corresponding pairs of sets, and mimic the proof of Theorem 2.6).

3.7 In the case of a family of two graph-directed sets with associated matrices $A^{(s)}$, show that the largest eigenvalues are given by

$$\rho(A^{(s)}) = \frac{1}{2}(A_{1,1}^{(s)} + A_{2,2}^{(s)}) + \frac{1}{2}((A_{1,1}^{(s)} - A_{2,2}^{(s)})^2 + 4A_{1,2}^{(s)}A_{2,1}^{(s)})^{1/2}.$$

By examining the effect on $\rho(A^{(s)})$ of decreasing each term $A_{i,j}^{(s)}$, show that $\rho(A^{(s)})$ is strictly decreasing in s.

3.8 Let E be the (compact) subset of $[0, 1]$ consisting of those numbers with decimal expansions containing only the digits 0, 2, 4, 6, 8. Deduce from Theorem 3.1 (or Theorem 3.2) that $\dim_H E = \dim_B E$. Deduce from Proposition 3.6 that $\dim_B E = \log 5/ \log 10$.

3.9 Let $V(r)$ be the area of the r-neighbourhood of the Sierpinski triangle of height 1. Show that $V(r) \asymp 3^k 2^{-2k}$ where $2^{-k} < r \leq 2^{-k-1}$, and deduce that the box dimension of the Sierpinski triangle is $\log 3/ \log 2$.

3.10 Let E be a compact subset of $\mathbb{R}$, and let E' be obtained from E by adding a single point inside each complementary interval of E. Show that $\dim_B E' = \dim_B E$ (assuming that $\dim_B E$ exists).

Chapter 4 Cookie-cutters and bounded distortion

In this chapter we introduce cookie-cutter sets, which may be thought of as 'non-linear Cantor sets'. This leads onto Chapter 5 where we shall use cookie-cutters to show how the theory of self-similar sets may be extended to non-linear analogues. Working with cookie-cutters allows the essential ideas of some very general theory to be presented in a relatively simple setting.

4.1 Cookie-cutter sets

We study a simple form of dynamical system called a cookie-cutter. A cookie-cutter has a fractal repeller, called a cookie-cutter set, that can be thought of as the attractor of a related iterated function system. For simplicity, we set this up so that the IFS has just two mappings, although extending the theory to more mappings requires few changes. We work in a subset X of $\mathbb{R}$, though the theory generalises to subsets of $\mathbb{R}^n$.

Let X be a bounded non-empty closed interval, and let X_1 and X_2 be disjoint subintervals of X. Let $f: X_1 \cup X_2 \to X$ be such that X_1 and X_2 are each mapped bijectively onto X (see Figure 4.1). We assume that f has a continuous derivative (later we will require a stronger differentiability condition) and is expanding, so that $|f'(x)| > 1$ for all $x \in X_1 \cup X_2$. (Often, f will be the restriction to a set X of a function defined on a larger domain, perhaps even the whole of $\mathbb{R}$. For example, f might be the restriction to X of a unimodal function as in Figure 4.2.)

We study the dynamical system given by iterating points by f. Of particular interest is the set

$$E = \{x \in X : f^k(x) \text{ is defined and in } X_1 \cup X_2 \text{ for all } k = 0, 1, 2, \ldots\} \quad (4.1)$$

where f^k is the k-th iterate of f. Thus E is the set of points that remain in $X_1 \cup X_2$ under iteration by f. Since $E = \cap_{k=0}^{\infty} f^{-k}(X)$, a decreasing sequence of compact sets, the set E is compact and non-empty.

Clearly, E is invariant under f, in that

$$f(E) = E = f^{-1}(E), \quad (4.2)$$

since $x \in E$ if and only if $f(x) \in E$. Moreover E is a *repeller*, in the sense that

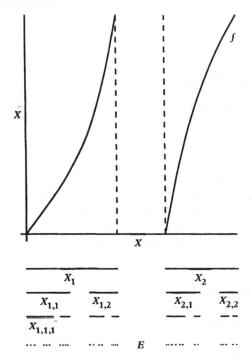

Figure 4.1   A cookie-cutter function $f: X_1 \cup X_2 \to X$ with repelling cookie-cutter set $E = \bigcap_{k=0}^{\infty} f^{-k}(X)$

points not in E (however close to E they may be) are eventually mapped outside $X_1 \cup X_2$ under iteration by f. Indeed, in examples such as Figure 4.2, $f^k(x) \to -\infty$ for all $x \notin E$, see FG, Section 13.1.

An equivalent way of viewing this situation is as the 'inverse' of an iterated function scheme. We define $F_1, F_2 : X \to X$ as the two branches of the inverse of f. Thus

$$F_1(x) = f^{-1}(x) \cap X_1$$
$$F_2(x) = f^{-1}(x) \cap X_2$$

so F_1 and F_2 map X bijectively onto X_1 and X_2 respectively. Then

$$f(x) = \begin{cases} F_1^{-1}(x) & (x \in X_1) \\ F_2^{-1}(x) & (x \in X_2). \end{cases} \qquad (4.3)$$

Since f has a continuous derivative with $|f'(x)| > 1$ on the compact set $X_1 \cup X_2$, there are numbers $0 < c_{min} \leq c_{max} < 1$ such that $1 < c_{max}^{-1} \leq |f'(x)| \leq c_{min}^{-1} < \infty$ for all $x \in X_1 \cup X_2$. It follows that the inverse functions F_1, F_2 are

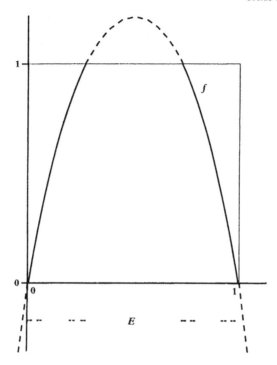

Figure 4.2 A unimodal function $f: \mathbb{R} \to \mathbb{R}$ gives rise to a cookie-cutter system. In this example the cookie-cutter set $E = \cap_{k=0}^{\infty} f^{-k}[0,1]$ is a repeller for f, with $\lim_{k \to \infty} f^k(x) = -\infty$ for all $x \notin E$

differentiable, with $c_{min} \leq |F_1'(x)|, |F_2'(x)| \leq c_{max}$ for $x \in X$. By the mean value theorem, for $i = 1, 2$

$$c_{min}|x - y| \leq |F_i(x) - F_i(y)| \leq c_{max}|x - y| \quad (x, y \in X). \tag{4.4}$$

From (4.1) the repeller E of f satisfies

$$E = F_1(E) \cup F_2(E). \tag{4.5}$$

Since F_1 and F_2 are contractions on X, equation (4.5) is satisfied by a unique non-empty compact set E, using the fundamental IFS property, Theorem 2.5. Thus the repeller E of f is the attractor of the IFS $\{F_1, F_2\}$.

As in Section 2.2, we index the intervals associated with the IFS by the sequences $I_k = \{(i_1, \ldots, i_k) : i_j = 1 \text{ or } 2\}$ formed by 1s and 2s, with $I = \cup_{k=0}^{\infty} I_k$. In particular, for each $\boldsymbol{i} = (i_1, \ldots, i_k)$ we write $X_{\boldsymbol{i}} = X_{i_1, \ldots, i_k} = F_{i_1} \circ \cdots \circ F_{i_k}(X)$. Thus $f^k : X_{\boldsymbol{i}} \to X$ is a bijection between closed intervals with inverse $F_{i_1} \circ \cdots \circ F_{i_k}$, and more generally, $f^k : X_{i_1, \ldots, i_m} \to X_{i_{k+1}, \ldots, i_m}$ is a bijection for each $k \leq m$. Since $X \supset X_1 \cup X_2$ with the union disjoint, we have $X_{\boldsymbol{i}} \supset X_{\boldsymbol{i},1} \cup X_{\boldsymbol{i},2}$

with the union disjoint for all i. Thus $E_k \equiv \cup_{i \in I_k} X_i$ consists of 2^k disjoint closed intervals, with $(E_k)_{k=0}^{\infty}$ a decreasing sequence of compact sets. As in (2.28) $E = \cap_{k=0}^{\infty} E_k$, and E is totally disconnected and topologically equivalent to a Cantor set.

We note, using (4.4), that

$$c_{\min}|X_i| \leq |X_{i,i}| \leq c_{\max}|X_i| \tag{4.6}$$

for all $i \in I$ and $i = 1, 2$.

In the simplest case, the IFS consisting of similarity transformations (2.24) corresponds to the map $f : [0, \frac{1}{3}] \cup [\frac{2}{3}, 1] \to [0, 1]$ given by $f(x) = 3x \pmod 1$. Then the repeller E is the middle-third Cantor set and E_k consists of 2^k intervals of length 3^{-k}. However, we shall be more interested in the case where F_1, F_2 are not similarity maps. For example, $F_1, F_2 = [0, 1] \to [0, 1]$ defined by

$$F_1(x) = \tfrac{1}{3}x + \tfrac{1}{10}x^2, \quad F_2(x) = \tfrac{1}{3}x + \tfrac{2}{3} - \tfrac{1}{10}x^2 \tag{4.7}$$

gives an invariant set E that is a 'non-linear perturbation' of the middle-third Cantor set. Equivalently, E is the repeller of a non-linear function f defined on a pair of subintervals of $[0, 1]$ in terms of F_1 and F_2 by (4.3).

A dynamical system $f : X_1 \cup X_2 \to X$ of this form, or the equivalent IFS $\{F_1, F_2\}$ on X, is termed a *cookie-cutter system* and the set E is called a *cookie-cutter set*. In general the mappings F_1 and F_2 are not similarity transformations, and E is a 'distorted' Cantor set, which nevertheless is 'approximately self-similar'.

4.2 Bounded distortion for cookie-cutters

The principle of bounded distortion makes precise the idea of a set being 'approximately self-similar', in that any sufficiently small neighbourhood may be mapped onto a large part of the set by a transformation that is not unduly distorting.

We first prove the 'principle of bounded variation' for a general function ϕ defined on a cookie-cutter set E, and then choose ϕ in a way that relates to the geometry of E to obtain the bounded distortion result. Let $f : X_1 \cup X_2 \to X$ be a cookie-cutter system with corresponding IFS $\{F_1, F_2\}$ and repeller E, as in Section 4.1. Let $\phi : X_1 \cup X_2 \to \mathbb{R}$ be a Lipschitz function, satisfying

$$|\phi(x) - \phi(y)| \leq a|x - y| \quad (x, y \in X_1 \cup X_2) \tag{4.8}$$

for some $a > 0$. (In fact the theory merely requires $\phi : E \to \mathbb{R}$, but in practice ϕ is usually defined naturally on a larger set than E.)

We are interested in the values of ϕ at successive iterates of points under f. In particular we shall estimate the sums

$$S_k\phi(x) \equiv \phi(x) + \phi(fx) + \phi(f^2x) + \cdots + \phi(f^{k-1}x)$$
$$= \sum_{j=0}^{k-1} \phi(f^jx) \tag{4.9}$$

for $k = 1, 2, \ldots$, where f^jx is the j-th iterate of x under f. (To avoid excessively clumsy notation we often write f^jx for $f^j(x)$ in this context.) Note that $S_k\phi(x)$ is defined provided $x \in X_i$ for some $i \in I_k$. If $x = F_{i_1} \circ \cdots \circ F_{i_k}w$ for $w \in X$, we have the alternative form

$$S_k\phi(F_{i_1} \circ \cdots \circ F_{i_k}w) = \sum_{j=1}^{k} \phi(F_{i_j} \circ \cdots \circ F_{i_k}w). \tag{4.10}$$

We may think of $\frac{1}{k}S_k\phi(x)$ as the average of ϕ at x and its first $k-1$ iterates.

The principle of bounded variation is a consequence of the Lipschitz condition on ϕ. The principle asserts that the sums $S_k\phi(x)$ do not vary too much with x in a sense that is uniform in k.

Proposition 4.1 (principle of bounded variation)

Let $\phi : X \to \mathbb{R}$ be a Lipschitz function.

(a) *There exists a number b such that for all $k = 1, 2, \ldots$ and all $(i_1, \ldots, i_k) \in I_k$ we have*

$$|S_k\phi(x) - S_k\phi(y)| \leq b \tag{4.11}$$

whenever $x, y \in X_{i_1,\ldots,i_k}$.

(b) *More generally for all $q \geq k$ and all $(i_1, \ldots, i_q) \in I_q$ we have*

$$|S_k\phi(x) - S_k\phi(y)| \leq b|X|^{-1}|X_{i_{k+1},\ldots,i_q}| \tag{4.12}$$

whenever $x, y \in X_{i_1,\ldots,i_q}$.

Proof By repeated application of (4.4) we have $|X_{i_1,\ldots,i_k}| = |F_{i_1} \circ \cdots \circ F_{i_k}(X)| \leq c_{\max}^k|X|$ for all $(i_1, \ldots, i_k) \in I$. If $x, y \in X_{i_1,\ldots,i_k}$ then $f^jx, f^jy \in X_{i_{j+1},\ldots,i_k}$ for $j = 0, 1, \ldots, k-1$, so by (4.8)

$$|\phi(f^jx) - \phi(f^jy)| \leq a|f^jx - f^jy|$$
$$\leq a|X_{i_{j+1},\ldots,i_k}|$$
$$\leq ac_{\max}^{k-j}|X|.$$

Hence

$$|S_k\phi(x) - S_k\phi(y)| = \left| \sum_{j=0}^{k-1} \phi(f^j x) - \sum_{j=0}^{k-1} \phi(f^j y) \right|$$

$$\le \sum_{j=0}^{k-1} |\phi(f^j x) - \phi(f^j y)|$$

$$\le \sum_{j=0}^{k-1} ac_{\max}^{k-j}|X|$$

$$\le ac_{\max}|X|/(1 - c_{\max}),$$

giving (4.11) with $b = ac_{\max}|X|/(1 - c_{\max})$.

The proof of (b) is very similar, noting that if $x, y \in X_{i_1,\ldots,i_q}$ then $f^j x, f^j y \in X_{i_{j+1},\ldots,i_q}$ so that

$$|\phi(f^j x) - \phi(f^j y)| \le ac_{\max}^{k-j}|X_{i_{k+1},\ldots,i_q}|. \qquad \square$$

It is sometimes convenient to write (4.11) in the form

$$e^{-b} \le \frac{\exp S_k\phi(x)}{\exp S_k\phi(y)} \le e^b. \tag{4.13}$$

We now assume that $f: X_1 \cup X_2 \to X$ is of differentiability class C^2, that is, twice differentiable with continuous second derivative (in the one-sided sense at the interval ends). Equivalently, F_1 and F_2 are of class C^2 on X. We choose

$$\phi(x) = -\log|f'(x)| \tag{4.14}$$

for $x \in X_1 \cup X_2$. Since $0 < |f'(x)|$, the function ϕ has a bounded continuous first derivative on $X_1 \cup X_2$. By the mean value theorem, ϕ satisfies a Lipschitz condition on both X_1 and X_2 and so on $X_1 \cup X_2$.

The sums $S_k\phi$ turn out to be just what we need to estimate the size of the X_i. Applying the chain rule for the derivative of a composition of functions to f^k we get

$$(f^k)'(x) = f'(f^{k-1}x) \times f'(f^{k-2}x) \times \cdots \times f'(x), \tag{4.15}$$

where dashes denote differentiation. Taking logarithms,

$$-\log|(f^k)'(x)| = \sum_{j=0}^{k-1} -\log|f'(f^j x)| \tag{4.16}$$

$$= \sum_{j=0}^{k-1} \phi(f^j x)$$

$$= S_k\phi(x). \tag{4.17}$$

using (4.14) and (4.9). (This is valid provided we have $f^j x \in X_1 \cup X_2$ for

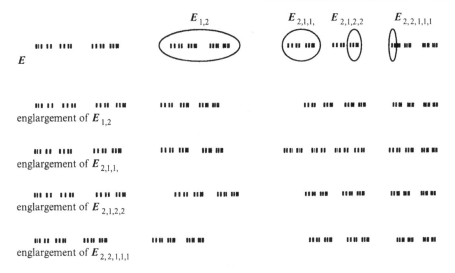

Figure 4.3 The principle of bounded distortion for a cookie-cutter set E. The figure shows similar copies of some of the components $E_{i_1,\dots,i_k}$, which look like 'distorted' versions of E. The 'amount' of distortion is bounded over all k and $(i_1,\cdots,i_k)$

$j = 0, 1, \dots, k-1$, that is for $x \in \cup_{i \in I_k} X_i$.) Thus, the sums $S_k\phi$ have a nice interpretation in terms of the derivative of iterates of f.

The mapping $f^k : X_{i_1,\dots,i_k} \to X$ is a bijection, but much more than this, it is a bi-Lipschitz mapping with constants not too different from $|X_{i_1,\dots,i_k}|^{-1}$; in particular $|X_{i_1,\dots,i_k}| \asymp |X||(f^k)'(x)|^{-1}$ for all $x \in X_{i_1,\dots,i_k}$, see Figure 4.3. This is made precise in the following proposition in which we apply the principle of bounded variation taking ϕ as in (4.14).

Proposition 4.2   (principle of bounded distortion)

There are numbers b_0 and b_1 such that for all $k = 0, 1, 2, \dots$ and for all $(i_1, \dots, i_k) \in I_k$, we have

$$b_0^{-1} \le |X_{i_1,\dots,i_k}||(f^k)'(x)| \le b_0 \tag{4.18}$$

for all $x \in X_{i_1,\dots,i_k}$. Moreover, $f^k : X_{i_1,\dots,i_k} \to X$ satisfies

$$b_1^{-1}|y - z| \le |f^k(y) - f^k(z)||X_{i_1,\dots,i_k}| \le b_1|y - z| \tag{4.19}$$

for all $y, z \in X_{i_1,\dots,i_k}$.

Proof We have $X_{i_1,\dots,i_k} = F_{i_1} \circ \cdots \circ F_{i_k}(X)$, so $f^k : X_{i_1,\dots,i_k} \to X$ is a differentiable bijection. Applying the mean value theorem to f^k gives that for

$y, z \in X_{i_1,...,i_k}$ there exists $w \in X_{i_1,...,i_k}$ such that

$$f^k(y) - f^k(z) = (y - z)(f^k)'(w). \tag{4.20}$$

Choosing y and z to be the end-points of $X_{i_1,...,i_k}$ then $f^k(y), f^k(z)$ are the end-points of X, so

$$|X| = |X_{i_1,...,i_k}||(f^k)'(w)| \tag{4.21}$$

for some $w \in X_{i_1,...,i_k}$. Using the bounded variation principle in the form (4.13) with (4.17) we see that for all $x, w \in X_{i_1,...,i_k}$

$$e^{-b} \leq \frac{|(f^k)'(x)|}{|(f^k)'(w)|} \leq e^b. \tag{4.22}$$

Combining this with (4.21) gives (4.18), and then (4.20) gives (4.19), where b_0 and b_1 depend only on $|X|$ and b. $\square$

In the form (4.19), the principle of bounded distortion says that the f^k may be uniformly approximated by similarity transformations.

It is sometimes useful to write (4.18) in the alternative form

$$b_0^{-1} \leq \frac{|X_{i_1,...,i_k}|}{|(F_{i_1} \circ \cdots \circ F_{i_k})'(x)|} \leq b_0$$

for all $x \in X$. Note that in the special case where F_1, F_2 are similarities with ratios c_1, c_2, then $f'(x)$ is constant on $X_{i_1,...,i_k}$, and the argument reduces to give $|X_{i_1,...,i_k}| = c_{i_1} c_{i_2} \ldots c_{i_k} |X|$, as would be expected for a self-similar set.

The crucial point of Proposition 4.2 is that b_0 and b_1 do not depend on k. Although the mean value estimate is applied to the composition of k functions, $f^j(x)$, and thus $f'(f^j(x))$, does not vary much as x ranges over $X_{i_1,...,i_k}$ except when j is close to k. This controls the variation of $(f^k)'(x)$ over $X_{i_1,...,i_k}$, using (4.15).

One useful consequence of the bounded distortion principle is that for each i, the sets $X_{i,1}$ and $X_{i,2}$ are reasonably well separated inside X_i. Furthermore the X_i are comparable with balls (intervals) in a uniform way.

Corollary 4.3

Let E be a cookie-cutter set and let $d = \text{dist}(X_1, X_2)$.

(a) For all $i \in I$

$$db_1^{-1}|X_i| \leq \text{dist}(X_{i,1}, X_{i,2}) \leq |X_i|. \tag{4.23}$$

(b) Let $\lambda = db_1^{-1} c_{\min}$. For all i, if $x \in X_i \cap E$ and $|X_i| \leq r < |X_i| c_{\min}^{-1}$, then

$$B(x, \lambda r) \cap E \subset X_i \cap E \subset B(x, r). \tag{4.24}$$

Proof The mapping $f^k : X_i \to X$ is a differentiable bijection satisfying (4.19). Taking $y \in X_{i,1}$ and $z \in X_{i,2}$ so that $f^k(y) \in X_1$ and $f^k(z) \in X_2$ satisfy $\mathrm{dist}(f^k(y), f^k(z)) = d$, the right-hand inequality of (4.19) yields the left-hand inequality of (4.23). The right-hand inequality of (4.23) is clear, since $X_i \subset X_{i,1} \cup X_{i,2}$.

For (b) we note that if $i \in I_k$ and $\lambda r < db_1^{-1}|X_i|$, then by (a) $B(x, \lambda r)$ is disjoint from X_j for all $j \in I_k$ with $j \neq i$. Thus the left-hand inclusion of (4.24) holds; the right-hand inclusion is immediate. $\square$

We now deduce that small parts of E may be mapped onto large parts 'without too much distortion'. More precisely, there exists a bi-Lipschitz mapping from each small ball with centre in E to a large part of E, with the Lipschitz constants comparable with the size of the ball. A set that satisfies the conclusion of Corollary 4.4 is sometimes called *approximately self-similar* or *quasi-self-similar*.

Corollary 4.4

Let E be a cookie-cutter set. Then there are numbers $c > 0$ and $r_0 > 0$ such that, for every ball B with centre in E and radius $r < r_0$, there exists a mapping $g : E \cap B \to E$ satisfying

$$c^{-1}r^{-1}|x - y| \leq |g(x) - g(y)| \leq cr^{-1}|x - y| \quad (x, y \in E \cap B). \qquad (4.25)$$

Proof Let $r < r_0 \equiv db_1^{-1}|X|$ and let $x \in E$. Then by (4.6) we may find k and $i = (i_1, \ldots, i_k) \in I_k$ such that $x \in X_i$ and $c_{\min}db_1^{-1}|X_i| \leq r < db_1^{-1}|X_i|$. By Corollary 4.3(b) $E \cap B(x, r) \subset X_i$, and using (4.19) $f^k : E \cap B(x, r) \to E$ satisfies

$$b_1^{-1}|X_i|^{-1}|y - z| \leq |f^k(y) - f^k(z)| \leq b_1|X_i|^{-1}|y - z|$$

so

$$c_{\min}db_1^{-2}r^{-1}|y - z| \leq |f^k(y) - f^k(z)| \leq dr^{-1}|y - z|,$$

which is (4.25), taking $g = f^k$. $\square$

A cookie-cutter set E is also approximately self-similar in the 'opposite' sense, in that the whole of E may be mapped into small neighbourhoods of E without too much distortion.

Corollary 4.5

Let E be a cookie-cutter set. Then there are numbers $c > 0$ and $r_0 > 0$ such that for every ball B with centre in E and radius $r < r_0$ there exists a mapping $g : E \to E \cap B$ satisfying

$$c^{-1}r|x - y| \leq |g(x) - g(y)| \leq cr|x - y| \quad (x, y \in E). \qquad (4.26)$$

Proof Let $x \in E$ and $r < r_0 = |X|$. We may find $\boldsymbol{i} = (i_1, \ldots, i_k)$ such that $x \in X_i$ and $c_{\min} r < |X_i| \leq r$, so $X_i \subset B(x, r)$. Using (4.19) with $y = F_{i_1} \circ \cdots \circ F_{i_k}(x)$ and $z = F_{i_1} \circ \cdots \circ F_{i_w}(w) \in X_i$ we have

$$b_1^{-1}|X_i||x - w| \leq |F_{i_1} \circ \cdots \circ F_{i_k}(x) - F_{i_1} \circ \cdots \circ F_{i_k}(w)| \leq b_1|X_i||x - w|$$

so

$$c_{\min} b_1^{-1} r|x - w| \leq |F_{i_1} \circ \cdots \circ F_{i_k}(x) - F_{i_1} \circ \cdots \circ F_{i_k}(w)| \leq b_1 r|x - w|,$$

which is (4.26), taking g as the restriction of $F_{i_1} \circ \cdots \circ F_{i_k}$ to E. □

These results on approximate self-similarity allow the implicit theorems of Section 3.1 to be applied to cookie-cutter sets.

Corollary 4.6

Let E be a cookie-cutter set with $\dim_H E = s$. Then $\underline{\dim}_B E = \overline{\dim}_B E = \dim_P E = s$ and $0 < \mathcal{H}^s(E) < \infty$.

Proof Corollary 4.4 together with Theorem 3.1 gives that these dimensions are equal and that $0 < \mathcal{H}^s(E)$, and Corollary 4.5 and Theorem 3.2 give that $\mathcal{H}^s(E) < \infty$. □

These dimensional properties will also follow, together with a formula for the dimension itself, from the more sophisticated approach of Chapter 5.

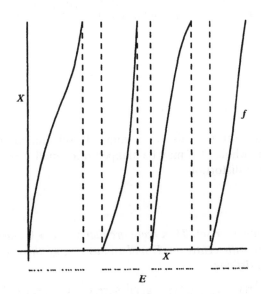

Figure 4.4 A 'four-part' cookie-cutter f with repeller $E = \bigcap_{k=0}^{\infty} f^{-k}(X)$

Just as in Example 3.4, we can also get information on subsets of cookie-cutter sets which satisfy $f(A) \subset A$ or $f(A) \supset A$, see Exercise 4.3.

Note that the theory of this chapter (and the next) applies in many other situations, which are summarised in Section 5.5. In particular only trivial modification is required for *m-part cookie-cutters* $f: X_1 \cup \cdots \cup X_m \to X$ where $f: X_i \to X$ is an expanding bijection for $i = 1, 2, \ldots, m$, see Figure 4.4.

4.3 Notes and references

The bounded distortion principle has been derived in various contexts, see, for example, Bedford (1986), Falconer (1989), Ruelle (1982) and Sullivan (1983) as well as other references listed in Section 5.7 in connection with the thermodynamic formalism.

Exercises

4.1 Find the function f corresponding to F_1, F_2 given by (4.7).

4.2 Show directly that the middle-third Cantor set is approximately self-similar, and find the least number c which can be used in (4.25) in this case.

4.3 Let A be a non-empty compact subset of a cookie-cutter set E with $s = \dim_H A$. Show analogously to Corollary 3.4, using (4.25) and (4.26), that if $A \subset f^{-1}(A)$ then $s = \dim_B A$ and $\mathcal{H}^s(A) > 0$, and that if $f(A) \supset A$ then $s = \dim_B A$ and $\mathcal{H}^s(A) < \infty$.

Chapter 5 The thermodynamic formalism

There are many problems in mathematics which may readily be solved in 'linear' cases, but which have non-linear counterparts that are much harder to analyse. For instance, the middle-third Cantor set E is the attractor of the IFS with the two similarity transformations on $\mathbb{R}$ given by (2.24). Many properties of the Cantor set follow quickly from this 'linear' description, for example its Hausdorff dimension is $\log 2 / \log 3$, using Theorem 2.7. However, we may wish to work with 'non-linear' sets such as the cookie-cutters described in Chapter 4, for example the subset of $[0,1]$ determined by the IFS (4.7). These non-linear constructions are much harder to analyse; there is no simple expression for the dimension of E, nor is it even clear how to obtain accurate dimension estimates.

This chapter describes a procedure which allows many results and ideas from the linear or piecewise-linear situation to be extended to non-linear cases. A major objective is to derive formulae for the dimension of fractals defined by non-linear systems, but many other aspects of dynamical systems and fractals may be treated using this approach.

For ease of exposition we present the thermodynamic formalism in the context of the cookie-cutter system described in Section 4.1. Nevertheless this illustrates the essential ideas of the thermodynamic formalism which are much more widely applicable.

Many of the notions in this chapter were first developed in the context of statistical mechanics, a subject which has remarkable parallels to dynamical systems theory, see Section 5.6. This is the reason for the name 'thermodynamic formalism' and terms such as 'pressure', 'Gibbs measure' and 'entropy'. It should be emphasised that a knowledge of statistical mechanics is not a prerequisite here!

5.1 Pressure and Gibbs measures

Two ingredients are required for the thermodynamic formalism: a suitable dynamical system or IFS and a Lipschitz function defined on an associated invariant set.

Here we treat the cookie-cutter system described in Section 4.1. Recall that X is a real closed interval with disjoint subintervals X_1 and X_2 and

$f: X_1 \cup X_2 \to X$ is an expanding mapping with a continuous second derivative that maps both X_1 and X_2 bijectively onto X. This dynamical system has a cookie-cutter repeller E which equivalently may be regarded as the attractor of the IFS given by a pair of contractions $\{F_1, F_2\}$ on X. We use the notation of Section 4.1; in particular we recall the hierarchy of nested intervals X_i indexed by sequences of 1s and 2s.

We also require a Lipschitz function $\phi: X_1 \cup X_2 \to \mathbb{R}$, satisfying

$$|\phi(x) - \phi(y)| \le a|x - y| \quad (x, y \in X_1 \cup X_2) \tag{5.1}$$

for some $a > 0$. The bounded variation theory of Section 4.2 applies, and as there we write

$$S_k\phi(x) = \sum_{j=0}^{k-1} \phi(f^j x) \tag{5.2}$$

for $x \in \cup_{i \in I_k} X_i$. We recall from Proposition 4.1 that there is a number b such that

$$|S_k\phi(x) - S_k\phi(y)| \le b \tag{5.3}$$

or, equivalently,

$$e^{-b} \le \frac{\exp(S_k\phi(x))}{\exp(S_k\phi(y))} \le e^b \tag{5.4}$$

for all $x, y \in X_i$, for all $i \in I_k$ and all k.

Initially, we work with a general Lipschitz function ϕ. In the next section appropriate choice of ϕ will lead to the dimension formula for E.

Our first objective is to find a measure μ supported by E and a number $P(\phi)$ such that

$$\mu(X_i) \asymp \exp(-kP(\phi)) \exp(S_k\phi(x))$$

for all $i \in I_k$ and $x \in X_i$. The number $P(\phi)$ is of considerable importance and is called the pressure of ϕ, and the measure μ is called a Gibbs measure for ϕ.

We prove part (a) of the following theorem now; this is enough to establish the formula for the dimension of E in the next section. Part (b) asserts that we can impose further stipulations on μ; the proof of this involves more sophisticated functional analysis and is deferred until Section 5.3.

Theorem 5.1

(a) *For all k and $i \in I_k$ let $x_i \in X_i$. Then the limit*

$$P(\phi) = \lim_{k \to \infty} \frac{1}{k} \log \sum_{i \in I_k} \exp S_k\phi(x_i) \tag{5.5}$$

exists and does not depend on the $x_i \in X_i$ chosen. Furthermore there exists a Borel probability measure μ supported by E and a number $a_0 > 0$ such that, *for all k and all $i = i_1, \ldots, i_k \in I_k$, we have*

$$a_0^{-1} \le \frac{\mu(X_{i_1,\ldots,i_k})}{\exp(-kP(\phi) + S_k\phi(x))} \le a_0 \tag{5.6}$$

for all $x \in X_{i_1,\ldots,i_k}$.

(b) It is possible to choose μ satisfying (5.6) so that in addition, either
(i) μ satisfies the transformation property

$$\mu(f(A)) = \exp P(\phi) \int_A \exp(-\phi(x))\mathrm{d}\mu(x)$$

for all Borel sets A, or
(ii) μ is invariant under f, that is for all $g \in C(E)$

$$\int g(x)\mathrm{d}\mu(x) = \int g(f(x))\mathrm{d}\mu(x).$$

Proof of (a)   Fix $w \in E$. From (5.2)

$$S_{k+m}\phi(x) = S_k\phi(x) + S_m\phi(f^k x) \tag{5.7}$$

for $k, m = 1, 2, \ldots$, so taking exponentials and summing,

$$
\begin{aligned}
\sum_{x: f^{k+m}x=w} \exp S_{k+m}\phi(x) &= \sum_{x: f^{k+m}x=w} \exp(S_k\phi(x))\exp(S_m\phi(f^k x))\\
&= \sum_{z: f^m z=w} \sum_{x: f^k x=z} \exp(S_k\phi(x))\exp(S_m\phi(f^k x))\\
&= \sum_{z: f^m z=w} \exp(S_m\phi(z)) \sum_{x: f^k x=z} \exp(S_k\phi(x))\\
&\le e^b \sum_{z: f^m z=w} \exp(S_m\phi(z)) \sum_{x: f^k x=w} \exp(S_k\phi(x))
\end{aligned}
$$

by (5.4). Writing

$$s_k = \sum_{x: f^k x=w} \exp(S_k\phi(x)) \tag{5.8}$$

this becomes the right-hand inequality of

$$e^{-b} s_k s_m \le s_{k+m} \le e^b s_k s_m. \tag{5.9}$$

The left-hand inequality follows in the same way, using the other inequality of (5.4). Taking logarithms and writing $a_k = \log s_k$ gives

$$a_k + a_m - b \le a_{k+m} \le a_k + a_m + b.$$

By Corollary 1.2 we have that $\lim_{k\to\infty} \frac{1}{k} a_k = \lim_{k\to\infty} \frac{1}{k}\log s_k$ exists, that is, the limit (5.5) exists in the case where $x_i = F_{i_1} \circ \cdots \circ F_{i_k} w$ for each $i = (i_1, \ldots, i_k)$. Using (5.4) it follows that the limit (5.5) exists and is the same for any choice of $x_i \in X_i$. We also note that Corollary 1.2 applied to the sequences (a_k) and $(-a_k)$ gives that $kP(\phi) - b \leq a_k \leq kP(\phi) + b$ for all k, that is

$$e^{-b} \exp(kP(\phi)) \leq s_k \leq e^b \exp(kP(\phi)). \tag{5.10}$$

We now construct a measure μ satisfying (5.6) by defining discrete measures μ_m on $\mathbb{R}$ and taking a limit as $m \to \infty$. For $m = 1, 2, \ldots$ and any set A, define

$$\mu_m(A) = \frac{1}{s_m} \sum_{x \in A: f^m x = w} \exp(S_m \phi(x))$$

(thus the sum is over the points $F_{i_1} \circ \cdots \circ F_{i_m} w$ that lie in A). Clearly μ_m is a discrete measure supported by E (since $w \in E$ we have $x \in E$ for all x with $f^m x = w$) and $\mu_m(E) = 1$. Thus there is a Borel measure μ supported by E that is a weak limit of a subsequence of the measures μ_m (see Proposition 1.9).

Certainly $\mu(E) = 1$. Moreover, if $i \in I_k$ and $k \leq m$

$$\mu_m(X_i) = \frac{1}{s_m} \sum_{x \in X_i: f^m x = w} \exp(S_m \phi(x))$$

$$= \frac{1}{s_m} \sum_{x \in X_i: f^m x = w} \exp(S_k \phi(x)) \exp(S_{m-k} \phi(f^k x))$$

using (5.7). Thus, if y is any point of X_i we have by (5.4) that

$$e^{-b} \mu_m(X_i) \leq s_m^{-1} \exp(S_k \phi(y)) \sum_{x \in X_i: f^m x = w} \exp(S_{m-k} \phi(f^k x)) \leq e^b \mu_m(X_i)$$

or

$$e^{-b} \mu_m(X_i) \leq s_m^{-1} \exp(S_k \phi(y)) \sum_{z \in X: f^{m-k} z = w} \exp(S_{m-k} \phi(z)) \leq e^b \mu_m(X_i),$$

since $f^k : X_i \to X$ is a bijection. Using (5.8) gives

$$e^{-b} \mu_m(X_i) \leq \exp\left(S_k \phi(y) \frac{s_{m-k}}{s_m}\right) \leq e^b \mu_m(X_i)$$

so by (5.9)

$$e^{-2b} \mu_m(X_i) \leq s_k^{-1} \exp(S_k \phi(y)) \leq e^{2b} \mu_m(X_i).$$

This is true for all $m \geq k$, so for the weak limit μ of a subsequence of (μ_m)

$$\frac{e^{-2b}}{s_k} \leq \frac{\mu(X_i)}{\exp(S_k \phi(y))} \leq \frac{e^{2b}}{s_k}.$$

Inequality (5.6) follows using (5.10).

Part (*b*) follows from Theorem 5.5 which will be proved in Section 5.3.   □

The number $P(\phi)$ defined by (5.5) is called the *topological pressure* or *pressure* of ϕ and a measure satisfying (5.6) for some $a_0 > 0$ is termed a *Gibbs measure* for ϕ. By definition any two Gibbs measures for ϕ are equivalent. We have shown that a cookie-cutter set supports a Gibbs measure.

In (5.5) we may take x_i to be any point of X_i. As $F_{i_1} \circ \cdots \circ F_{i_k} : X \to X_i \subset X$ is a contraction, there is a unique $x_i \in X_i$ with $F_{i_1} \circ \cdots \circ F_{i_k} x_i = x_i$, or equivalently with $f^k x_i = x_i$. Thus, we may choose the points $x_i \in X_i$ for $i \in I_k$ to be the set of 2^k fixed points of f^k, that is the periodic points of f which have periods dividing k. This leads to the following expression for the pressure, which avoids reference to the X_i:

$$P(\phi) = \lim_{k \to \infty} \frac{1}{k} \log \sum_{x \in \text{Fix} f^k} \exp(S_k \phi(x)) \tag{5.11}$$

$$= \lim_{k \to \infty} \frac{1}{k} \log \sum_{x \in \text{Fix} f^k} \exp(\phi(x) + \phi(fx) + \cdots + \phi(f^{k-1}x)),$$

where $\text{Fix} f^k$ denotes the set of fixed points of f^k.

5.2 The dimension formula

By appropriate choice of the Lipschitz function ϕ, the theory of the previous section specialises to give an elegant formula for the Hausdorff dimension of the cookie-cutter set E in terms of pressure.

To find the Hausdorff measure and dimension of a set we need to estimate the sums in (2.7). For a cookie-cutter set E it is natural to utilise the coverings of E provided by the intervals $\{X_i : i \in I_k\}$. Certainly

$$\mathcal{H}_\delta^s(E) \le \sum_{i \in I_k} |X_i|^s \tag{5.12}$$

provided that k is chosen so that $\max_{i \in I_k} |X_i| \le \delta$. Letting $\delta \to 0$ gives

$$\mathcal{H}^s(E) \le \liminf_{k \to \infty} \sum_{i \in I_k} |X_i|^s. \tag{5.13}$$

It follows that $\dim_H E \le s$ for any s for which this lower limit is finite. However, as we shall see, much more is true. Under very general conditions on f these sums are very well behaved, with $\sum_{i \in I_k} |X_i|^s \asymp \exp(kP_s)$ where P_s is the pressure of a certain function. Thus the pressure P_s is the exponential growth rate of the sums of the s-th powers of the interval lengths at the k-th level of the hierachy. Moreover, the limit in (5.13) provides a *lower* as well as an upper estimate for $\mathcal{H}^s(E)$, so the Hausdorff dimension is given by the number s which makes $P_s = 0$, and furthermore the restriction of $\mathcal{H}^s$ to E is a Gibbs measure.

For $s \in \mathbb{R}$ we take

$$\phi(x) = -s \log |f'(x)| \tag{5.14}$$

in (5.1) and consider the pressure $P(-s \log |f'|)$ as s varies, where f' is the derivative of f. Then (5.5) becomes, for any choice of $x_i \in X_i$,

$$P(-s \log |f'|) = \lim_{k \to \infty} \frac{1}{k} \log \sum_{i \in I_k} \exp \left(\sum_{j=0}^{k-1} -s \log |f'(f^j x_i)| \right) \tag{5.15}$$

$$= \lim_{k \to \infty} \frac{1}{k} \log \sum_{i \in I_k} |(f^k)'(x_i)|^{-s} \tag{5.16}$$

$$= \lim_{k \to \infty} \frac{1}{k} \log \sum_{i \in I_k} |X_i|^s \tag{5.17}$$

using (5.2), (4.16) and (4.18). (Observe the resemblance between this and (5.13).) We examine the behaviour of $P(-s \log |f'|)$ as s varies.

Lemma 5.2

For $s \in \mathbb{R}$ and $\delta > 0$ we have

$$-\delta m_2 \leq P(-(s+\delta) \log |f'|) - P(-s \log |f'|) \leq -\delta m_1 \tag{5.18}$$

where

$$0 < m_1 \equiv \inf_{x \in X_1 \cup X_2} \log |f'(x)| \leq \sup_{x \in X_1 \cup X_2} \log |f'(x)| \equiv m_2 < \infty.$$

In particular, $P(-s \log |f'|)$ is strictly decreasing and continuous in s, with $\lim_{s \to -\infty} P(-s \log |f'|) = \infty$ and $\lim_{s \to \infty} P(-s \log |f'|) = -\infty$.

Proof   For $\delta > 0$

$$\frac{1}{k} \log \sum_{i \in I_k} \exp \left(\sum_{j=0}^{k-1} -(s+\delta) \log |f'(f^j x_i)| \right)$$

$$\leq \frac{1}{k} \log \sum_{i \in I_k} \left(\exp \left(\sum_{j=0}^{k-1} -s \log |f'(f^j x_i)| \right) \exp(-\delta k m_1) \right)$$

$$\leq \frac{1}{k} \log \left(\sum_{i \in I_k} \exp \left(\sum_{j=0}^{k-1} -s \log |f'(f^j x_i)| \right) \right) - \delta m_1.$$

Letting $k \to \infty$ and using (5.15) gives the right-hand inequality of (5.18). The left-hand inequality follows in a similar way. $\square$

Thus the graph of $P(-s \log |f'|)$ has the form indicated in Figure 5.1 (the function is convex, see Exercise 5.5). In particular, there is a unique number s

such that $P(-s\log|f'|) = 0$. This number turns out to be the Hausdorff dimension of the cookie-cutter set E.

Theorem 5.3

Let s be the unique real number satisfying

$$P(-s\log|f'|) = 0. \tag{5.19}$$

Then $\dim_H E = s$ *and* $0 < \mathcal{H}^s(E) < \infty$. *Moreover, the restriction of* $\mathcal{H}^s$ *to E is a Gibbs measure, and in particular there is a number* $a_1 > 0$ *such that*

$$a_1^{-1}|X_i|^s \le \mathcal{H}^s(E \cap X_i) \le a_1|X_i|^s \tag{5.20}$$

for all $i \in I_k$ *and all k.*

Proof Let s be given by (5.19), take $\phi(x) = -s\log|f'(x)|$ for this s and let μ be an associated Gibbs measure given by Theorem 5.1. Since $P(\phi) = 0$, (5.6) becomes

$$a_0^{-1} \le \mu(X_i)/\exp(S_k\phi(x)) \le a_0$$

for all $x \in X_i$, for all $i \in I_k$, and all k, so

$$a_0^{-1} \le \mu(X_i) \Big/ \exp\left(-s\sum_{j=0}^{k-1}\log|f'(f^jx)|\right) = \mu(X_i)/|(f^k)'(x)|^{-s} \le a_0$$

using the chain rule (4.16). Combining this with (4.18) there is a number a_2

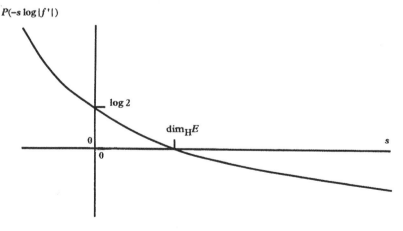

Figure 5.1 Form of the pressure function $P(-s\log|f'|)$ for a (two-part) cookie-cutter system

such that

$$a_2^{-1} \leq \frac{\mu(X_i)}{|X_i|^s} \leq a_2 \tag{5.21}$$

for all i. Thus, μ is a probability measure supported by E with the measure of every interval X_i comparable to $|X_i|^s$.

The measure μ may be related to Hausdorff measure on E as a consequence of the bounded distortion principle. Let $x \in E$ and $r < c_{\min}^{-1}|X|$ (in the notation of Section 4.2); we estimate $\mu(B(x,r))$ where $B(x,r)$ is the interval (one-dimensional ball) with centre x and length $2r$. Using (4.6) we may find an integer k and $i \in I_k$, such that $x \in X_i$ and

$$|X_i| \leq r < c_{\min}^{-1}|X_i|.$$

From (4.24)

$$\mu(B(x, \lambda r)) \leq \mu(X_i) \leq \mu(B(x, r)),$$

where $\lambda = d b_1^{-1} c_{\min}$ is independent of x and r, so from (5.21)

$$a_2^{-1} \mu(B(x, \lambda r)) \leq |X_i|^s \leq a_2 \mu(B(x, r)).$$

Thus, for some $b_2 > 0$,

$$b_2^{-1} r^s \leq \mu(B(x, r)) \leq b_2 r^s \tag{5.22}$$

for all $x \in E$ and r sufficiently small. It follows from Proposition 2.2 that $b_2^{-1} \leq \mathcal{H}^s(E) \leq 2^s b_2$ and $\dim_H E = \dim_B E = s$. Similarly, it follows that μ is equivalent to the restriction of s-dimensional Hausdorff measure to E, that is for every Borel set A we have $b_2^{-1} \mathcal{H}^s(E \cap A) \leq \mu(A) \leq 2^s b_2 \mathcal{H}^s(E \cap A)$. Taking $A = X_i$ and combining this with (5.21) gives (5.20). $\square$

The dimension formula (5.19) may be viewed in several ways. Choosing the x_i in (5.16) to be the fixed points of f^k, as in (5.11), $\dim_H E$ is given in dynamical terms as the value of s such that

$$0 = \lim_{k \to \infty} \frac{1}{k} \log \sum_{x \in \mathrm{Fix} f^k} |(f^k)'(x)|^{-s}.$$

Alternatively, using (5.17), $\dim_H E$ is the value of s such that

$$\lim_{k \to \infty} \left(\sum_{i \in I_k} |X_i|^s \right)^{1/k} = 1, \tag{5.23}$$

and furthermore $0 < a_1^{-1} \mathcal{H}^s(E) \leq \sum_{i \in I_k} |X_i|^s \leq a_1 \mathcal{H}^s(E) < \infty$ for all k, by summing (5.20) over $i \in I_k$. Thus these natural sums give bounds for Hausdorff measures; compare this with (5.13).

In the special case when E is a self-similar set and F_1 and F_2 are similarity transformations of ratios r_1 and r_2, $\dim_H E$ is the number s such that $r_1^s + r_2^s = 1$, see (2.42). Here $f^k : X_{i_1,\ldots,i_k} \to X$ is a similarity of ratio

$(r_{i_1} \cdots r_{i_k})^{-1}$, so

$$\sum_{i \in I_k} |(f^k)'(x_i)|^{-s} = \sum_{i \in I_k} (r_{i_1} \cdots r_{i_k})^s = (r_1^s + r_2^s)^k = 1.$$

Using (5.16) s satisfies $P(-s \log |f'|) = 0$ so the pressure formula does indeed generalise the dimension formula (2.42) for self-similar sets to a non-linear setting.

It is worth noting that, by inspecting the derivation of the pressure formula, there is control over the rate of convergence in (5.16). In fact, if b is such that $s| \log |(f^k)'(x)| - \log |(f^k)'(y)|| \le b$ for $x, y \in X_i$ and $i \in I_k$, then

$$\left| P(-s \log |f'|) - \frac{1}{k} \log \sum_{i \in I_k} |(f^k)'(x_i)|^{-s} \right| \le \frac{2b}{k} \tag{5.24}$$

for all k, see Exercise 5.2, and this together with (5.18) enables the accuracy of corresponding approximations to $\dim_H E$ to be gauged.

*5.3 Invariant measures and the transfer operator

This section presents some slightly more sophisticated aspects of the thermodynamic formalism, leading to an alternative characterisation of pressure. A major rôle is played by measures that are *invariant* under the cookie-cutter function f, that is measures μ supported by E which satisfy

$$\int g(f(x)) \mathrm{d}\mu(x) = \int g(x) \mathrm{d}\mu(x) \tag{5.25}$$

for every continuous $g : E \to \mathbb{R}$. This is equivalent to $\mu(f^{-1}(A)) = \mu(A)$ for every Borel set A. In the next section we will derive the *variational principle*, that the pressure is the supremum of a certain expression over all invariant probability measures μ on E.

In order to proceed, we must show that the measure μ in Theorem 5.1 may be chosen to be invariant. This extra, rather delicate, requirement is usually achieved by a functional analytic approach utilizing properties of an operator L_ϕ.

Let E be the cookie-cutter repeller for f and let and $\phi : E \to \mathbb{R}$ be a Lipschitz function as above. Write $C(E)$ for the space of real valued continuous functions on E. We define the *transfer operator* or *Sinai–Bowen–Ruelle operator* $L_\phi : C(E) \to C(E)$ by

$$(L_\phi g)(x) = g(F_1 x) \mathrm{e}^{\phi(F_1 x)} + g(F_2 x) \mathrm{e}^{\phi(F_2 x)} \tag{5.26}$$

$$= \sum_{y: f(y) = x} g(y) \mathrm{e}^{\phi(y)}. \tag{5.27}$$

The operator L_ϕ is linear (that is $L_\phi(g_1 + g_2) = L_\phi g_1 + L_\phi g_2$ and $L_\phi(\lambda g) = \lambda(L_\phi g)$ for scalars λ) and is positive (that is if $g(x) > 0$ for all $x \in E$ then $(L_\phi g)(x) > 0$ for all $x \in E$).

We will sometimes need to apply the operator L_ϕ repeatedly. Writing L_ϕ^k for the k-th iterate of L_ϕ (so that $L_\phi^2 g = L_\phi(L_\phi g)$, etc), we get by repeated substitution of (5.26) into itself that

$$(L_\phi^k g)(x) = \sum_{(i_1,\dots,i_k)\in I_k} g(F_{i_1} \circ \cdots \circ F_{i_k} x) \exp[\phi(F_{i_1} \circ \cdots \circ F_{i_k} x)$$

$$+ \phi(F_{i_2} \circ \cdots \circ F_{i_k} x) + \cdots + \phi(F_{i_k} x)]$$

$$= \sum_{(i_1,\dots,i_k)\in I_k} g(F_{i_1} \circ \cdots \circ F_{i_k} x) \exp(S_k \phi(F_{i_1} \circ \cdots \circ F_{i_k} x)). \qquad (5.28)$$

The following identity which relates L_ϕ to f will be useful. Let $g_1, g_2 \in C(E)$. Then

$$(L_\phi((g_1 \circ f) \times g_2))(x) = \sum_{i=1,2} g_1(f(F_i x)) g_2(F_i x) e^{\phi(F_i x)}$$

$$= \sum_{i=1,2} g_1(x) g_2(F_i x) e^{\phi(F_i x)}$$

$$= g_1(x)(L_\phi g_2)(x). \qquad (5.29)$$

The principal properties of the transfer operator are given by the following theorem, which is a version of the Ruelle–Perron–Frobenius theorem.

Theorem 5.4

(a) *There exists $\lambda > 0$ and $w \in C(E)$ with $w(x) > 0$ for all $x \in E$ such that w is an eigenfunction of L_ϕ with eigenvalue λ, that is*

$$L_\phi w = \lambda w. \qquad (5.30)$$

(b) *There exists a Borel probability measure μ supported by E such that*

$$\int (L_\phi g) \mathrm{d}\mu = \lambda \int g \mathrm{d}\mu \qquad (5.31)$$

for all $g \in C(E)$.

(c) *The measure ν on E defined by*

$$\int g \mathrm{d}\nu = \int g w \mathrm{d}\mu \qquad (5.32)$$

for all $g \in C(E)$ is invariant under f. (We assume that w is normalised so that $\nu(E) = 1$.)

Proof Since ϕ is Lipschitz, there is a number $a > 0$ large enough so that $e^{|\phi(F_i x) - \phi(F_i y)|} \leq e^{a|x-y|}$ for $x, y \in E$ and $i = 1, 2$. Let $c_{max} < 1$ be as in (4.4), and fix $\alpha > 0$ such that $\alpha c_{max} + a \leq \alpha$. Let $\beta = e^{-\alpha|E|} > 0$ where $|E|$ is the diameter of E. Define

$$B = \{g \in C(E) : \beta \leq g(x) \leq 1 \text{ and } g(x) \leq g(y)e^{\alpha|x-y|} \text{ for all } x, y \in E\}.$$

Then B is a convex set (that is $tg_1 + (1 - t)g_2 \in B$ if $g_1, g_2 \in B$ and $0 \leq t \leq 1$) and B is an equicontinuous subset of $C(E)$, so B is a $\| \, \|_\infty$−compact subset of $C(E)$ by the Arzela–Ascoli theorem. We show that a normalised version of L_ϕ maps B into itself and so has a fixed point.

Let g satisfy

$$0 \leq g(x) \leq g(y)e^{\alpha|x-y|} \quad (x, y \in E). \tag{5.33}$$

Then, if $x, y \in E$, we have $|F_i x - F_i y| \leq c_{max}|x - y|$, so by (5.26)

$$(L_\phi g)(x) \leq \sum_{i=1,2} g(F_i y)e^{\alpha c_{max}|x-y|} e^{a|x-y|} e^{\phi(F_i y)}$$

$$\leq e^{\alpha|x-y|}(L_\phi g)(y). \tag{5.34}$$

Define the normalised mapping T_ϕ on B by $T_\phi g(x) = L_\phi g(x)/ \| L_\phi g \|_\infty$. By the above $(T_\phi g)(x) \leq e^{\alpha|x-y|}(T_\phi g)(y)$ for $x, y \in E$, and so, since $\| T_\phi g \|_\infty = 1$, we have $\beta = e^{-\alpha|E|} \leq (T_\phi g)(y) \leq 1$ for all $y \in E$. Thus T_ϕ maps the convex compact set B into itself, so by the Schauder fixed point theorem, there exists $w \in B$ with $T_\phi w = w$, or $L_\phi w = \lambda w$ where $\lambda = \| L_\phi w \|_\infty$. Since $w \in B$ it follows that $w(x) > 0$ and $\lambda > 0$, completing the proof of (*a*).

Now define a set of measures $\mathcal{M} \equiv \{\mu : \text{spt}\mu \subset E \text{ and } \int w d\mu = 1\}$, where w is as in (*a*). We regard $\mathcal{M}$ as a subset of the space $C(E)^*$ of continuous linear functionals on E which may be identified with the signed measures on E. Let L_ϕ^* denote the dual mapping to L_ϕ defined on $C(E)^*$ by

$$\int g d(L_\phi^* \mu) = \int (L_\phi g) d\mu \tag{5.35}$$

for $g \in C(E)$. Then for $\mu \in \mathcal{M}$

$$\int w d\left(\frac{1}{\lambda}L_\phi^* \mu\right) = \int \frac{1}{\lambda}(L_\phi w)d\mu = \int w d\mu = 1.$$

Hence $\frac{1}{\lambda}L_\phi^*$ maps $\mathcal{M}$ into itself. Since $\mathcal{M}$ is a convex and a compact subset of $C(E)^*$ in the weak-* topology, the Schauder fixed point theorem gives a measure $\mu \in \mathcal{M}$ such that $\frac{1}{\lambda}L_\phi^* \mu = \mu$. Thus (5.31) holds using (5.35); multiplying μ by a constant to get $\mu(E) = 1$ then ensures that μ is a probability measure as required for (*b*).

To check that ν given by (5.32) is invariant under f, let $g \in C(E)$. Then

$$\int g(x)\mathrm{d}\nu(x) = \int g(x)w(x)\mathrm{d}\mu(x) \qquad\qquad \text{by (5.32)}$$

$$= \lambda^{-1} \int g(x)(L_\phi w)(x)\mathrm{d}\mu(x) \qquad\qquad \text{by (5.30)}$$

$$= \lambda^{-1} \int (L_\phi((g \circ f) \times w))(x)\mathrm{d}\mu(x) \qquad\qquad \text{by (5.29)}$$

$$= \lambda^{-1}\lambda \int (g \circ f)(x)w(x)\mathrm{d}\mu(x) \qquad\qquad \text{by (5.31)}$$

$$= \int g(f(x))\mathrm{d}\nu(x) \qquad\qquad \text{by (5.32).} \quad \square$$

The transfer operator has many other important spectral properties which we do not pursue here, for example λ is an eigenvalue of multiplicity one with the remainder of the spectrum of L_ϕ inside a disc of radius strictly less than λ.

The transfer operator is intimately related to the pressure $P(\phi)$. The eigenvalue λ of Theorem 5.4 turns out to be $\exp P(\phi)$, and μ, and thus ν, are Gibbs measures for ϕ. This is expressed in the next theorem, which extends Theorem 5.1 by showing that the Gibbs measure may be chosen to be invariant.

Theorem 5.5

With λ, μ and ν as in Theorem 5.4 we have that $\log \lambda = P(\phi)$ and that μ and ν are Gibbs measures for ϕ. Thus μ and ν are Borel probability measures on E such that for some $a_0 > 0$

$$a_0^{-1} \le \frac{\mu(X_i)}{\exp(-kP(\phi) + S_k\phi(x))}, \quad \frac{\nu(X_i)}{\exp(-kP(\phi) + S_k\phi(x))} \le a_0 \qquad (5.36)$$

for all k and $i \in I_k$ and $x \in X_i$. The measure ν is invariant under f. The measure μ satisfies

$$\mu(f^k(A)) = \exp kP(\phi) \int_A \exp(-S_k\phi(x))\mathrm{d}\mu(x) \qquad (5.37)$$

for every Borel set $A \subset E$ and $k = 1, 2, \ldots$.

Proof We write 1_A for the indicator function of the set A. Let $i = (i_1, \ldots, i_k) \in I_k$. Since $F_{j_1} \circ \cdots \circ F_{j_k} x \in X_{i_1, \ldots, i_k}$ if and only if $j_1 = i_1, \ldots, j_k = i_k$, we get from (5.28) that for $A \subset X_{i_1, \ldots, i_k} \cap E$

$$L_\phi^k(e^{-S_k\phi(x)}1_A(x)) = \exp(-S_k\phi(F_{i_1} \circ \cdots \circ F_{i_k}x))1_A(F_{i_1} \circ \cdots \circ F_{i_k}x)$$
$$\times \exp(S_k\phi(F_{i_1} \circ \cdots \circ F_{i_k}x))$$
$$= 1_{f^k(A)}(x),$$

since $x \in f^k(A)$ if and only if $F_{i_1} \circ \cdots \circ F_{i_k}(x) \in A$ for $A \subset X_{i_1,\ldots,i_k}$. Integrating and using (5.31) k times

$$\mu(f^k(A)) = \int 1_{f^k(A)}(x) d\mu(x)$$

$$= \int L_\phi^k(e^{-S_k\phi(x)} 1_A(x)) d\mu(x)$$

$$= \lambda^k \int e^{-S_k\phi(x)} 1_A(x) d\mu(x)$$

$$= \lambda^k \int_A e^{-S_k\phi(x)} d\mu(x). \tag{5.38}$$

This holds for any Borel set A contained in $X_i \cap E$ for $i \in I_k$, so by addition (5.38) follows for any Borel set $A \subset E$. Putting $A = X_i \cap E$ in (5.38) and using (5.4),

$$e^{-b} \le \lambda^k e^{-S_k\phi(y)} \mu(X_i) \le e^b$$

for all $y \in X_i$. By summing over $i \in I_k$ and comparing with (5.5) it is clear that $P(\phi) = \log \lambda$ so (5.38) gives (5.37). The inequality (5.36) for μ also follows, and holds for the equivalent measure ν, since from (5.32)

$$\left(\inf_{x \in E} w(x)\right) \mu(X_i) \le \nu(X_i) \le \left(\sup_{x \in E} w(x)\right) \mu(X_i)$$

where $0 < \inf_{x \in E} w(x) \le \sup_{x \in E} w(x) < \infty$. $\square$

Note that part (*b*) of Theorem 5.1 follows immediately from Theorem 5.5.

It is now easy to deduce a further important property of Gibbs measures, namely ergodicity. We say that a measure μ is *ergodic* for f if every measurable set $A \subset X$ which is invariant (in the sense that $f^{-1}(A) = A$) has either $\mu(A) = 0$ or $\mu(X \setminus A) = 0$. Thus in the ergodic situation, the only invariant sets are the ones that are trivial when measured by μ.

Observe that if A is an invariant set, then $A = \cap_{k=0}^\infty f^{-k}(A) \subset \cap_{k=0}^\infty f^{-k}(X) = E$ so $A \subset E$. Moreover, $A = \cup_{i \in I_k} F_{i_1} \circ \cdots \circ F_{i_k}(A)$, so that $A \cap X_i = F_{i_1} \circ \cdots \circ F_{i_k}(A)$ and

$$f^k(A \cap X_i) = A. \tag{5.39}$$

Corollary 5.6

Any Gibbs measure μ satisfying (5.6) is ergodic for f.

Proof First let μ be the measure of Theorem 5.5 which satisfies (5.37). If A is an invariant set then $A \subset E$, and by (5.39) and (5.37)

$$\mu(A) = \mu(f^k(A \cap X_i)) = \exp kP(\phi) \int_{A \cap X_i} \exp(-S_k\phi(x)) d\mu(x)$$

for all $i \in I_k$. Thus by (5.4)

$$e^{-b}\mu(A) \leq \exp(kP(\phi) - S_k\phi(x))\mu(A \cap X_i)$$

for any $x \in X_i$. In exactly the same way, replacing A by the invariant set E,

$$\exp(kP(\phi) - S_k\phi(x))\mu(E \cap X_i) \leq e^b\mu(E).$$

But $\mu(E) = 1$ and μ is supported by E, so combining these inequalities gives

$$\mu(A)\mu(X_i) = \mu(A)\mu(E \cap X_i) \leq e^{2b}\mu(A \cap X_i)$$

for all X_i. Since $\{X_i \cap E : i \in I_k, k = 0, 1, \ldots\}$ generates the Borel subsets of E, we have that

$$\mu(A)\mu(B) \leq e^{2b}\mu(A \cap B)$$

for every Borel set $B \subset E$. Taking $B = E \setminus A$ gives $\mu(A)\mu(E \setminus A) \leq e^{2b}\mu(A \cap (E \setminus A)) = 0$ so that either $\mu(A) = 0$ or $\mu(E \setminus A) = 0$, as required.

Any other Gibbs measure ν corresponding to a given ϕ is, by definition (5.6), equivalent to this μ in the sense of (1.22). Hence if A is an invariant set, then either $\nu(A) = 0$ or $\nu(E \setminus A) = 0$. $\square$

One conclusion of this section is that Gibbs measures may be chosen to be both invariant and ergodic under f. Such measures play a particularly important rôle in ergodic theory, as we shall see in Chapter 6.

5.4 Entropy and the variational principle

Entropy quantifies the rate at which information can be gleaned about a dynamical system from a sequence of repeated observations. Systems that are equivalent in a certain sense have the same entropy, and consequently entropy is an important invariant in dynamical systems theory. We first define entropy and then obtain the variational principle: that pressure is the maximum of an expression involving entropy.

We continue with the cookie-cutter system introduced in Section 4.1, with $f: X_1 \cup X_2 \to X$ and repeller E. Let μ be a probability measure on E which we assume to be invariant under f. Consider the following 'experiment' to determine the position of a point $x \in E$ by observing whether its iterates lie in X_1 or X_2. For $j = 0, 1, 2, \ldots$ let i_j be the integer 1 or 2 such that $f^j x \in X_{i_j}$. Thus, regarding $f^j x$ as the position of a particle, originally at x, after time j, we think of X_{i_j} as an observation of whether the particle is to the left (in X_1) or right (in X_2) at time j. A natural question to ask is how accurately a sequence of k observations $(X_{i_0}, X_{i_1}, \ldots, X_{i_{k-1}})$ determines the initial position x of the particle. In particular, what is the measure

$$\mu\{x : f^j x \in X_{i_j} \text{ for } j = 0, 1, \ldots, k - 1\} = \mu(X_{i_0,\ldots,i_{k-1}}) \qquad (5.40)$$

of the set of x which corresponds to a given set of observations? If this measure is small, then x is closely determined (according to the measure μ) by these k observations.

If for a given $x \in E$ we find that $\mu(X_{i_0,\ldots,i_{k-1}}) \simeq c^k$ for large k, where $0 < c < 1$, then with each successive observation we improve our knowledge of the location of x by a factor of about c. Thus the rate of acquisition of information about x is roughly

$$-\log c = -\tfrac{1}{k}\log \mu(X_{i_0,\ldots,i_{k-1}}) = -\tfrac{1}{k}\log \mu(X_i), \qquad (5.41)$$

where here $i = i_0, i_1, \ldots, i_{k-1}$. Of course this value depends on x which is not known *a priori* in our experiment. Nevertheless we can average (5.41) over all $x \in E$ with respect to the measure μ by forming the sums

$$-\frac{1}{k}\sum_{i \in I_k} \mu(X_i) \log \mu(X_i).$$

This represents the average rate of information gain over time $0 \le j \le k - 1$ and taking the limit as $k \to \infty$ gives an average over all time.

This heuristic argument leads us to define the *entropy of f with respect to the measure μ* to be

$$h_\mu \equiv h_\mu(f) = \lim_{k \to \infty} -\frac{1}{k}\sum_{i \in I_k} \mu(X_i) \log \mu(X_i). \qquad (5.42)$$

Of course, this presupposes that the limit exists; to prove this we use the subadditive inequality.

Proposition 5.7

Let μ be an invariant probability measure supported by the repeller E of f. Then the entropy $h_\mu(f)$ given by (5.42) exists.

Proof We show that the sequence $\sum_{i \in I_k} -\mu(X_i) \log \mu(X_i)$ is subadditive so that the limit (5.42) exists by Proposition 1.1.

For brevity, we write $\psi(t) = -t \log t (t > 0)$, $\psi(0) = 0$; then ψ is a concave function on $[0, \infty)$, that is, $-\psi$ is a convex function. For positive integers i and j and each $i \in I_i$ and $j \in I_j$ we have, assuming that $\mu(X_j) > 0$ for all $j \in I_j$,

$$\psi(\mu(X_i)) = \psi\left(\sum_{j \in I_j} \mu(X_{ij})\right)$$

$$= \psi\left(\sum_{j \in I_j} \mu(X_j)\mu(X_{ij})/\mu(X_j)\right)$$

$$\geq \sum_{j \in I_j} \mu(X_j)\psi(\mu(X_{ij})/\mu(X_j))$$

using the concavity of ψ (see Proposition 1.4) and that $\sum_{j \in I_j} \mu(X_j) = 1$. From the definition of ψ

$$\psi(\mu(X_i)) \geq \sum_{j \in I_j} \mu(X_j)\frac{\mu(X_{ij})}{\mu(X_j)}(\log \mu(X_j) - \log \mu(X_{ij}))$$

$$\geq \sum_{j \in I_j} \mu(X_{ij})\log \mu(X_j) + \sum_{j \in I_j} \psi(\mu(X_{ij})).$$

If $\mu(X_j) = 0$ for some $j \in I_j$, the same inequality may be obtained by summing over those $j \in I_j$ with $\mu(X_j) \neq 0$. Now summing over $i \in I_i$ we have

$$\sum_{i \in I_i} \psi(\mu(X_i)) \geq \sum_{j \in I_j} \mu(X_j)\log \mu(X_j) + \sum_{ij \in I_{i+j}} \psi(\mu(X_{ij})),$$

noting that $\sum_{i \in I_i} \mu(X_{ij}) = \mu(X_j)$ since μ is invariant. Thus,

$$\sum_{i \in I_{i+j}} \psi(\mu(X_i)) \leq \sum_{i \in I_i} \psi(\mu(X_i)) + \sum_{i \in I_j} \psi(\mu(X_i)).$$

In other words, the sequence $a_i \equiv \sum_{i \in I_i} \psi(\mu(X_i)) = \sum_{i \in I_i} -\mu(X_i)\log \mu(X_i)$ satisfies the subadditive property (1.2) so the limit (5.42) exists by Proposition 1.1. □

In fact, entropy is much more general than suggested by (5.42). If $X = Y_1 \cup \ldots \cup Y_m$ is *any* partition of X into a finite number of (measurable) 'boxes', we can observe the sequence of boxes $(Y_{i_0}, \ldots, Y_{i_{k-1}})$ occupied by the iterates of x, so that $f^j x \in Y_{i_j}$ for each j, and estimate the rate of information gain as

$$\lim_{k \to \infty} -\frac{1}{k}\sum \mu(Y_{i_0,\ldots,i_{k-1}})\log \mu(Y_{i_0,\ldots,i_{k-1}})$$

where $Y_{i_0,\ldots,i_{k-1}} = \{x : f^j x \in Y_{i_j} \text{ for } j = 0, 1, \ldots, k-1\}$. Remarkably, this limit exists and equals $h_\mu(f)$ for *any* reasonable partition $Y_1 \cup \ldots \cup Y_m$ (reasonable in the sense that the observations $(Y_{i_0}, \ldots, Y_{i_k})$ distinguish between almost all $x \in E$ if k is sufficiently large). For our purposes, however, it is adequate to take (5.42) as the definition of entropy.

The variational principle characterises pressure as the maximum of a certain expression over all invariant probability measures. Preparatory to proving this, we need the following expression for an integral with respect to an invariant measure.

Lemma 5.8

Let E be a cookie-cutter set as above, let $\phi : E \to \mathbb{R}$ be a Lipschitz function and let μ be an invariant probability measure on E. Then for any choice of $x_i \in X_i$ we have

$$\int \phi(x)\mathrm{d}\mu = \lim_{k \to \infty} \frac{1}{k} \sum_{i \in I_k} S_k \phi(x_i) \mu(X_i). \tag{5.43}$$

Proof First observe that if μ is invariant then $\int \phi(x)\mathrm{d}\mu = \int \phi(f^j x)\mathrm{d}\mu$ for all j (applying (5.25) repeatedly), so that

$$\int \phi(x)\mathrm{d}\mu = \frac{1}{k} \int \sum_{j=0}^{k-1} \phi(f^j x)\mathrm{d}\mu = \frac{1}{k} \int S_k \phi(x)\mathrm{d}\mu. \tag{5.44}$$

Hence, if $x_i \in X_i$ for each $i \in I_k$, we have on splitting the range of integration

$$\left| \int \phi(x)\mathrm{d}\mu - \frac{1}{k} \sum_{i \in I_k} S_k \phi(x_i) \mu(X_i) \right|$$

$$= \left| \frac{1}{k} \sum_{i \in I_k} \left(\int_{x \in X_i} S_k \phi(x)\mathrm{d}\mu - S_k \phi(x_i) \mu(X_i) \right) \right|$$

$$\leq \frac{1}{k} \sum_{i \in I_k} \mu(X_i) \max_{x \in X_i} |S_k \phi(x) - S_k \phi(x_i)|$$

$$\leq b/k$$

using the bounded variation estimate (5.3) and that $\mu(E) = 1$. Letting $k \to \infty$ gives (5.43). $\square$

Theorem 5.9   (variational principle)

Let $\phi : E \to \mathbb{R}$ be Lipschitz. Then

$$P(\phi) = \sup\{h_\mu + \int \phi \mathrm{d}\mu : \mu \text{ is an invariant probability measure on } E\}.$$

The supremum is attained by the invariant Gibbs measure ν of Theorem 5.5.

Proof Choose $x_i \in X_i$ for all $i \in I$. For $k = 0, 1, 2, \ldots$ and μ an invariant probability measure, define

$$p_k(\mu) \equiv \frac{1}{k} \sum_{i \in I_k} \mu(X_i)[-\log \mu(X_i) + S_k \phi(x_i)] \tag{5.45}$$

$$\leq \frac{1}{k} \log \sum_{i \in I_k} \exp S_k \phi(x_i)$$

by Corollary 1.5. Letting $k \to \infty$, and noting that the limits of the three parts of this last inequality are given by (5.42), (5.43) and (5.5) respectively, we get

$$h_\mu + \int \phi \, d\mu \leq P(\phi).$$

To see that equality holds for the invariant Gibbs measure ν of Theorem 5.5, note that from (5.36)

$$p_k(\nu) \geq \frac{1}{k} \sum_{i \in I_k} \nu(X_i)[-\log(a_0 \exp(-kP(\phi) + S_k \phi(x_i))) + S_k \phi(x_i)]$$

$$= \frac{1}{k} \sum_{i \in I_k} \nu(X_i)[-\log a_0 + kP(\phi)]$$

$$= P(\phi) - \tfrac{1}{k} \log a_0.$$

Combining with (5.45) (with μ replaced by ν) and letting $k \to \infty$, again using (5.42) and (5.43), gives $h_\nu + \int \phi \, d\nu \geq P(\phi)$. $\quad\square$

Taking $\psi = -s \log |f'|$ and using Theorem 5.3 we get the variational formula for the Hausdorff dimension of the repeller E. It is the unique real number s such that

$$0 = \sup\{h_\mu - s \int \log |f'| d\mu : \mu \text{ is an invariant probability measure on } E\}.$$

5.5 Further applications

The methods of this chapter, and the underlying bounded variation principle of Section 4.2, adapt to many other dynamical systems and iterated function systems. We list some further situations for which the dimension of the attractor or repeller may be found in terms of a pressure, with an invariant Gibbs measure supported by the set concerned, and satisfying a variational condition.

Weaker differentiability conditions

The thermodynamic theory holds with little modification under weaker conditions on ϕ and f. For the bounded variation principle to hold it is enough that ϕ is a Hölder continuous function of exponent ϵ for some $\epsilon > 0$ (so that

$$|\phi(x) - \phi(y)| \leq a|x - y|^\epsilon \qquad (5.46)$$

rather than the Lipschitz condition (5.1)), see Exercise 5.1. Consequently, for the bounded distortion principle and the dimension formula, it is enough for $f'(x)$ to satisfy an ϵ-Hölder condition; in this case f is said to be of differentiability class $C^{1+\epsilon}$.

More general iterated function systems

Let X be a closed subset of $\mathbb{R}^n$ and let $F_1, \ldots, F_m$ be C^2 contractions (or $C^{1+\epsilon}$ contractions for some $\epsilon > 0$), with non-empty compact attractor E satisfying $E = \cup_{i=1}^m F_i(E)$. If $n = 1$ and X can be chosen so that $\{F_i(X)\}_{i=1}$ are disjoint sets, a trivial modification of the analysis of Chapters 4 and 5 shows that the 'm-part cookie-cutter E' is approximately self-similar and has dimension given by $P(-s \log |f'|) = 0$, where the sums in the definition (5.5) of P are now over index sets $I_k = \{(i_1, \ldots, i_k) : 1 \leq i_j \leq m\}$, see Figure 4.4.

In higher dimensions the situation is more complicated. Let X be a 'nice' domain in $\mathbb{R}^n$, for example $\mathbb{R}^n$ itself or a convex set or a connected domain with a smooth boundary. We say that a differentiable mapping $F : X \to X$ is *conformal* if its derivative $F'(x)$ (regarded as a linear mapping) is a similarity transformation. Thus a conformal mapping is a 'local similarity' that transforms small regions to nearly similar images. Let $F_1, \ldots, F_m : X \to X$ be injections that are C^2 conformal contractions. If $X_i = F_i(X)$ for $i = 1, \ldots, m$, and $X_1, \ldots, X_m$ are disjoint, we may regard the attractor E of this IFS as the repeller of a function $f : X_1 \cup \ldots \cup X_m \to X$ where $f(x) = F_i^{-1}(x)$ if $x \in X_i$. The analysis proceeds more or less as in Chapters 4 and 5, but using higher-dimensional versions of the chain rule and mean value theorem, to give bounded variation and distortion principles and approximate self-similarity of E. The dimension of E is found by solving $P(-s \log \|f'\|) = 0$, where $\|f'\|$ is the expansion ratio of the derivative f', with E supporting an invariant Gibbs measure μ with $\mu(X_i) \asymp |X_i|^s$.

If $X_1, \ldots, X_m$ are not disjoint but E satisfies the open set condition, the IFS attractor E cannot always be identified with the repeller of a dynamical system. Nevertheless, by working in terms of conformal mappings $F_1, \ldots, F_m$ throughout rather than with f (so that occurrences of $f^k : X_{i_1,\ldots,i_k} \to X$ are replaced by $(F_{i_1} \circ \cdots \circ F_{i_k})^{-1}$ with $S_k \phi$ defined by (4.10)) the theory proceeds in the same way.

Graph-directed systems

The thermodynamic formalism allows a non-linear version of the graph-directed systems defined by (3.12) to be treated. For $e \in \mathcal{E}_{i,j}$ we take

$F_e : X_j \to X_i$ to be C^2 contractions on appropriate sets $X_1, \ldots, X_q \subset \mathbb{R}^n$ and obtain a family of sets $E_1, \ldots, E_q$ satisfying (3.12). Under certain conditions the aggregate of these sets may be the repeller of a certain dynamical system f with local inverses given by the F_e. In the case of $X_i \subset \mathbb{R}^1$, or for $X_i \subset \mathbb{R}^n$ with the mappings F_e conformal, the thermodynamic formalism leads to a 'pressure analogue' of the formula $\rho(A^{(s)}) = 1$ of Corollary 3.5 together with Gibbs measures on the E_i.

Julia sets

Julia sets of certain complex analytic functions $f : \mathbb{C} \to \mathbb{C}$ give a very important class of approximately self-similar sets. (Note that f being analytic implies the corresponding real function $f : \mathbb{R}^2 \to \mathbb{R}^2$ is conformal). Then f may have a repeller E, so that there is a neighbourhood X of E (with $E \subset \text{int}(X)$) such that $E = \{z \in X : f^k(z) \in X \text{ for all } k = 1, 2, \ldots\}$, with f expanding on X, in that $|f'(z)| > 1$ for $z \in X$. Such an E is the *Julia set* of f. This situation occurs, for example, for $f(z) = z^2 + c$ for 'most' complex numbers c; in particular if $|c|$ is small when E is homeomorphic to a circle, or if $|c|$ is large when E is totally disconnected, see FG, Section 14.3. In the simplest case, with $f(z) = z^2 + c$ and $|c|$ large, the set X may be chosen so that the inverse f^{-1} has two branches $F_1, F_2 : X \to X$ which are conformal contractions, see Figure 5.2. Then the Julia set E is the attractor of the conformal IFS $\{F_1, F_2\}$. Thus E is

Figure 5.2   The Julia set of $f(z) = z^2 + 0.5 + 0.3i$, which is a self-conformal set under the IFS consisting of the two branches of the inverse of f

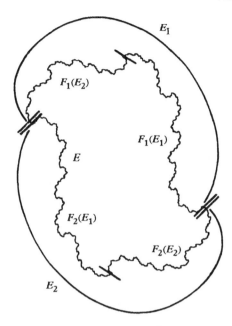

Figure 5.3 The Julia set E of $f(z) = z^2 + 0.279 + 0.3i$. Such a set may be regarded as a graph-directed set using a Markov partition $E = E_1 \cup E_2$. With F_1 and F_2 as the 'north-east' and 'south-west' branches of f^{-1}, we have $E_1 = F_1(E_1) \cup F_1(E_2)$ and $E_2 = F_2(E_1) \cup F_2(E_2)$

approximately self-similar and has Hausdorff and box dimensions given by $P(-s\log|f'|) = 0$.

For other analytic functions $f: \mathbb{C} \to \mathbb{C}$ it is often possible to find a decomposition $E = E_1 \cup \ldots \cup E_m$ (called a *Markov partition* of E), a matrix $(\mathcal{E}_{i,j})$ of 0s and 1s, and functions $F_{i,j}: E_j \to E_i$, defined when $\mathcal{E}_{i,j} = 1$, such that for each i the branches of f^{-1} near E_i are given by $F_{i,j}$. This is essentially a graph-directed system of conformal mappings, see Figure 5.3, and as before the thermodynamic theory gives the dimension of the Julia set E as the solution of $P(-s\log|f'|) = 0$, with $0 < \mathcal{H}^s(E) < \infty$.

The thermodynamic approach leads to many further properties of Julia sets. For instance, $\dim_{\mathrm{H}}$(Julia set of $z^2 + c$) is asymptotic to $1 + |c|^2/4\log 2$ for small c, and to $2\log 2/\log|c|$ for large c. The theory extends to more general polynomials and other analytic functions.

Non-conformal repellers in $\mathbb{R}^n$

The thermodynamic formalism does not generalise so easily to mappings that are not conformal, though some progress has been made in a few special cases.

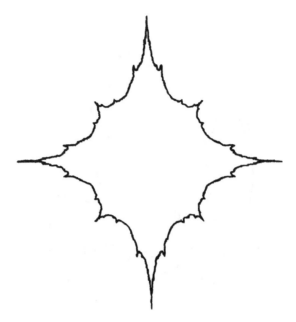

Figure 5.4   Repeller of the non-conformal mapping $(x, y) \rightarrow (x^2 - y^2,\ 6xy)$

For instance, let S denote the interval $[0, 1]$ made into a circle by identifying 0 and 1, and let $f \colon S \times \mathbb{R} \to S \times \mathbb{R}$ be an expanding map that preserves vertical lines, that is with $f(x, y) = (f_1(x), f_2(x, y))$ for suitable f_1 and f_2. Then f has a repelling curve that encircles the cylinder $S \times \mathbb{R}$. Under certain conditions this repeller is a fractal curve with dimension given by $P(s\phi_1 + \phi_2) = 0$ where P is a pressure and ϕ_1 and ϕ_2 are functions involving the expansion rates of f in the x and y directions.

For certain diffeomorphisms of two-dimensional regions where there is a local decomposition of the mapping into one expanding and one contracting component, there is a formula involving pressure that gives the dimension of the invariant horseshoe that occurs.

For functions f without some sort of decomposition into 'independent' directions, for example that of Figure 5.4, the situation is much more complex. A certain amount of progress towards a dimension formula has been made; one approach that leads to upper bounds for dimension is sketched in Section 12.1.

5.6 Why 'thermodynamic' formalism ?

A question frequently asked by mathematicians involved with dynamical systems is why the name 'thermodynamic' is given to this approach. Although

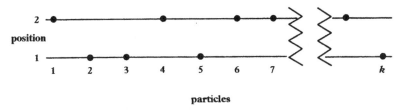

position

particles

Figure 5.5 The configuration of the particle system $(i_1, \ldots, i_k)$ has energy $E_{i_1, \ldots, i_k}$ and probability of occurrence proportional to $\exp(-sE_{i_1, \ldots, i_k})$

the analogues are formal rather than physical, many of the ideas, such as the existence of Gibbs measures, were originally developed in statistical mechanics, and translated to dynamical systems many years later.

For simplicity, we consider a one-dimensional particle chain with particles situated at integer points $(1, 2, 3, \ldots, k)$. Each particle is in one of two possible positions, which we label 1 or 2 (see Figure 5.5). Thus a configuration of the system is specified by a k-term sequence $i = (i_1, \ldots, i_k)$ where $i_j = 1$ or 2 for $j = 1, \ldots, k$. Each configuration has an associated *energy*, denoted by $E_{i_1, \ldots, i_k}$, and for real numbers s we define the *partition function*

$$Z_k^s = \sum_{i_1, \ldots, i_k} \exp(-sE_{i_1, \ldots, i_k}),$$

where the sum is over all 2^k possible state configurations.

When the number of particles is large, the particles interact and exchange energy in the process. The fundamental principle governing such interactions is Boltzmann's Law, that the probability of a small part of the system being in a particular configuration is proportional to a power of the energy of that configuration. Thus

$$\text{Probability of configuration } (i_1, \ldots, i_k) = \frac{\exp(-sE_{i_1, \ldots, i_k})}{Z_k^s}$$

for a number s which may be identified with the reciprocal of the absolute temperature of the system (multiplied by Boltzmann's constant).

In the simplest situation, the energy of each particle depends only on its own state, so that $E_{i_1, \ldots, i_k}$ is the sum of the individual energies of the particles. Thus, writing $\phi(i_1, \ldots, i_k)$ for the energy of the first particle (depending only on i_1) and f for the 'shift' $f(i_1, i_2, \ldots, i_k) = (i_2, \ldots, i_k, j_1)$, where j_1 is chosen arbitrarily, we have that $E_{i_1, \ldots, i_k} = \phi(i) + \phi(f(i)) + \ldots + \phi(f^{k-1}(i)) \equiv S_k \phi(i)$.

Thus

$$Z_k^s = \sum_{i \in I_k} \exp(-sS_k\phi(i))$$

and

$$\text{Probability of configuration}(i) = \frac{\exp(-sS_k\phi(i))}{Z_k^s}$$

for $i \in I_k$.

We allow the particle chain to become infinite, by letting $k \to \infty$; this is known as taking the *thermodynamic limit*. Provided that the energies satisfy certain reasonable physical conditions, such as some form of translation invariance, an argument parallel to that used to establish the bounded variation principle and existence of Gibbs measures for dynamical systems shows that

$$Z_k^s \asymp \exp(kP_s)$$

and that

probability (a configuration $(j_1, j_2, \ldots)$ with $j_1 = i_1, \cdots, j_k = i_k$ occurs)

$$\asymp \exp(-sS_k\phi(i) - kP_s)$$

for all finite sequences $i = (i_1, \ldots, i_k)$. The number P_s is the pressure of the physical system and this probability is the classical Gibbs distribution for states.

The parallels with the formulae and relationships that occur in the cookie-cutter dynamical systems are now apparent. We have the following correspondences:

statistical mechanics	*dynamical systems*
finite particle configuration	interval X_i
infinite particle configuration	point of E
energy E_i	$S_k\phi(x)$
inverse temperature	dimension
partition function	the sum $\sum_{i \in I_k} \exp(S_k\phi(x_i))$
pressure	topological pressure
Gibbs distribution on configurations	Gibbs measure on E.

Whilst it may be inappropriate to say that 'the dimension of a repeller is the reciprocal of the temperature that renders the pressure zero', many features of dynamical systems may be studied to advantage using parallels from statistical mechanics.

5.7 Notes and references

The remarkable insight that thermodynamic methods could be applied to dynamical systems began with Sinai (1972) and was developed by Bowen (1975) and Ruelle (1978). These last two books provide quite technical accounts as do Parry and Pollicott (1990) who extend the work via zeta functions to study periodic orbits of continuous systems. A nice survey of this material is given by Bedford (1991) and a variety of related survey articles are included in Bedford *et al.* (1991). For a very general treatment of entropy and pressure see Walters (1982). For the functional analytic properties used in proving Theorem 5.4 see Dunford and Schwartz (1958).

Ruelle (1982) treats the case of conformal maps in regions of $\mathbb{R}^n$ with special reference to Julia sets. Bedford (1986) and Mauldin and Williams (1988) analyse graph-directed or recurrent IFSs, and Bedford and Urbanski (1990) study maps preserving vertical lines. McClusky and Manning (1983) obtain the formula for horseshoe attractors, and Falconer (1994) considers general non-conformal repellers.

Exercises

5.1 Prove the bounded variation principle, Proposition 4.1, with the Lipschitz condition (4.8) replaced by the ϵ-Hölder condition (5.46).

5.2 Verify (5.24) which gives a bound on the rate of convergence in the pressure definition.

5.3 Let $f: X_1 \cup X_2 \to X$ be a cookie-cutter system of the usual form, so that f is a 2–1 function on E. Show that if μ is any invariant probability measure on E then $h_\mu \leq \log 2$, and that we can always find μ with $h_\mu = \log 2$.

5.4 Consider the system (2.24) so that $f(x) = 3x$ (mod 1) and E is the middle-third Cantor set. Find $P(-s \log |f'|)$ for all real numbers s. Let μ be the 'uniformly distributed' probability measure on E (so that $\mu(X_i) = 2^{-k}$ if $i \in I_k$). Find h_μ and verify directly that $h_\mu + \int \phi \, d\mu = P(\phi)$ when $\phi = -s \log |f'|$.

5.5 Show that $P(-s \log |f'|)$ given by (5.15) is a convex function of s.

Chapter 6   The ergodic theorem and fractals

The ergodic theorem is one of the most fundamental and useful results in probability theory and dynamical systems. In this chapter we prove the ergodic theorem and show how it may be applied to fractal geometry, in particular to local properties of fractals such as densities and average densities. Other applications of the ergodic theorem will be encountered later in the book.

6.1 The ergodic theorem

The setting for the ergodic theorem is as follows. There is a set X, a mapping $f: X \to X$, and a finite measure μ on X. We assume for every measurable set $A \subset X$ the inverse image $f^{-1}(A)$ is measurable and

$$\mu(f^{-1}(A)) = \mu(A), \tag{6.1}$$

that is, μ is *invariant under f* or *f is measure preserving* for μ. Condition (6.1) is equivalent to

$$\int g(x)\mathrm{d}\mu(x) = \int g(f(x))\mathrm{d}\mu \tag{6.2}$$

for all measurable $g : X \to \mathbb{R}$. We assume, as always, that μ is a Borel regular measure so (6.1) is equivalent to (6.2) holding for all continuous $g : X \to \mathbb{R}$.

Recall that the measure μ is *ergodic* for f if every measurable set A such that $A = f^{-1}(A)$ has $\mu(A) = 0$ or $\mu(X \setminus A) = 0$. Essentially, this means that the system is indecomposable: if there is a set A with $A = f^{-1}(A)$ and $0 < \mu(A) < \mu(X)$, then we can consider separately the two independent systems obtained by restricting f to A and to $X \setminus A$.

If μ is ergodic and $\phi : X \to \mathbb{R}$ is a measurable function such that $\phi(x) = \phi(f(x))$ for all x, then there is a number λ such that

$$\phi(x) = \lambda \tag{6.3}$$

for almost all x. To see this, note that for each $\lambda \in \mathbb{R}$, the set $A = \{x \in X : \phi(x) < \lambda\}$ is measurable and satisfies $f^{-1}(A) = A$, so by ergodicity either $\mu(A) = 0$ or $\mu(X \setminus A) = 0$, that is either $\phi(x) < \lambda$ for almost all x or $\phi(x) \geq \lambda$ for almost all x.

We regard $f: X \to X$ as a dynamical system, so that the k-th iterate $f^k x$ represents the position at time k of a particle at x at time 0. For $\phi: X \to \mathbb{R}$, we think of $\frac{1}{k}\sum_{j=0}^{k-1}\phi(f^j x)$ as the *time average* of ϕ evaluated at the first k iterates of x. The ergodic theorem asserts that for almost all x these averages approach a limit as $k \to \infty$. Furthermore, if μ is ergodic, the limit is independent of x and equals the *space average* of ϕ, that is $\int_X \phi(y)d\mu(y)$. Thus in the ergodic case, the time average of ϕ for almost all initial points x equals the space average of ϕ.

Theorem 6.1   (ergodic theorem)

Let $f: X \to X$, let μ be a finite measure on X that is invariant under f, and let $\phi \in L^1(\mu)$. Then the limit

$$\Phi(x) \equiv \lim_{k \to \infty} \frac{1}{k}\sum_{j=0}^{k-1}\phi(f^j x) \tag{6.4}$$

exists for μ-almost all x. Moreover, if μ is ergodic then

$$\Phi(x) = \frac{1}{\mu(X)}\int_X \phi(y)d\mu(y) \tag{6.5}$$

for μ-almost all x.

Proof   For simplicity we give the proof on the assumption that for some M we have $|\phi(x)| \leq M$ for all $x \in X$. For the extension to $\phi \in L^1(x)$, see Exercise 6.1. Write

$$\alpha_k(x) = \frac{1}{k}\sum_{j=0}^{k-1}\phi(f^j x) \tag{6.6}$$

for the average of ϕ over the first k iterates, and $\bar{\alpha}(x) = \lim\sup_{k \to \infty}\alpha_k(x)$. The crux of the proof is to show that for all $\epsilon > 0$

$$\int \bar{\alpha}(x)d\mu(x) \leq \int \phi(x)d\mu(x) + \epsilon. \tag{6.7}$$

To verify this define

$$\tau(x) = \min\{k > 0 : \alpha_k(x) \geq \bar{\alpha}(x) - \epsilon\},$$

so $\tau(x) < \infty$ for all x by definition of $\bar{\alpha}$. Should we be fortunate enough for there to exist $T < \infty$ such that $\tau(x) \leq T$ for all x, then the sum (6.6) may be broken into blocks of length at most T such that the average of the $\phi(f^j x)$ over the j in each block is at least $\bar{\alpha}(x) - \epsilon$. More precisely, for each x we define a sequence $(k_1, k_2, \ldots)$ inductively, by taking $k_1 = \tau(x)$ and $k_i = \tau(f^{k_1 + \cdots + k_{i-1}}x)$

for $i = 2, 3, \ldots$, so that

$$\sum_{j=k_1+\cdots+k_{i-1}}^{k_1+\cdots+k_i-1} \phi(f^j x) = k_i \alpha_{k_i}(f^{k_1+\cdots+k_{i-1}} x)$$

$$\geq k_i(\bar{\alpha}(f^{k_1+\cdots+k_{i-1}} x) - \epsilon)$$

$$= k_i(\bar{\alpha}(x) - \epsilon), \tag{6.8}$$

since $\bar{\alpha}(x) = \bar{\alpha}(f^k x)$ for all k. Summing over i,

$$\sum_{j=0}^{k-1} \phi(f^j x) \geq k(\bar{\alpha}(x) - \epsilon)$$

whenever k is of the form $k_1 + \cdots + k_i$. For an arbitrary integer k, by comparing with the sum to $k_1 + \cdots + k_i$ terms for the largest i such that $k_1 + \cdots + k_i \leq k$, we get

$$\sum_{j=0}^{k-1} \phi(f^j x) \geq (k - T)(\bar{\alpha}(x) - \epsilon) - TM,$$

since $0 < k - (k_1 + \cdots + k_i) \leq T$. Integrating (noting that $\int \phi(f^j x) \mathrm{d}\mu(x) = \int \phi(x) \mathrm{d}\mu(x)$ by (6.2)), dividing by k, and letting $k \to \infty$ gives (6.7) in this case.

Now suppose we are less fortunate and $\tau(x)$ is unbounded. We may choose T large enough to make $\mu(A) < \epsilon$, where $A \equiv \{x : \tau(x) > T\}$, and modify ϕ on A by defining $\phi^* : X \to \mathbb{R}$ by

$$\phi^*(x) = \begin{cases} \phi(x) & (x \notin A) \\ M & (x \in A). \end{cases}$$

We define α_k^* as in (6.6) but with ϕ replaced by ϕ^*. Letting

$$\tau^*(x) = \min\{k > 0 : \alpha_k^*(x) \geq \bar{\alpha}(x) - \epsilon\},$$

we now have $\tau^*(x) \leq T$ for all x (since $\tau^*(x) = 1$ for $x \in A$). Proceeding just as before we obtain

$$\int \bar{\alpha}(x) \mathrm{d}\mu(x) \leq \int \phi^*(x) \mathrm{d}\mu(x) + \epsilon$$

$$= \int \phi(x) \mathrm{d}\mu(x) + \int (\phi^*(x) - \phi(x)) \mathrm{d}\mu(x) + \epsilon$$

$$\leq \int \phi(x) \mathrm{d}\mu(x) + 2M\epsilon + \epsilon$$

which gives (6.7) since ϵ may be chosen arbitrarily small.

Since ϵ is arbitrary (6.7) implies $\int \bar{\alpha}(x) \mathrm{d}\mu(x) \leq \int \phi(x) \mathrm{d}\mu(x)$. A symmetrical argument gives $\int \phi(x) \mathrm{d}\mu(x) \leq \int \underline{\alpha}(x) \mathrm{d}\mu(x)$, where $\underline{\alpha}(x) = \liminf_{k \to \infty} \alpha_k(x)$, and combining these inequalities $\int(\bar{\alpha}(x) - \underline{\alpha}(x))d\mu(x) \le 0$. As $0 \le \bar{\alpha}(x) - \underline{\alpha}(x)$ it follows that $\underline{\alpha}(x) = \bar{\alpha}(x)$ for almost all x, so the common value equals the limit (6.4).

Clearly, $\alpha_k(fx) - \alpha_k(x) = (\phi(f^k x) - \phi(x))/k$, so $\Phi(fx) = \Phi(x)$ whenever either limit exists. Thus if μ is ergodic, (6.3) implies that $\Phi(x) = \lambda$ for almost all x for some λ. Using the dominated convergence theorem and the invariance of μ,

$$\lambda\mu(X) = \int \Phi(x)d\mu(x) = \lim_{k\to\infty} \int \alpha_k(x)d\mu(x) = \int \phi(x)d\mu(x),$$

which is (6.5). $\square$

Whilst this form of the ergodic theorem is suited to our applications to self-similar sets, we need the following generalisation, which may be thought of as an 'approximate ergodic theorem', for studying non-linear cookie-cutter sets.

Corollary 6.2

Let $f : X \to X$, let μ be a finite invariant measure on X, and let $\phi_n \in L^1(\mu)$ for $n = 1, 2, \ldots$. Suppose that for all positive integers k and n and all $x \in X$

$$|\phi_n(f^k x) - \phi_{n+k}(x)| < \epsilon_n \tag{6.9}$$

where $\epsilon_n \searrow 0$. Then the limit $\Phi(x) = \lim_{m\to\infty}(1/m)\sum_{k=0}^{m-1} \phi_k(x)$ exists for μ-almost all x. If μ is ergodic, then $\Phi(x)$ is almost everywhere constant.

Proof   For $m \ge 1$ and $n \ge 1$ we have identically that

$$\frac{1}{m+n} \sum_{k=0}^{m+n-1} \phi_k(x) = \frac{1}{m+n} \sum_{k=0}^{n-1} \phi_k(x) \tag{6.10}$$

$$+ \frac{1}{m+n} \sum_{k=0}^{m-1} [\phi_{n+k}(x) - \phi_n(f^k x)] \tag{6.11}$$

$$+ \frac{m}{m+n} \frac{1}{m} \sum_{k=0}^{m-1} \phi_n(f^k x). \tag{6.12}$$

Fixing n and letting $m \to \infty$, (6.10) converges to 0 for almost all x, (6.11) is bounded in modulus by ϵ_n and (6.12) converges for almost all x to a number, $\Phi_n^*(x)$ say, using Theorem 6.1. Thus for almost all x,

$$\Phi_n^*(x) - \epsilon_n \le \liminf_{m\to\infty} \frac{1}{m} \sum_{k=0}^{m-1} \phi_k(x)$$

$$\le \limsup_{m\to\infty} \frac{1}{m} \sum_{k=0}^{m-1} \phi_k(x) \le \Phi_n^*(x) + \epsilon_n.$$

Letting $n \to \infty$, it follows that

$$\lim_{m \to \infty} \frac{1}{m} \sum_{k=0}^{m-1} \phi_k(x) = \lim_{n \to \infty} \Phi_n^*(x) \equiv \Phi(x),$$

for almost all x.

If μ is ergodic, then Φ_n^* is almost everywhere constant, by Theorem 6.1, so Φ is almost everywhere constant. $\square$

A simple application of the ergodic theorem concerns the Liapounov exponents of a dynamical system. Let $X \subset \mathbb{R}$ be a closed interval and let $f: X \to X$ be a C^1 mapping. The Liapounov exponent reflects the local rate of expansion or contraction under iteration by f. We define the *Liapounov* or *characteristic exponent* $\lambda(x)$ of f at x by

$$\lambda(x) \equiv \lim_{k \to \infty} \frac{1}{k} \log |(f^k)'(x)| \tag{6.13}$$

$$= \lim_{k \to \infty} \frac{1}{k} \sum_{j=0}^{k-1} \log |f'(f^j x)| \tag{6.14}$$

using the chain rule (4.16). Thus for a small interval J centred at x, we have $|f^k(J)| \simeq \exp(k\lambda(x))|J|$. Of course, this definition assumes that the limit (6.13) exists; under reasonable conditions this follows from the ergodic theorem.

Proposition 6.3

Let μ be an invariant ergodic measure for a C^1 mapping $f: X \to X$, and suppose that $\int \log |f'(x)| d\mu(x) > -\infty$. Then there is a number λ such that the Liapounov exponent $\lambda(x)$ exists and equals λ for μ-almost all x.

Proof This is immediate from Theorem 6.1, on taking $\phi(x) = \log |f'(x)|$. $\square$

Thus under these conditions we refer to *the* Liapounov exponent λ of f.

Liapounov exponents may be generalised to differentiable mappings $f: X \to X$ where X is a suitable subset of $\mathbb{R}^n$. The derivative $(f^k)'(x)$ is a linear mapping on $\mathbb{R}^n$, and we may write $a_1^k(x) \geq a_2^k(x) \geq \ldots \geq a_n^k(x)$ for the lengths of the principal semi-axes of the ellipsoid $(f^k)'(B)$ where B is the unit ball in $\mathbb{R}^n$. The Liapounov exponents are defined as the logarithmic rates of growth with k of these semi-axis lengths:

$$\lambda_j(x) = \lim_{k \to \infty} \frac{1}{k} \log a_j^k(x) \quad (j = 1, \ldots, n).$$

Thus the Liapounov exponents describe the distortion of an infinitesimal ball under iteration by f. Using a more sophisticated version of the ergodic theorem, it may be shown that if μ is invariant and ergodic with respect to f

then there are numbers $\lambda_1 \geq \lambda_2 \geq \ldots \geq \lambda_n \geq 0$ such that $\lambda_j(x) = \lambda_j$ for μ-almost all x for all j.

6.2 Densities and average densities

Self-similar sets and cookie-cutter sets support natural measures that are invariant and ergodic with respect to their defining transformations. It is this that makes the ergodic theorem such a useful tool in their study. We first summarise the properties of these measures.

Lemma 6.4

(a) *Let $E \subset \mathbb{R}^n$ be a self-similar IFS attractor of dimension s satisfying the strong separation condition (see Section 2.2) and let $\mu = \mathcal{H}^s|_E$ be the restriction of s-dimensional Hausdorff measure $\mathcal{H}^s$ to E. Then μ is invariant and ergodic with respect to $f: E \to E$ (where f is given by the inverse map (2.41)).*

(b) *More generally, let μ be the self-similar measure on the self-similar set E given by (2.43)–(2.44), where E satisfies the strong separation condition. Then μ is invariant and ergodic with respect to $f: E \to E$.*

(c) *A cookie-cutter set E of dimension s supports an invariant ergodic probability measure μ that is equivalent to $\mathcal{H}^s|_E$.*

Proof

(a) With the usual notation for IFSs (see Section 2.2) we have that

$$f^{-1}(A) = \bigcup_{i=1}^{m} F_i(A)$$

for $A \subset E$, with this union disjoint. By the scaling property (2.13) of $\mathcal{H}^s$, and hence of μ, and by (2.42)

$$\mu(f^{-1}(A)) = \sum_{i=1}^{m} \mu(F_i(A)) = \sum_{i=1}^{m} r_i^s \mu(A) = \mu(A)$$

so that μ is invariant under f.

Now suppose that $A \subset E$ is measurable and $A = f^{-1}(A)$, so that $A = f^{-k}(A) = \bigcup_{i \in I_k} F_{i_1} \circ \cdots \circ F_{i_k}(A)$ for all k and $i = (i_1, \ldots, i_k)$. Then $A \cap E_{i_1, \ldots, i_k} = F_{i_1} \circ \cdots \circ F_{i_k}(A)$ and by the scaling property

$$\mu(A \cap E_i) = (r_{i_1} \ldots r_{i_k})^s \mu(A) = \mu(E_i)\mu(E)^{-1}\mu(A).$$

Let $\mathcal{C}$ be the class of sets $U \subset E$ such that

$$\mu(A \cap U) = \mu(U)\mu(A)\mu(E)^{-1}. \tag{6.15}$$

We have shown that $E_i \in \mathcal{C}$ for all $i \in I$, and by additivity any countable union of such sets is in $\mathcal{C}$. By the regularity of μ, given any measurable set $U \subset E$, we may find a sequence of these sets decreasing to U, so that (6.15) holds for any such U. In particular, taking $U = A$, we have $\mu(A) = \mu(A \cap A) = \mu(A)\mu(A)\mu(E)^{-1}$ giving $\mu(A) = 0$ or $\mu(A) = \mu(E)$, so μ is ergodic.

(b) We note that we have a scaling property of the form $\mu(F_i(A)) = p_i\mu(A)$ for any $A \subset E$. The proof proceeds as in (a) with r_i^s replaced by p_i.

(c) This non-linear analogue of (a) was proved in Chapter 5: Theorem 5.3 shows that the restriction of $\mathcal{H}^s$ to E is a Gibbs measure, Theorem 5.5 shows that there is an equivalent invariant Gibbs measure, and Corollary 5.6 shows that these measures are ergodic. $\square$

It follows from Lemma 6.4(c) that the Liapounov exponents of a cookie-cutter system exist and are constant $\mathcal{H}^s$-almost everywhere on E.

We now examine some consequences of ergodicity for densities of certain fractals. Let $E \subset \mathbb{R}^n$ be a Borel set of Hausdorff dimension s with $0 < \mathcal{H}^s(E) < \infty$. Again, for convenience, we write $\mu = \mathcal{H}^s|_E$ for the restriction of s-dimensional Hausdorff measure to E, so that

$$\mu(A) = \mathcal{H}^s(E \cap A). \tag{6.16}$$

Recall from (2.17)–(2.18) (see also FG, Chapter 5) that the (s-dimensional) *lower* and *upper densities* of E at x are defined as

$$\underline{D}^s(x) = \underline{D}^s(E, x) = \lim_{r \to 0} \inf \mathcal{H}^s(E \cap B(x, r))/(2r)^s = \lim_{r \to 0} \inf \mu(B(x, r))/(2r)^s$$

$$\tag{6.17}$$

and

$$\overline{D}^s(x) = \overline{D}^s(E, x) = \lim_{r \to 0} \sup \mathcal{H}^s(E \cap B(x, r))/(2r)^s = \lim_{r \to 0} \sup \mu(B(x, r))/(2r)^s$$

$$\tag{6.18}$$

for $x \in \mathbb{R}^n$. These densities indicate the concentration of the set E around the point x. Clearly, $\underline{D}^s(x) \leq \overline{D}^s(x)$ for all x, but for irregular 'fractal' sets, this inequality is strict for μ-almost all $x \in E$ (see Section 2.1). Nevertheless, a consequence of ergodicity is that, in the case of self-similar sets and cookie-cutter sets, the lower densities and the upper densities are almost everywhere constant.

Proposition 6.5

Let E be either a self-similar set satisfying the strong separation condition or a cookie-cutter set, and let $s = \dim_H E$. Then there exist numbers $\underline{d}$ and $\overline{d}$ with

$0 < \underline{d} \leq \bar{d} \leq 1$ *such that*

$$\underline{D}^s(x) = \underline{d} \quad and \quad \bar{D}^s(x) = \bar{d}$$

for $\mathcal{H}^s$-almost all $x \in E$.

Proof Let $f: X \to X$ be the function defining the self-similar set or cookie-cutter set E. We assume that X is such that E is in the interior of X, so $\underline{D}^s(x) = \underline{D}^s(f(x))$ for all $x \in E$ by (2.19).

By Lemma 6.4 (a) or (c) there is an ergodic measure μ equivalent to the restriction of $\mathcal{H}^s$ to E. Since $\underline{D}^s$ is a measurable function of x (see Exercise 6.4), it follows from (6.3) that $\underline{D}^s(x)$ is constant μ-almost everywhere and thus for $\mathcal{H}^s$-almost all $x \in E$. A similar argument applies to $\bar{D}^s(x)$.

We note that $0 < \underline{D}^s(x)$ for all $x \in E$ (see (5.22)) and that $\bar{D}^s(x) \leq 1$ for $\mathcal{H}^s$-almost all x (indeed this is so for all x, see FG, Proposition 5.1), so $0 < \underline{d} \leq \bar{d} \leq 1$. $\square$

A classical result in the theory of densities (FG, Section 5.1) states that if $0 < \mathcal{H}^s(E) < \infty$ and s is non-integral, then $\underline{D}^s(x) < \bar{D}^s(x)$ for $\mathcal{H}^s$-almost all x, that is, the density fails to exist, so $\underline{d} < \bar{d}$ in Proposition 6.5. This means that the ratio

$$\mu(B(x,r))/(2r)^s = \mathcal{H}^s(E \cap B(x,r))/(2r)^s \tag{6.19}$$

'oscillates' more or less between $\underline{d}$ and $\bar{d}$ when r is small. It is natural to try and describe this oscillation, and in particular to find the 'average' value of (6.19) for small r. Since self-similar sets exhibit similarities at scales that approach 0 at a geometric rate (for example, the middle-third Cantor set has self-similarity ratios $\frac{1}{3}, \frac{1}{9}, \frac{1}{27}, \ldots$) it is appropriate to use a form of averaging that assigns equal weight to each such scaling step.

Thus, we introduce the *logarithmic averages*

$$A^s(x, T) = \frac{1}{T} \int_{t=0}^{T} \mu(B(x, e^{-t}))(2e^{-t})^{-s} dt$$

$$= 2^{-s} \frac{1}{T} \int_{t=0}^{T} \mu(B(x, e^{-t})) e^{st} dt. \tag{6.20}$$

We define the *lower* and *upper average densities* of E or μ at x by

$$\underline{A}^s(x) = \liminf_{T \to \infty} A^s(x, T)$$

and

$$\bar{A}^s(x) = \limsup_{T \to \infty} A^s(x, T).$$

If $\underline{A}^s(x) = \bar{A}^s(x)$, we term the common value the *average density* or *order-two density*, which we denote by $A^s(x)$. Then

$$A^s(x) = \lim_{T \to \infty} \frac{1}{T} \int_{t=0}^{T} \frac{\mu(B(x, e^{-t}))}{(2e^{-t})^s} dt. \tag{6.21}$$

(Of course, these average densities may be defined for more general measures μ than just the restriction of Hausdorff measures.)

It is easy to see that for all x

$$\underline{D}^s(x) \leq \underline{A}^s(x) \leq \bar{A}^s(x) \leq \bar{D}^s(x). \tag{6.22}$$

The densities and average densities are defined locally in that they are determined by the restriction of μ to $B(x, r)$ for any $r > 0$; thus they reflect the local structure of E at x.

We have remarked that for a fractal set E the density $D^s(x)$ does not in general exist. However, for many fractals, including self-similar sets and cookie-cutter sets, the average density $A^s(x)$ *does* exist and takes the same value at almost all x. We demonstrate this using the ergodic theorem. We first illustrate the method in the particularly simple setting when E is the middle-third Cantor set.

Theorem 6.6

Let E be the middle-third Cantor set, let $s = \log 2 / \log 3$, and let $\mu = \mathcal{H}^s|_E$ (that is the natural uniformly distributed measure on E). Then for μ-almost all $x \in E$ the average density $A^s(x)$ exists with

$$A^s(x) = \frac{1}{2^s \log 2} \iint_{|x-y| \geq 1/3} |x - y|^{-s} d\mu(x) d\mu(y) = 0.62344\dots. \tag{6.23}$$

Proof For $k = 0, 1, 2, \dots$ write

$$\phi_k(x) = \int_{t=k}^{k+1} \frac{\mu(B(x, 3^{-t}))}{2^{-t}} dt. \tag{6.24}$$

(Of course, $B(x, r)$ is just the interval $[x - r, x + r]$ in this case.) Let $f : E \to E$ be given by $f(x) = 3x \pmod 1$; then f reflects the self-similarity of E in a natural way (for $r \leq \frac{1}{3}$, the 'picture' of $f(E \cap B(x, r))$ is just that of $E \cap B(x, r)$ enlarged by a factor 3). Using this and the scaling property of μ, we note that replacing x by $f(x)$ and t by $t - 1$ in (6.24) has the effect of doubling both the numerator and denominator of the integrand. Thus $\phi_k(x) = \phi_{k-1}(f(x))$, and iterating

$$\phi_k(x) = \phi_{k-1}(fx) = \phi_{k-2}(f^2 x) = \dots = \phi_0(f^k x).$$

Since μ is invariant and ergodic for f (by Lemma 6.4(a) or by checking directly), the ergodic theorem, Theorem 6.1, tells us that for μ-almost all $x \in E$,

$$\int_E \phi_0(y)d\mu(y) = \lim_{k\to\infty} \frac{1}{k}\sum_{j=0}^{k-1}\phi_0(f^jx)$$

$$= \lim_{k\to\infty} \frac{1}{k}\sum_{j=0}^{k-1}\phi_j(x)$$

$$= \lim_{k\to\infty} \frac{1}{k}\int_{t=0}^k \frac{\mu(B(x,3^{-t}))}{2^{-t}}dt$$

$$= 2^s \lim_{T\to\infty} \frac{1}{T}\int_{t=0}^T \frac{\mu(B(x,e^{-t}))}{(2e^{-t})^s}dt$$

$$= 2^s A^s(x)$$

using (6.24); the boundedness of the integrand enables us to pass from the discrete to the continuous limit. Thus for μ-almost all x, the average density $A^s(x)$ exists and equals a, where

$$a = 2^{-s}\int_E \phi_0(y)d\mu(y).$$

It is convenient to put a in the form (6.23). Substituting (6.24), and writing 1_X for the indicator function of the set X,

$$2^s a = \int_E \int_{t=0}^1 2^t \mu(B(x,3^{-t}))dt\,d\mu(x)$$

$$= \int_E \int_{t=0}^1 \int_E 2^t 1_{\{|x-y|\le 3^{-t}\}}d\mu(y)dt\,d\mu(x)$$

$$= \int_E \int_E \int_{t=0}^{\min\{1,-\log|x-y|/\log 3\}} e^{t\log 2}dt\,d\mu(x)d\mu(y)$$

$$= \frac{1}{\log 2}\iint_{|x-y|<1/3} 1d\mu(x)d\mu(y)$$

$$+ \frac{1}{\log 2}\iint_{|x-y|\ge 1/3}(|x-y|^{-s}-1)d\mu(x)d\mu(y)$$

$$= \frac{1}{\log 2}\iint_{|x-y|\ge 1/3}|x-y|^{-s}d\mu(x)d\mu(y),$$

for (6.23). (Here we have split the integral with respect to t and used that $(\mu \times \mu)\{(x,y) : |x-y| < 1/3\} = (\mu \times \mu)\{(x,y) : |x-y| \ge 1/3\}$ by virtue of the similarity of the left and right parts of the Cantor set.)

The integrand (6.23) is non-singular over the domain of integration, and is not too awkward to evaluate numerically. The integrand may be thought of as the expectation of a random variable with distribution given in terms of the measure $\mu \times \mu$, and may be evaluated roughly using a Monte Carlo method, or more precisely as a series involving the moments of the random variable. $\square$

The above proof adapts without difficulty to the case where E is an 'm-part Cantor set' obtained from the unit interval by repeated subdivision into equal and equally spaced subintervals of λ times the length of the parent interval. Thus E is the invariant set of the IFS consisting of the similarities $F_i(x) = \lambda x + (i - 1)(1 - \lambda)/(m - 1)$ for $i = 1, \ldots, m$. Then, $s = \dim_H E = -\log m/\log \lambda$, with $\mathcal{H}^s(E) = 1$ and

$$A^s(x) = \frac{1}{2^s \log m} \int \int |x - y|^{-s} d\mu(x) d\mu(y)$$

for μ-almost all x, where μ is the restriction of $\mathcal{H}^s$ to E and where the double integral is evaluated over $(x, y) \in \bigcup_{i \neq j} E_i \times E_j$.

Average densities for sets of this form are plotted in Figure 6.1. In particular, it can be seen that sets of the same dimension have different average densities; thus the average density is a parameter that can distinguish between sets of equal dimension.

For sets that are self-similar under similarities of several different ratios the proof that $A^s(x)$ exists and is constant almost everywhere is more involved, and for general cookie-cutter sets there are further complications, requiring the

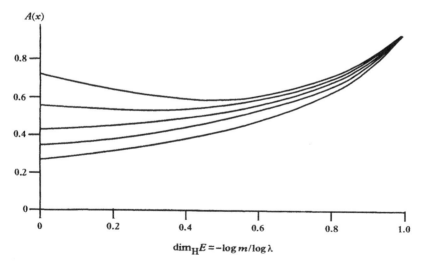

Figure 6.1   The average density of an m-part Cantor set E for various values of m and $\dim_H E$

approximate ergodic theorem, Corollary 6.2. Nevertheless, the proof of Theorem 6.6 is the prototype for these generalisations.

Let $f : X_1 \cup X_2 \to X$ be a C^2 cookie-cutter system with repeller E, which we assume to lie in the interior of X. Let $s = \dim_H E$, so that $0 < \mathcal{H}^s(E) < \infty$, and let $\mu = \mathcal{H}^s|_E$. For convenience, we collect together the facts we need to study average densities of E.

We may find r_0 so that if $0 < r \leq r_0$ and $x \in X$ then

$$B(x, r) \subset X \tag{6.25}$$

so, in particular, $F_{i_1} \circ \cdots \circ F_{i_k}$ is defined on all such $B(x, r)$ for all $(i_1, \ldots, i_k) \in I_k$ for all k. From (5.20) we have that for some number $d > 0$

$$\mu(B(x, r))r^{-s} \leq d \tag{6.26}$$

for all $x \in E$ and $r > 0$. We note from (4.12) and (4.17) that for some $b > 0$

$$|\log|(f^k)'(x)| - \log|(f^k)'(y)|| \leq b|X_{i_{k+1}, \ldots, i_q}|$$

for all $x, y \in X_{i_1, \ldots, i_q}$, for all $q \geq k$ and all $(i_1, \ldots, i_q)$. In particular, if $x, y \in X_{i_1, \ldots, i_k}$ and q is the greatest integer such that $f^k(x), f^k(y) \in X_{i_{k+1}, \ldots, i_q}$, then

$$|\log|(f^k)'(x)| - \log|(f^k)'(y)|| \leq b_1 d^{-1}|f^k(x) - f^k(y)| \tag{6.27}$$

by virtue of (4.23).

Theorem 6.7

Let E be a cookie-cutter set and let ν be any invariant ergodic measure on E. Then there is a number a such that the average density $A^s(x)$ exists and equals a for ν-almost all $x \in E$.

*Proof The basic argument is parallel to that of Theorem 6.6, but the non-linearity of the cookie-cutter system requires the extension of the ergodic theorem, Corollary 6.2.

For $n = 0, 1, 2, \ldots$ and $x \in E$ let

$$\phi_n(x) = \int_{\log|(f^n)'(x)|}^{\log|(f^{n+1})'(x)|} e^{st}\mu(B(x, e^{-t}))dt, \tag{6.28}$$

where $\mu = \mathcal{H}^s|_E$. We show that condition (6.9) of Corollary 6.2 is satisfied by the ϕ_n.

For the time being fix $x \in E$ and integers n and k. Write

$$t_n = \log|(f^n)'(f^k x)| \tag{6.29}$$

(so the derivative of f^n is evaluated at $f^k x$) and assume that n is large enough to ensure $e^{-t_n} \leq r_0$. Let $F_{i_1} \circ \ldots \circ F_{i_k}$ be the branch of f^{-k} such that $F_{i_1} \circ \ldots \circ F_{i_k}(f^k x) = x$.

Let x_-, x_+ be points of the interval

$$F_{i_1} \circ \cdots \circ F_{i_k}(B(f^k x, e^{-t_n})) \tag{6.30}$$

such that

$$|(f^k)'(x_-)| = \inf|(f^k)'(x)|$$

and

$$|(f^k)'(x_+)| = \sup|(f^k)'(x)|$$

where the infimum and supremum are over the interval (6.30). By the Lipschitz property of Hausdorff measure (2.11), with the Lipschitz constant obtained from the mean value theorem in the usual way, we have for $t \geq t_n$,

$$\mu(B(f^k x, e^{-t})) \leq |(f^k)'(x_+)|^s \mu(F_{i_1} \circ \cdots \circ F_{i_k}(B(f^k x, e^{-t})))$$
$$\leq |(f^k)'(x_+)|^s \mu(B(x, e^{-t}|(f^k)'(x_-)|^{-1})) \tag{6.31}$$

(the inclusion in the last step follows from an application of the mean value theorem to $F_{i_1} \circ \cdots \circ F_{i_k}$, noting that $(F_{i_1} \circ \cdots \circ F_{i_k})'(f^k(x)) = (f^k)'(x)^{-1})$.
From (6.28), (6.31) and then substituting $u = t + \log|(f^k)'(x_-)|$

$$\phi_n(f^k x) = \int_{t_n}^{t_{n+1}} e^{st} \mu(B(f^k x, e^{-t})) dt$$

$$\leq \int_{t_n}^{t_{n+1}} e^{st} \mu(B(x, e^{-t}|(f^k)'(x_-)|^{-1})) |(f^k)'(x_+)|^s dt$$

$$= \int_{u=t_n+\log|(f^k)'(x_-)|}^{t_{n+1}+\log|(f^k)'(x_-)|} e^{su} \mu(B(x, e^{-u})) \left|\frac{(f^k)'(x_+)}{(f^k)'(x_-)}\right|^s du$$

$$\leq \int_{t_n+\log|(f^k)'(x)|}^{t_{n+1}+\log|(f^k)'(x)|} e^{su} \mu(B(x, e^{-u})) du + \epsilon_n,$$

where $\epsilon_n \to 0$ as $n \to \infty$ uniformly in k and x, using (6.26) and (6.27) and noting that $|f^k x - f^k x_-|, |f^k x - f^k x_+| \to 0$ uniformly as $n \to \infty$, by virtue of (6.30). Since $t_n + \log|(f^k)'(x)| = \log|(f^{n+k})'(x)|$ by the chain rule, this is

$$\phi_n(f^k x) \leq \phi_{n+k}(x) + \epsilon_n$$

where $\epsilon_n \to 0$ as $n \to \infty$. This is half of the inequality

$$|\phi_n(f^k x) - \phi_{n+k}(x)| \leq \epsilon_n \quad (x \in E),$$

and the other half follows in exactly the same way.
By Corollary 6.2

$$\frac{1}{m} \int_0^{\log|(f^m)'(x)|} e^{st} \mu(B(x, e^{-t})) dt = \frac{1}{m} \sum_{k=0}^{m-1} \phi_k(x) \to a_0$$

for some $a_0 > 0$ for ν-almost all x, since ν is an invariant ergodic measure. By Proposition 6.3 $\frac{1}{m}\log|(f^m)'(x)| \to \lambda$ for μ-almost all x, where λ is the Liapounov exponent, so

$$\frac{1}{\log|(f^m)'(x)|}\int_0^{\log|(f^m)'(x)|} e^{st}\mu(B(x,e^{-t}))dt \to a_0\lambda^{-1}$$

as $m \to \infty$. Using that the integrand is bounded (see (6.26)) we may change from the discrete limit as $m \to \infty$ to the continuous limit as $T \to \infty$ of (6.21) to get that $A^s(x) = a_0\lambda^{-1}2^{-s}$ for ν-almost all x. $\square$

By choosing ν appropriately, we get the natural result on the existence of average densities for cookie-cutters.

Corollary 6.8

Let E be a cookie-cutter set of Hausdorff dimension s. Then there is a number $a > 0$ such that

$$A^s(x) \equiv \lim_{T\to\infty}\int_{t=0}^T \frac{\mathcal{H}^s(E \cap B(x,e^{-t}))}{(2e^{-t})^s}dt = a \tag{6.32}$$

for $\mathcal{H}^s$-almost all $x \in E$.

Proof By Lemma 6.4(c) there is an invariant ergodic measure ν on E equivalent to $\mathcal{H}^s|_E$, so (6.32) is a restatement of Proposition 6.7 in this case. $\square$

Unlike density, average density is a well-defined and natural parameter for describing a wide class of fractals. Unfortunately, average densities are usually difficult to calculate or even to estimate numerically. It might be possible to estimate the average densities of certain cookie-cutters in a way analogous to that indicated in Theorem 6.6 for the Cantor set, but this would be quite involved.

An alternative way of expressing the average density $A^s(x)$ is in terms of the behaviour of the singular integral $\int|x-y|^{-s}d\mu(y)$.

Proposition 6.9

Let μ be a finite measure on $\mathbb{R}^n$, and let $x \in \mathbb{R}^n$ be such that

$$\mu(B(x,r)) \leq dr^s \tag{6.33}$$

for $x \in E$ and $r > 0$ and such that the average density $A^s(x)$ of μ exists at x. Then

$$A^s(x) = \lim_{\epsilon\to 0}\frac{1}{2^s s|\log \epsilon|}\int_{|x-y|\geq\epsilon}\frac{d\mu(y)}{|x-y|^s}. \tag{6.34}$$

Proof We have, after writing $m(r) = \mu(B(x,r))$ in (6.20),

$$A^s(x,T) = 2^{-s}T^{-1} \int_0^T e^{st}m(e^{-t})dt.$$

Substituting $r = e^{-t}$ and $\epsilon = e^{-T}$ and then integrating by parts gives

$$A^s(x, -\log \epsilon) = 2^{-s}|\log \epsilon|^{-1} \int_{r=\epsilon}^1 r^{-s-1}m(r)dr$$

$$= 2^{-s}|\log \epsilon|^{-1}s^{-1}\left[m(\epsilon)\epsilon^{-s} - m(1) + \int_{r=\epsilon}^1 r^{-s}dm(r)\right]$$

$$= 2^{-s}|\log \epsilon|^{-1}s^{-1}\left[O(1) + \int_{|x-y|\geq\epsilon} |x - y|^{-s}d\mu(y)\right],$$

using (6.33) and that $\int_1^\infty r^{-s}dm(r) < \infty$. Thus (6.34) follows.   □

Conversely, the singular integrals may be expressed in terms of average densities, so that (6.34) becomes

$$\int_{|x-y|\geq\epsilon} \frac{d\mu(y)}{|x - y|^s} \sim 2^s s|\log \epsilon|A^s(x). \tag{6.35}$$

as $\epsilon \to 0$. Since the average density, or equivalently, this singular integral, depends only on μ in arbitrarily small neighbourhoods of x, it is easy to see that for any continuous $f: E \to \mathbb{R}$ we have that

$$\int_{|x-y|\geq\epsilon} \frac{f(y)d\mu(y)}{|x - y|^s} \sim 2^s s|\log \epsilon|f(x)A^s(x). \tag{6.36}$$

In fact (6.36) holds for μ-almost all x if $f \in L^1(\mu)$ (that is, if f is measurable with $\int |f|d\mu < \infty$). This result, which lies in the realms of harmonic analysis, may be proved using a version of the Hardy–Littlewood maximal theorem to transfer the formula from continuous to integrable functions. In the case when $s = \log 2/\log 3$, E is the middle-third Cantor set and $\mu = \mathcal{H}^s|_E$, (6.36) becomes

$$\int_{|x-y|\geq\epsilon} \frac{f(y)d\mu(y)}{|x - y|^s} \sim |\log \epsilon|f(x) \times 0.60912\ldots$$

for almost all x, using (6.23).

6.3 Notes and references

For a full treatment of the ergodic theorem and its variants see the books by Parry (1981) and Petersen (1983). The ergodic theorem was first proved by Birkhoff (1931); the proof given here is essentially that of Katznelson and Weiss (1982).

Applications of the ergodic theorem to dynamical systems, and in particular to Liapounov exponents, are described in Pollicott (1992).

Salli (1985) gave a non-ergodic proof of Proposition 6.5. Average densities were introduced by Bedford and Fisher (1992) who demonstrated their existence for cookie-cutter sets. The approach described here was given by Falconer (1992b). Values of average density were calculated by Patzschke and Zähle (1993) in the case of the middle-third Cantor set and by Leistritz (1994) for the m-part Cantor sets. Average densities of Brownian paths and other random sets are considered in Bedford and Fisher (1992) and in Falconer and Xiao (1995). Patzschke and Zähle (1993) study local properties of fractal functions using similar techniques.

Exercises

6.1 Deduce the ergodic theorem in the case $\phi \in L^1(\mu)$ (rather than with the restriction $|\phi(x)| \leq M$ for all x). To do this, first assume that $\phi(x) \geq 0$ for all x, and apply Theorem 6.1 to the function $\phi'(x) = \max\{\phi(x), M\}$. Alternatively, modify the proof of Theorem 6.1 to allow for ϕ to be unbounded.

6.2 Verify the inequalities (6.22).

6.3 Let E be a compact subset of a real interval $[a, b]$ with $0 < \mathcal{H}^s(E) < \infty$ and assume that the average density $A^s(x)$ exists at some $x \in E$. Let $f : [a, b] \to \mathbb{R}$. Show that if f is an injection with continuous derivative, then $A^s(x)$ equals the average density of $f(E)$ at $f(x)$, but that this need not be true if $f : [a, b] \to f([a, b])$ is merely bi-Lipschitz.

6.4 Verify that the densities $\underline{D}^s(x)$ and $\overline{D}^s(x)$ given by (6.17) and (6.18) are μ-measurable functions.

6.5 Let $s = \log 2/\log 3$, let E be the middle-third Cantor set and $\mu = \mathcal{H}^s|_E$, and let $p > 0$. Show, similarily to Theorem 6.6, that the p-th power average densities $\lim_{T \to \infty} T^{-1} \int_0^T \mu(B(x, e^{-t}))^p \, e^{pst} dt$ exist and are constant for μ-almost all x.

6.6 Define the *(right) one-sided average density* of μ by $A_R^s(x) = \lim_{T \to \infty} \int_{t=0}^T \mu([x, x + e^{-t}])/e^{-ts} dt$. Show, with μ as the usual measure on the Cantor set, that $A_R^s(x)$ exists and is almost everywhere constant. How are this value, and the corresponding almost sure left one-sided density, related to the average density of μ?

6.7 Verify (6.36) in the case of a continuous function f.

Chapter 7 The renewal theorem and fractals

The renewal theorem is another major theorem of probabilistic analysis that has been applied with advantage to fractal geometry. The self-similarity of certain fractals is reflected in relationships equivalent to the renewal equation, and the conclusions of the renewal theorem then translate to give information about the structure of the fractals.

7.1 The renewal theorem

Let $g : \mathbb{R} \to \mathbb{R}$ be a given function and let μ be a given Borel probability measure with support in $[0, \infty)$. The integral equation

$$f(t) = g(t) + \int_0^\infty f(t - y) \mathrm{d}\mu(y) \quad (t \in \mathbb{R}) \tag{7.1}$$

is called the *renewal equation*. We shall be interested in 'solutions' f of this equation; in particular the renewal theorem will tell us about the behaviour of $f(t)$ as $t \to \infty$. We often think of the variable t as 'time', so that (7.1) relates the value of f at time t to its earlier values.

The renewal equation has been studied extensively as an integral equation in its own right, but it is also of fundamental importance in probability theory. The example usually quoted concerns the renewal or replacement of light bulbs. At time 0, a new light bulb is installed, the instant it blows it is replaced by another bulb, and so on. With μ as the probability measure giving the distribution of lifetimes of the bulbs (so that $\mu([t_1, t_2])$ is the probability of a failure in the time interval $[t_1, t_2]$) and taking $g(t) = 0$ $(t < 0)$ and $g(t) = 1$ $(t \geq 0)$, equation (7.1) is satisfied by $f(t)$, the expected number of replacements up to time t. To see this note that, conditional on the first replacement occurring at time $y > 0$, we have for $t > y$ that

$$\#(\text{replacements up to time } t) = 1 + \#(\text{replacements up to time } t - y).$$

The renewal equation may conveniently be expressed in the language of convolutions. Recall that, if μ is a Borel measure on $[0, \infty)$ and f a Borel measurable function on $\mathbb{R}$, the *convolution* $f * \mu$ is defined by

$$(f * \mu)(t) = \int_0^\infty f(t - y) d\mu(y), \tag{7.2}$$

assuming that this integral exists. In this notation (7.1) becomes

$$f = g + f * \mu. \tag{7.3}$$

We will need to take repeated convolutions of such equations with μ. The k-th order convolution, μ^{*k}, is defined inductively, taking μ^{*0} as the unit point mass at 0 (so $f * \mu^{*0} = f$), with $\mu^{*1} = \mu$, and, in general for $k = 1, 2, \ldots$,

$$(f * \mu^{*k})(t) = (f * \mu^{*(k-1)} * \mu)(t) \tag{7.4}$$

$$= \int_0^\infty (f * \mu^{*(k-1)})(t - y) d\mu(y)$$

$$= \int_0^\infty \cdots \int_0^\infty f(t - y_1 - \cdots - y_k) d\mu(y_1) \ldots d\mu(y_k). \tag{7.5}$$

(Convolution is easily seen to satisfy the associative law, so brackets are not required in (7.4).) The form (7.5) is obtained by substituting (7.2) in itself $(k - 1)$ times; the distribution of the sums $y_1 + \cdots + y_k$ is central to renewal theory.

We first show that, under suitable conditions, the renewal equation has a unique solution. We take μ to be a Borel probability measure with support contained in $[0, \infty)$ such that

$$\lambda \equiv \int_0^\infty t \, d\mu(t) < \infty. \tag{7.6}$$

To avoid trivial cases we assume that μ is not concentrated at 0, that is $\mu(\{0\}) < 1$, which clearly implies that for every $a > 0$

$$\gamma_a \equiv \int_0^\infty e^{-at} d\mu(t) < 1. \tag{7.7}$$

We assume throughout that $g : \mathbb{R} \to \mathbb{R}$ is a function with a discrete set of discontinuities, and such that for some $c > 0$ and $\alpha > 0$

$$|g(t)| \le c e^{-\alpha |t|} \quad (t \in \mathbb{R}). \tag{7.8}$$

In particular, g is bounded and integrable. (In fact the theory goes through under rather weaker conditions on g: it is enough for g to be 'directly Riemann integrable'.)

Given conditions (7.6)–(7.8) we may exhibit the solution of the renewal equation. We write $\mathcal{F}$ for the space of Borel measurable functions $f : \mathbb{R} \to \mathbb{R}$ such that $\lim_{t \to -\infty} f(t) = 0$ and such that f is bounded on the half-line $(-\infty, a]$ for every $a \in \mathbb{R}$. The space $\mathcal{F}$ is a natural one in which to seek solutions of (7.1).

Proposition 7.1

Let g and μ satisfy (7.6)–(7.8). There is a unique $f \in \mathcal{F}$ that satisfies the renewal equation (7.1), given by

$$f = \sum_{k=0}^{\infty} g * \mu^{*k}, \tag{7.9}$$

that is

$$f(t) = \sum_{k=0}^{\infty} \int_0^{\infty} \cdots \int_0^{\infty} g(t - y_1 - \cdots - y_k) d\mu(y_1) \ldots d\mu(y_k). \tag{7.10}$$

Moreover, $f(t)$ is bounded for $t \in \mathbb{R}$, and if g is continuous, then f is uniformly continuous on $\mathbb{R}$.

Proof We prove this result under the simplifying assumption that there exists $\tau > 0$ such that

$$\mu[0, \tau] = 0. \tag{7.11}$$

(This is true in all our applications, for the proof without this extra assumption see Exercise 7.1.)

Using that μ is a probability measure and (7.8), we have for $t \in \mathbb{R}$

$$\sum_{k=0}^{m} \int_0^{\infty} \cdots \int_0^{\infty} |g(t - y_1 - \cdots - y_k)| d\mu(y_1) \cdots d\mu(y_k)$$

$$= \int_0^{\infty} \cdots \int_0^{\infty} \sum_{k=0}^{m} |g(t - y_1 - \cdots - y_k)| d\mu(y_1) \cdots d\mu(y_m)$$

$$\leq c \int_0^{\infty} \cdots \int_0^{\infty} \sum_{k=0}^{m} e^{-\alpha|t - y_1 - \cdots - y_k|} d\mu(y_1) \cdots d\mu(y_m) \tag{7.12}$$

$$\leq c \int_0^{\infty} \cdots \int_0^{\infty} 2/(1 - e^{-\alpha\tau}) d\mu(y_1) \cdots d\mu(y_m) \tag{7.13}$$

$$= 2c/(1 - e^{-\alpha\tau}). \tag{7.14}$$

(In summing to get (7.13) we note that, by (7.11), the numbers $y_1, \ldots, y_m$ are all at least τ, except for a set of $(y_1, \ldots, y_m)$ of $\mu \times \cdots \times \mu$ measure 0. Thus the series (7.10) is absolutely uniformly convergent with

$$|f(t)| \leq 2c/(1 - e^{-\alpha\tau}) \tag{7.15}$$

for all $t \in \mathbb{R}$.

For $t < 0$, the bound (7.12) becomes

$$c \int_0^{\infty} \cdots \int_0^{\infty} \sum_{k=0}^{m} e^{\alpha(t - y_1 - \cdots - y_k)} d\mu(y_1) \cdots d\mu(y_m) \leq c\, e^{\alpha t}/(1 - e^{-\alpha\tau}), \tag{7.16}$$

again using (7.11). Applying this to (7.10) gives

$$|f(t)| \leq c\,e^{\alpha t}/(1 - e^{-\alpha \tau}) \tag{7.17}$$

for $t < 0$, so $\lim_{t \to -\infty} f(t) = 0$, giving $f \in \mathcal{F}$.

For all t, we have identically that

$$\sum_{k=0}^{m} (g * \mu^{*k})(t) = g(t) + \sum_{k=0}^{m-1} ((g * \mu^{*k}) * \mu)(t)$$

$$= g(t) + \int_{0}^{\infty} \sum_{k=0}^{m-1} (g * \mu^{*k})(t - y)d\mu(y).$$

Letting $m \to \infty$ and noting (7.14), the dominated convergence theorem gives that f satisfies (7.1).

Suppose now that $f_1, f_2 \in \mathcal{F}$ are both solutions of (7.1). Then $f_0 \equiv f_1 - f_2 \in \mathcal{F}$ satisfies $f_0 = f_0 * \mu$, so by repeated convolution with μ we have $f_0 = f_0 * \mu^{*k}$ for all k. Hence for all $u \in \mathbb{R}$ and $k = 1, 2, \ldots$

$$|f_0(t)| = \left| \int \cdots \int_{y_1 + \cdots + y_k > u} f_0(t - y_1 - \cdots - y_k)d\mu(y_1) \ldots d\mu(y_k) \right.$$

$$\left. + \int \cdots \int_{u - y_1 - \cdots - y_k \geq 0} f_0(t - y_1 - \cdots - y_k)d\mu(y_1) \ldots d\mu(y_k) \right|$$

$$\leq \sup_{v \leq t-u} |f_0(v)| + \int_{0}^{\infty} \cdots \int_{0}^{\infty} \sup_{v \leq t} |f_0(v)| e^{u - y_1 - \cdots - y_k} d\mu(y_1) \ldots d\mu(y_k)$$

$$\leq \sup_{v \leq t-u} |f_0(v)| + \sup_{v \leq t} |f_0(v)| e^{u} \left(\int_{0}^{\infty} e^{-y} d\mu(y) \right)^{k}.$$

Given $\varepsilon > 0$ we may choose u large enough to make the first term less than $\frac{1}{2}\varepsilon$ (since $\lim_{v \to -\infty} f_0(v) = 0$), and then choose k large enough to make the second term less than $\frac{1}{2}\varepsilon$ (using (7.7) and that f_0 is bounded). Thus $|f_0(t)| < \varepsilon$ for all $\varepsilon > 0$, so $f_0(t) = 0$ giving $f_1(t) = f_2(t)$ for all t.

Finally we show that if g is continuous then f is uniformly continuous. Given $\varepsilon > 0$, it follows from (7.8) that $|g(t)| \leq \varepsilon e^{-\frac{1}{2}\alpha|t|}$ for all $|t| \geq T$, if T is chosen large enough. Together with uniform continuity of g on the interval $[-T - 1, T + 1]$, this implies that there is a number $h_0 > 0$ such that for all $t \in \mathbb{R}$ and $0 < h \leq h_0$

$$|g(t + h) - g(t)| \leq 2\varepsilon e^{-\frac{1}{2}\alpha|t|}. \tag{7.18}$$

Replacing $g(\cdot)$ by $g(h + \cdot) - g(\cdot)$ in (7.9) or (7.10) we get

$$f(t + h) - f(t) = \sum_{k=0}^{\infty} ((g(h + \cdot) - g(\cdot)) * \mu^{*k})(t).$$

Thus applying the estimate (7.15) to $f(t+h)-f(t)$, using the bounds on $|g(t+h)-g(t)|$ given by (7.18) in place of (7.8), gives

$$|f(t+h)-f(t)| \le 4\varepsilon/(1-\mathrm{e}^{-\frac{1}{2}\alpha\tau})$$

for all $t \in \mathbb{R}$ and $0 < h \le h_0$. Since ε may be chosen arbitrarily small, it follows that f is uniformly continuous on $\mathbb{R}$. $\square$

Two contrasting situations arise in renewal theory. On the one hand, μ may be supported by a discrete set consisting of integer multiples of a number $\tau > 0$, in which case μ^{*k} is supported by the same set for all k. Then by (7.10) $f(t)$ depends only on $g(t-k\tau)$ for $k = 0, 1, 2, \ldots$. On the other hand there may be no such number τ. To see the significance of this distinction, consider again the light bulb replacement example. If the bulbs happen to be manufactured in such a way that they always blow after some exact multiple of 24 hours, then, if the first bulb is fitted at midnight, renewals will only occur at midnight. On the other hand, if the bulb lifetime is not distributed with some such underlying periodicity then in the long term a bulb may require changing at any time of the day or night.

Thus we define a measure μ to be τ-*arithmetic* if $\tau > 0$ is the greatest positive number such that the support of μ is contained in the additive group $\tau\mathbb{Z} \equiv \{\tau k : k \in \mathbb{Z}\}$. If there is no such $\tau > 0$ then we call μ *non-arithmetic*. For a simple example, suppose that μ is supported by the two point set $\{y_1, y_2\}$, where $y_1, y_2 > 0$. Then μ is non-arithmetic if y_1/y_2 is irrational, and τ-arithmetic if y_1/y_2 is rational with $y_1 = k_1\tau$ and $y_2 = k_2\tau$ where k_1, k_2 are co-prime integers. Similarly, μ is non-arithmetic if, for example, the support of μ contains an interval.

The renewal theorem concerns the limiting behaviour of solutions f of the renewal equation as $t \to \infty$. The main conclusion is that in the non-arithmetic case $\lim_{t\to\infty} f(t)$ exists, and in the τ-arithmetic case f is asymptotic to a function that is periodic with period τ.

Various proofs of the renewal theorem have been devised, none particularly elementary. Here we describe two approaches: the first proof uses ideas from probability theory and has an intuitive interpretation in terms of a 'game', and the second proof uses Fourier transforms and requires the power of a Tauberian theorem.

Theorem 7.2 (renewal theorem)

Let g and μ satisfy (7.6)–(7.8), and let $f \in \mathcal{F}$ satisfy the renewal equation (7.1). If μ is non-arithmetic then

$$\lim_{t\to\infty} f(t) = \lambda^{-1} \int_{-\infty}^{\infty} g(y)\,\mathrm{d}y. \tag{7.19}$$

If μ is τ-arithmetic then for all $y \in [0, \tau)$

$$\lim_{k \to \infty} f(k\tau + y) = \lambda^{-1} \sum_{j=-\infty}^{\infty} g(j\tau + y). \tag{7.20}$$

First proof This proof uses probabilistic ideas. We give the proof in the arithmetic case, though it may be adapted to the non-arithmetic situation. By scaling the time coordinates, we may assume that μ is 1-arithmetic and by translating the origin we may assume that $y = 0$ in (7.20).

We express the solution of the renewal equation (7.10) in probabilistic terms. We let $(Y_1, Y_2, \ldots)$ be a sequence of independent identically distributed random variables each with probability measure μ (so for all $j, \mu(A)$ is the probability that $Y_j \in A$). We write P for the product probability measure on the set of sequences $(Y_1, Y_2, \ldots)$ and E for expectation with respect to P. In this notation we may write (7.10) as

$$f(t) = \sum_{k=0}^{\infty} \mathsf{E}(g(t - Y_1 - \cdots - Y_k))$$

$$= \sum_{j=-\infty}^{\infty} g(j) \mathsf{P}(Y_1 + \cdots + Y_k = t - j \text{ for some } k). \tag{7.21}$$

Fix $m \in \mathbb{Z}^+$ and consider the following 'game' played by two players A and B on a 'board' consisting of 'squares' numbered by the integers $\{-m, -m+1, \ldots, -1, 0, 1, 2, \ldots\}$, see Figure 7.1. Player A starts at square $-m$ and player B starts at 0. The players repeatedly throw a 'die' to determine the number of places moved: there is a probability $\mu(\{x\})$ of throwing an x and thus advancing x integers. It is intuitively obvious (at least to anyone that plays board games!) that, if y is a large integer, then A and B have virtually the same chance of landing on square y at some stage of the game, despite starting on different squares. To see this, suppose for simplicity that the probability measure μ is supported by a finite set of positive integers $\{1, \ldots, q\}$ with $\mu(\{1\}) > 0$, so the players always move between 1 and q

Player A

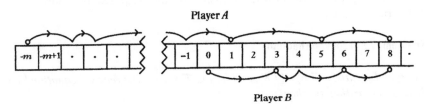

Player B

Figure 7.1 The 'game' in the proof of the renewal theorem. With probability one the players eventually land on the same square

places and have a positive probability of moving exactly one place. The players take consecutive turns as follows. Player A's first turn consists of repeated throws and moves until he lands on or passes B's position. Then player B has a turn of several throws until he lands on or passes A, then A has another such turn, and so on. Then, with probability 1, one player will eventually land on the square actually occupied by the other. This is because at the start of each turn, the players must be within q places of each other, so there is a probability of at least $\varepsilon = \mu(\{1\})^q > 0$ of the players coinciding at the end of that turn, independently of what has happened before. Thus the probability of the players not coinciding before n turns have passed is at most $(1 - \varepsilon)^n \to 0$ as $n \to \infty$. This argument adapts to the case of general μ; the 1-arithmetic condition means that there is a positive probability of the players coinciding within a finite number of turns, with condition (7.6) ensuring that the expected number of squares moved on any throw is finite.

Clearly, once A and B have landed on the same square the probability of visiting any given subsequent square is the same for both players. Thus, if S_n is the event 'the first square visited by both A and B is square n' and $r \geq n$, then

$$P(A \text{ lands on } r \,|S_n) = P(B \text{ lands on } r \,|S_n),$$

where these probabilities are conditional on S_n. By the 'translation invariance' of the game, for $r \geq m \geq 0$,

$$\begin{aligned}
|P(B \text{ lands on } r + m) &- P(B \text{ lands on } r)| \\
&= |P(A \text{ lands on } r) - P(B \text{ lands on } r)| \\
&= \left| \sum_{n=0}^{\infty} (P(A \text{ lands on } r|S_n) - P(B \text{ lands on } r|S_n))P(S_n) \right| \\
&\leq \sum_{n=r+1}^{\infty} P(S_n) \\
&= P\left(\bigcup_{n=r+1}^{\infty} S_n \right) \to 0
\end{aligned}$$

uniformly in m as $r \to \infty$, by the preceding paragraph. Thus we have that $(P(B \text{ lands on } r))_{r=1}^{\infty}$ is a Cauchy sequence, so

$$P(B \text{ lands on } r) \to \eta \tag{7.22}$$

as $r \to \infty$, for some $\eta \geq 0$. But the average number of squares moved by B on each throw is $\sum_{t=0}^{\infty} t\mu(\{t\}) = \lambda$ by (7.6), so clearly $\eta = \lambda^{-1}$. Interpreting (7.22) in terms of the random variables Y_i,

$$P(Y_1 + \cdots + Y_k = r \text{ for some } k) \to \eta = \lambda^{-1}$$

as $r \to \infty$. Thus the 'j-th' term of the sum (7.21) converges to $g(j)\lambda^{-1}$ as $t \to \infty$, so since $\sum_{j=-\infty}^{\infty} |g(j)| < \infty$, by (7.8), and μ is a probability measure, the dominated convergence theorem implies that $f(t) \to \sum_{j=-\infty}^{\infty} g(j)\lambda^{-1}$.

*Second proof This method depends on properties of Fourier transforms and Wiener's Tauberian theorem.

This concerns the existence of limits as $x \to \infty$ of convolutions of functions $(f * \psi)(x) \equiv \int_{-\infty}^{\infty} f(x - y)\psi(y)\mathrm{d}y$. Wiener's Tauberian theorem states that if f is a bounded function on $\mathbb{R}$ and there is a function $\psi \in L^1(\mathbb{R})$ such that $(f * \psi)(x) \to l$ as $x \to \infty$, then

$$(f * \phi)(x) \to l \int_{-\infty}^{\infty} \phi(y)\mathrm{d}y \Big/ \int_{-\infty}^{\infty} \psi(y)\mathrm{d}y \tag{7.23}$$

for *every* $\phi \in L^1(\mathbb{R})$, provided that the Fourier transform of ψ does not vanish, that is provided that

$$\hat{\psi}(u) \neq 0 \text{ for all } u \in \mathbb{R}, \tag{7.24}$$

where $\hat{\psi}(u) = \int_{-\infty}^{\infty} e^{iux}\psi(x)\mathrm{d}x$ is the Fourier transform of ψ.

[An intuitive way of thinking of this theorem for 'sufficiently nice' functions is that, if $\hat{\psi}(u) \neq 0$ and ϕ is a function with rapidly decreasing Fourier transform $\hat{\phi}$, then we may define $\hat{h}(u)$ by $\hat{\phi}(u) = \hat{\psi}(u)\hat{h}(u)$, so that by the convolution theorem, $\phi = \psi * h$, where h is the function with Fourier transform $\hat{h}$. Then, formally,

$$(f * \phi)(x) = ((f * \psi) * h)(x) = \int_{-\infty}^{\infty} (f * \psi)(x - y)h(y)\mathrm{d}y. \tag{7.25}$$

Provided that $h \in L^1(\mathbb{R})$, the major contribution to this integral comes from 'relatively small' values of y, so if x is very large, $(f * \psi)(x - y)$ is close to l for y where $h(y)$ is significant, and thus the integral is close to $l \int_{-\infty}^{\infty} h(y)\mathrm{d}y$. Since $\phi = h * \psi$ we have $\int_{-\infty}^{\infty} \phi(y)\mathrm{d}y = \int_{-\infty}^{\infty} \psi(y)\mathrm{d}y \int_{-\infty}^{\infty} h(y)\mathrm{d}y$ so (7.23) follows.]

We present the Tauberian proof of the renewal theorem in the case where μ is non-arithmetic. We first assume that g is continuous so that f is uniformly continuous by Proposition 7.1. Rearranging (7.1) and integrating gives

$$\int_{t=-\infty}^{x} g(t)\mathrm{d}t = \int_{y=0}^{\infty} \int_{t=-\infty}^{x} [f(t) - f(t - y)]\mathrm{d}t\,\mathrm{d}\mu(y)$$

$$= \int_{y=0}^{\infty} \int_{t=x-y}^{x} f(t)\mathrm{d}t\,\mathrm{d}\mu(y)$$

$$= \int_{t=-\infty}^{x} \int_{y=x-t}^{\infty} f(t)\mathrm{d}\mu(y)\mathrm{d}t$$

$$= \int_{t=-\infty}^{x} f(t)\mu[x - t, \infty)\mathrm{d}t.$$

Hence, defining $\psi(t) = 0$ for $t < 0$ and $\psi(t) = \mu([t, \infty))$ for $t \geq 0$, we have

$$\int_{-\infty}^{x} g(t)dt = \int_{-\infty}^{\infty} f(t)\psi(x - t)dt = (f * \psi)(x), \qquad (7.26)$$

so in particular

$$(f * \psi)(x) \rightarrow \int_{-\infty}^{\infty} g(t)dt \qquad (7.27)$$

as $x \rightarrow \infty$. Wiener's theorem (7.23) tells us that, provided that the Fourier transform condition (7.24) is satisfied, then for every $\phi \in L^1(\mathbb{R})$

$$(f * \phi)(x) \rightarrow \int_{-\infty}^{\infty} g(t)dt \int_{-\infty}^{\infty} \phi(y)dy \bigg/ \int_{-\infty}^{\infty} \psi(y)dy$$

$$= \lambda^{-1} \int_{-\infty}^{\infty} g(t)dt \int_{-\infty}^{\infty} \phi(y)dy, \qquad (7.28)$$

since by (7.6) $\int_{-\infty}^{\infty} \psi(y)dy = \int_{0}^{\infty} \mu([t, \infty))dt = \int_{0}^{\infty} t d\mu(t) = \lambda$. To check (7.24) we use that μ is non-arithmetic. For $u = 0$ we have $\hat{\psi}(0) = \int_{-\infty}^{\infty} \psi(y)dy = \lambda > 0$, as above. For $u \neq 0$, integration by parts gives

$$\hat{\psi}(u) = \int_{0}^{\infty} e^{iut} \mu([t, \infty))dt$$

$$= \left(\mu([0, \infty)) - \int_{0}^{\infty} e^{iut} d\mu(t) \right) \bigg/ iu$$

$$= \left(1 - \int_{0}^{\infty} e^{iut} d\mu(t) \right) \bigg/ iu.$$

Since $\mu([0, \infty)) = 1$, it is only possible to have $1 = \int_{0}^{\infty} e^{iut} d\mu(t)$ for some $u \neq 0$ if $e^{iut} = 1$ for almost all t. This would require that for μ-almost all t, there is an integer n such that $t = 2n\pi/u$, so that μ would be $(2\pi/u)$-arithmetic. We conclude that $\hat{\psi}(u) \neq 0$ for all $u \in \mathbb{R}$, so (7.24) and thus (7.28) holds.

To replace $(f * \phi)(x)$ by $f(x)$ in (7.28) we choose ϕ to be close to a 'Dirac delta function'. By Proposition 7.1, f is uniformly continuous on $\mathbb{R}$, so given $\varepsilon > 0$ there exists $\delta > 0$ such that $|f(x + h) - f(x)| < \varepsilon$ for all $x \in \mathbb{R}$ and $0 < h \leq \delta$. Choosing $\phi \geq 0$ such that $\int_{-\infty}^{\infty} \phi(y)dy = 1$ and such that $\phi(y) = 0$ if $|y| \geq \delta$, we have $|f(x) - (f * \phi)(x)| < \varepsilon$ for all x. Since ε may be chosen arbitrarily small, (7.28) implies that $f(x) \rightarrow \lambda^{-1} \int_{-\infty}^{\infty} g(t)dt$, as $x \rightarrow \infty$, which is (7.19).

Finally we extend the theorem to the case where g has a discrete set of discontinuities. Given $\varepsilon > 0$ we may find $g_0, g_1 : \mathbb{R} \rightarrow \mathbb{R}$ with $g = g_0 + g_1$, such that g_0 is continuous and such that $g_1 \geq 0$ and $\int_{-\infty}^{\infty} g_1(t)dt < \varepsilon$. Similarly, we may find a continuous $g_2 : \mathbb{R} \rightarrow \mathbb{R}$ such that $g_2 \geq g_1$ and $\int_{-\infty}^{\infty} g_2(t)dt < 2\varepsilon$. We may choose g_0, g_1 and g_2 so that they all satisfy the bound (7.8) for some constant c. Let f_0, f_1 and f_2 be the solutions to the renewal equation (given by (7.9)) corresponding to g_0, g_1 and g_2. We have established that the renewal theorem holds for the continuous functions g_0 and g_2, so

$$f_0(x) \to \lambda^{-1} \int_{-\infty}^{\infty} g_0(t)dt$$

as $x \to \infty$, and

$$0 \leq \limsup_{x\to\infty} f_1(x) \leq \limsup_{x\to\infty} f_2(x) = \lambda^{-1} \int_{-\infty}^{\infty} g_2(t)dt < 2\lambda^{-1}\varepsilon.$$

Thus

$$\left| f(x) - \lambda^{-1} \int_{-\infty}^{\infty} g(t)dt \right|$$

$$\leq |f(x) - f_0(x)| + \left| f_0(x) - \lambda^{-1} \int_{-\infty}^{\infty} g_0(t)dt \right| + \lambda^{-1} \int_{-\infty}^{\infty} |g_0(t) - g(t)|dt$$

$$= |f_1(x)| + \left| f_0(x) - \lambda^{-1} \int_{-\infty}^{\infty} g_0(t)dt \right| + \lambda^{-1} \int_{-\infty}^{\infty} g_1(t)dt$$

$$< 3\lambda^{-1}\varepsilon$$

if x is large enough, so the conclusion (7.19) holds in this case.   □

For our applications, μ will be of a rather special form, with support on a finite set, allowing the renewal theorem to be expressed in the following way. We say that a set $\{y_1, \ldots, y_m\}$ of positive real numbers is τ-arithmetic if $\tau > 0$ is the greatest number such that each y_j is an integer multiple of τ, and non-arithmetic if no such $\tau > 0$ exists.

Corollary 7.3

Let $m \geq 2$, let $y_1, \ldots, y_m > 0$ be 'times', and let $p_1, \ldots, p_m > 0$ be 'probabilities', so that $\sum_{j=1}^{m} p_j = 1$. Let g be as in (7.8) and let $f \in \mathcal{F}$ satisfy the renewal equation in the form

$$f(t) = g(t) + \sum_{j=1}^{m} p_j f(t - y_j). \tag{7.29}$$

If $\{y_1, \ldots, y_m\}$ is non-arithmetic then

$$\lim_{t\to\infty} f(t) = \lambda^{-1} \int_{-\infty}^{\infty} g(y)dy,$$

and if $\{y_1, \ldots, y_m\}$ is τ-arithmetic then

$$\lim_{k\to\infty} f(k\tau + y) = \lambda^{-1} \sum_{j=-\infty}^{\infty} g(j\tau + y)$$

for all $y \in [0, \tau)$, where $\lambda = \sum_{j=1}^{m} y_j p_j$.

Proof This is just a restatement of Theorem 7.2, taking μ to be the probability measure supported by $\{y_1, \ldots, y_m\}$ such that $\mu(\{y_j\}) = p_j$ for $j = 1, \ldots, m$. The definitions of $\{y_1, \ldots, y_m\}$ and μ being τ-arithmetic or non-arithmetic coincide. □

7.2 Applications to fractals

Let $q(r)$ be some measurement of a self-similar fractal E at scale r. It may be possible to use self-similarity to write down a relationship expressing $q(r)$ in terms of $q(r')$ for $r' > r$, and a substitution may convert this relationship into a renewal equation. The limiting behaviour of the solution assured by the renewal theorem may then correspond to the behaviour of $q(r)$ as $r \to 0$.

This is best illustrated by an example. Let $\{F_1, \ldots, F_m\}$ be an iterated function system, where F_i is a similarity of ratio r_i, with attractor $E \subset \mathbb{R}^n$, see Section 2.2. Thus $E = \cup_{i=1}^m F_i(E)$, where for convenience we assume that this union is disjoint. As in definitions (2.1)–(2.3), $N(r) \equiv N_r(E)$ will denote the *covering number function* of E, that is the least number of sets of diameter r that can cover E. We know that the box dimensions $\underline{\dim}_B E = \overline{\dim}_B E = s$, where $\sum_{i=1}^m r_i^s = 1$, see Theorem 2.7, so that $\lim_{r \to 0} \log N(r) / - \log r = s$, and it is not hard to show that there are $c_1, c_2 > 0$ such that $c_1 r^s \leq N(r) \leq c_2 r^s$ for all $r \leq 1$. The renewal theorem enables us to obtain much more precise information about $N(r)$ for small r. Recall that $f(r) \sim g(r)$ means that $\lim_{r \to 0} f(r) / g(r) = 1$.

Proposition 7.4

Let E be a self-similar set as above, with covering number function $N(r)$. If $\{\log r_1^{-1}, \ldots, \log r_m^{-1}\}$ is a non-arithmetic set, then for some $c > 0$

$$N(r) \sim cr^{-s} \tag{7.30}$$

as $r \to 0$, and if $\{\log r_1^{-1}, \ldots, \log r_m^{-1}\}$ is τ-arithmetic then

$$N(r) \sim r^{-s} p(-\log r) \tag{7.31}$$

as $r \to 0$, where p is a positive function with period τ.

Proof Let $d = \min_{i \neq j} \text{dist}(F_i(E), F_j(E))$ and let $N_i(r)$ be the minimum number of sets of diameter r that can cover $F_i(E)$. We observe that if $r < d$ a set of diameter r cannot intersect more than one of the $F_i(E)$, but if $r \geq d$ such a set may cover parts of several of the $F_i(E)$. Thus counting the number of sets of diameter r needed to cover E,

$$N(r) = \sum_{i=1}^m N_i(r) - h(r),$$

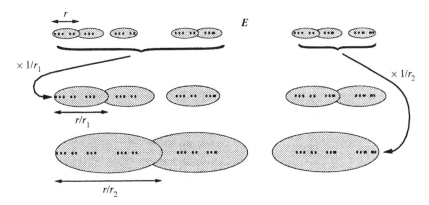

Figure 7.2 The set E shown is a self-similar set constructed using ratios r_1 and r_2. Scaling the left and right parts of E by $1/r_1$ and $1/r_2$ relates an r-cover of E to r/r_1- and r/r_2-covers of E, and this gives relation (7.32)

where $h(r) = 0$ for $0 < r < d$ and $h(r) \geq 0$ for $r \geq d$. Since $F_i(E)$ is similar to E at scale r_i, we have $N_i(r) = N(r/r_i)$, so

$$N(r) = \sum_{i=1}^{m} N(r/r_i) - h(r); \tag{7.32}$$

an equation that reflects the self-similarity of E, see Figure 7.2. We transform (7.32) using the substitutions

$$r = e^{-t}; \quad f(t) = e^{-st} N(e^{-t}); \quad g(t) = e^{-st} h(e^{-t}). \tag{7.33}$$

so

$$N(e^{-t}) = \sum_{i=1}^{m} e^{s \log r_i} e^{-s \log r_i} N(e^{-t - \log r_i}) - h(e^{-t});$$

thus after multiplying by e^{-st} we get,

$$f(t) = \sum_{i=1}^{m} r_i^s f(t - \log r_i^{-1}) - g(t). \tag{7.34}$$

With $s = \dim_B E$ we have $\sum_{i=1}^{m} r_i^s = 1$, so this is the renewal equation (7.29) with 'times' $y_i = \log r_i^{-1}$ and 'probabilities' $p_i = r_i^s$.

Since $N(r)$ is integer valued and increasing, f, and thus h, has discrete discontinuities. Since $h(r) = 0$ for $0 < r < d$, so $g(t) = 0$ for $t \geq -\log d$. Unfortunately, $g(t)$ is unbounded as $t \to -\infty$ so we cannot apply the renewal theorem to (7.34) as it stands. To get round this problem, we modify f and g by defining

$$f_0(t) = \begin{cases} 0 & (t < 0) \\ f(t) & (t \geq 0) \end{cases}$$

and

$$g_0(t) = \begin{cases} 0 & (t < 0) \\ g(t) - \sum_{\{i : r_i < e^{-t}\}} r_i^s f(t - \log r_i^{-1}) & (t \geq 0). \end{cases}$$

Then

$$f_0(t) = \sum_{i=1}^{m} r_i^s f_0(t - \log r_i^{-1}) - g_0(t) \qquad (7.35)$$

for $t \in \mathbb{R}$, where $g_0(t) = 0$ if $t \notin [0, -\log \min\{r_1, \ldots, r_m, d\}]$.

Corollary 7.3 now implies that $f(t) = f_0(t) \to c$ for some $c > 0$ in the non-arithmetic case, and that $f(t)$ is asymptotic to a positive periodic function in the arithmetic case; inverting the substitutions (7.33) leads to (7.30) and (7.31). □

Note that the constant c in (7.30) is given by the renewal theorem in terms of g_0, so that the limit of $N(r)$ as $r \to 0$ may be expressed in terms of the values of $N(r)$ over the range $1 \geq r \geq \min\{r_1, \ldots, r_m, d\}$, and similarly for the function p in (7.31). Since $\{\log r_1^{-1}, \ldots, \log r_m^{-1}\}$ is 'usually' a non-arithmetic set (as it will be if $\log r_i^{-1} / \log r_j^{-1}$ is irrational for some $i \neq j$), the 'usual' conclusion is that $r^s N(r)$ approaches a limit. Asymptotic periodicity (7.31) is the 'exceptional' possibility, although it occurs for self-similar sets with all the similarity ratios r_i equal, such as for the middle-third Cantor set.

To obtain more precise asymptotic forms, we specialise to subsets of the line. Let E be a self-similar subset of $[0,1]$ constructed using similarities of ratios $r_1, \ldots, r_m$ and assume that at the first stage of the construction, the gaps between the intervals have lengths $b_1, \ldots, b_{m-1}$. We define the *gap-counting function* of E by

$$G(r) = \#\{\text{complementary intervals of } E \text{ with length } \geq r\}.$$

(Thus for the middle-third Cantor set $r_1 = r_2 = b_1 = 1/3$ and $G(r) = 2^k - 1$ if $3^{-(k+1)} < r \leq 3^{-k}$.) The renewal theorem method is well suited to estimating $G(r)$ for small r.

Proposition 7.5

Let E be a self-similar subset of $[0, 1]$ as above. If $\{\log r_1^{-1}, \ldots, \log r_m^{-1}\}$ is a non-arithmetic set, then

$$G(r) \sim r^{-s} s^{-1} \sum_{i=1}^{m} b_i^s \bigg/ \sum_{i=1}^{m} r_i^s \log r_i^{-1}, \qquad (7.36)$$

as $r \to 0$. If $\{\log r_1^{-1}, \ldots, \log r_m^{-1}\}$ is τ-arithmetic, then, for each $\alpha > 0$,

$$G(\alpha \rho^k) \sim (\alpha \rho^k)^{-s}(1 - e^{-s\tau})^{-1} \sum_{i=1}^{m} \exp(s \log \alpha - s\tau \lfloor (\log b_i^{-1} + \log \alpha)/\tau \rfloor)$$

$$\times \left(\sum_{i=1}^{m} r_i^s \log r_i^{-1} \right)^{-1} \tag{7.37}$$

as $k \to \infty$, where $\rho = e^{-\tau}$ and here $\lfloor \ \rfloor$ denotes 'the greatest integer less than or equal to'.

Proof We note that for each i the gap lengths within the components $F_i(E)$ are those of E multiplied by r_i, so counting the gaps in E by counting those within each $F_i(E)$ together with those between the $F_i(E)$, we have

$$G(r) = \sum_{i=1}^{m} G(r/r_i) + \#\{i : b_i \geq r\}.$$

This formula is valid for all $r > 0$, and the right-hand term vanishes for $r \geq 1$. Substituting

$$r = e^{-t}; \quad f(t) = e^{-st}G(e^{-t}); \quad g(t) = e^{-st}\#\{i : b_i \geq e^{-t}\}$$

we get, just as in (7.34),

$$f(t) = \sum_{i=1}^{m} r_i^s f(t - \log r_i^{-1}) + g(t), \tag{7.38}$$

with $f(t) = g(t) = 0$ for $t \leq 0$. Taking $s = \dim_B E = \dim_H E$, we have $\sum_{i=1}^{m} r_i^s = 1$, and $|g(t)| \leq (m-1)e^{-st}$, so we may apply the renewal theorem directly to (7.38). Since

$$\int_{-\infty}^{\infty} g(t)dt = \sum_{i=1}^{m} \int_{-\log b_i}^{\infty} e^{-st}dt$$

$$= \sum_{i=1}^{m} s^{-1}b_i^s,$$

Corollary 7.3 gives

$$\lim_{t \to \infty} f(t) = s^{-1} \sum_{i=1}^{m} b_i^s \bigg/ \sum_{i=1}^{m} r_i^s \log r_i^{-1}$$

and substituting back gives (7.36). The τ-arithmetic formula follows in a similar manner. $\square$

Proposition 7.5 enables us to deduce the asymptotic behaviour of $V(r)$, the Lebesgue measure of the r-neighbourhood of E, see Section 3.2.

Corollary 7.6

Let E be a self-similar subset of $[0, 1]$ as above, and assume that $\{\log r_1^{-1}, \ldots, \log r_m^{-1}\}$ is non-arithmetic. Then as $r \to 0$

$$V(r) \sim r^{1-s} 2^{1-s} (1-s)^{-1} s^{-1} \sum_{i=1}^{m} b_i^s \bigg/ \sum_{i=1}^{m} r_i^s \log r_i^{-1}.$$

Proof We sketch the proof; the exact estimates required are easily completed. As in (3.17)

$$V(r) = 2r(G(2r) + 1) + \sum \{|A| : A \text{ is a gap of length} < 2r\}, \qquad (7.39)$$

where G is the gap-counting function. With $G(r) \sim cr^{-s}$, where c is the coefficient of r^{-s} in (7.36), we have

$$\sum \{|A| : A \text{ is a gap of length} < 2r\} = -\int_0^{2r} t \, dG(t)$$

$$\sim cs \int_0^{2r} t^{-s} dt$$

$$= cs(1-s)^{-1}(2r)^{1-s}.$$

From (7.39)

$$V(r) \sim c(2r)^{1-s} + cs(1-s)^{-1}(2r)^{1-s},$$
$$= c(2r)^{1-s}(1-s)^{-1}$$

as required. $\square$

Of course there is an analogue of Corollary 7.6 (with an even messier formula!) in the arithmetic case.

The renewal theory method may be applied to many other sets and quantities. The basic idea is always to use self-similarity to write down a recursion relation and to transform this into a renewal equation to enable the renewal theorem to be applied. The method may be applied to sets with a weaker separation condition, such as the open set condition, see Exercise 7.4. A more sophisticated version of the renewal theorem has been developed to study approximately self-similar sets such as cookie-cutter sets or attractors of conformal iterated function systems.

The method may be used to examine the infinitesimal limits of a variety of quantities reflecting the fractality of self-similar sets. An application to the asymptotic form of solutions to the heat equation on fractal domains is given in Section 12.3.

7.3 Notes and references

Accounts of renewal theory may be found in many probability texts, see for example Feller (1966) or Grimmett and Stirziker (1992). The Tauberian proof of the renewal theorem is given in Rudin (1973) and a version of the probabilistic proof in Lalley (1991). For other proofs see Lindvall (1977) and Levitin and Vassiliev (1996).

Renewal theory was applied to covering functions of fractals by Lalley (1988, 1991); the first of these papers treats self-similar sets with the open set separation condition. Renewal methods were applied to the gap counting functions and the length of the r-neighbourhood of self-similar sets in Kigami and Lapidus (1993) and Falconer (1995b). A generalisation of the renewal theorem suited to attractors of non-linear IFSs, such as cookie-cutter sets, was proved in Lalley (1989) where many applications are given.

Exercises

7.1 Prove Proposition 7.1 without the assumption (7.11). (Hint: A little more work is needed to estimate the sum $\sum_{k=0}^{m} e^{-\alpha|t-y_1-\cdots-y_k|}$ in (7.12) to get estimates like (7.15) and (7.17). This may be achieved by using (7.7) to bound $\int \cdots \int \#\{k : a \leq y_1 + \cdots + y_k \leq b\} \, d\mu(y_1) \ldots d\mu(y_m)$ for $a < b$.)

7.2 Adapt the first proof of Theorem 7.2 to the arithmetic case. (The 'game' now needs to be played on the real line, with the distance moved along the line determined by the probability measure μ.)

7.3 Find $N(r)$ explicitly for the middle-third Cantor set, and verify that (7.31) holds where p is a (non-constant) function with period $\log 3$.

7.4 Let E be the von Koch curve, and let $N(r)$ be the covering number function for E. Show that $N(r) = 4N(3r) - h(r)$, where $|h(r)| \leq 6$ for $r < 1/6$. Deduce that $N(r) \sim r^{-\log 4/\log 3} p(-\log r)$ where p has period $\log 3$. Thus the renewal theory method may be used for self-similar sets without strict separation of the parts.

7.5 Let E be the middle-third Cantor set and μ the restriction of $\mathcal{H}^s$ to E, where $s = \log 2/\log 3$. Define $M(r) \equiv (\mu \times \mu)\{(x, y) : |x - y| \leq r\} = \int \mu(B(x, r)) d\mu(x)$. Show that $M(r) = \frac{1}{2} M(3r) + q(r)$, where $q(r) = (\mu \times \mu)\{(x, y) : \frac{1}{3} \leq |x - y| \leq r\}$. Hence investigate the behaviour of $M(r)$ as $r \to 0$.

7.6 Let $E \subset \mathbb{R}^n$ be the self-similar attractor of the IFS with similarities $\{F_1, \ldots, F_m\}$ of ratios $r_1, \ldots, r_m$ with the components $F_i(E)$ disjoint. Fix $x \notin E$, and define $N(r) = \#\{(i_1, \ldots, i_k) : \text{dist}(F_{i_1} \circ \cdots \circ F_{i_k} x, E) \geq r\}$. Study the asymptotic behaviour of $N(r)$ as $r \to 0$.

Chapter 8 Martingales and fractals

The martingale convergence theorem gives general conditions that guarantee convergence of sequences of random variables or functions. In this chapter we prove the theorem and give two rather different applications to fractal geometry.

8.1 Martingales and the convergence theorem

Although the martingale convergence theorem can be formulated as a result in mathematical analysis, and indeed one of our applications will be in an analytic context, martingales are very naturally thought of in probabilistic terms. The word 'martingale' comes from the name of a classical gambling system (involving doubling one's stake after every lost game), and it is natural to think intuitively of martingales in the context of gambling.

A gambler plays a sequence of games against a casino, the games being fair in the sense that whatever amount is betted, the expected or average gain is zero. (Thus a game might involve throwing a die, with the gambler losing his stake unless a 6 is thrown, in which case he gets back 6 times his stake.) If the gambler's capital after the k-th game is denoted by the random variable Y_k then the fairness of the game requires that the expected value of Y_{k+1} equals Y_k, regardless of how much is betted and the outcomes of earlier games. One version of the martingale convergence theorem states that if $Y_k \geq 0$ for all k (that is if the player is not allowed to run into debt) then, with probability one, Y_k converges to a random variable Y satisfying $\mathsf{E}(Y) \leq \mathsf{E}(Y_0)$, where Y_0 is the initial capital and E denotes expectation. (In the gambling example Y has a high probability of being zero, unless the gambler is extremely cautious, in which case there might just be a positive probability that $Y > 0$!) Whatever 'system' the gambler uses to determine his stake for the k-th game (which may depend on the outcome of the first $k - 1$ games) there is almost sure convergence to an ultimate capital Y which has expectation no more than the initial capital. In particular this means that there is no gambling system that yields an expected profit for the gambler.

Analogues of this idea occur in a wide variety of situations in probabilistic analysis. We work in a sample space Ω with $\mathcal{F}$ a σ-field of events (so $\mathcal{F}$ is closed under countable unions and intersections and under taking complements) on which a probability measure P is defined. Let $\mathcal{F}_0 \subset \mathcal{F}_1 \subset \cdots \subset \mathcal{F}$ be an increasing sequence of σ-fields of events, and assume that $\mathcal{F}$ is the smallest σ-field that contains $\mathcal{F}_k$ for all k. For $k = 0, 1, 2, \ldots$ let Y_k be a random variable on $\mathcal{F}_k$. We call Y_k or, more precisely, $(Y_k, \mathcal{F}_k)$ a *martingale* if for all $k = 0, 1, 2, \ldots$

$$E(|Y_k|) < \infty \tag{8.1}$$

and

$$E(Y_{k+1} | \mathcal{F}_k) = Y_k. \tag{8.2}$$

Condition (8.2), that the expectation of Y_{k+1} conditional on $\mathcal{F}_k$ equals Y_k, means essentially that, whatever happens in the first k steps of the process, the expectation of Y_{k+1} nevertheless equals Y_k.

[Note: the technical measure theoretic definition of conditional expectation with respect to a σ-field is quite complicated. For our purposes it is enough to think of $E(Y_{k+1} | \mathcal{F}_k)$ as the mean value of Y_{k+1} calculated as though $Y_0, \ldots, Y_k$ are already known. The properties of conditional expectation that we shall use are very natural in terms of this interpretation.]

In the gambling example, $\mathcal{F}_k$ represents the set of all possible outcomes of the first k games. Then (8.2) says that regardless of what happens in the first k games, the expected value of the gambler's capital Y_{k+1} after the $(k + 1)$-th game equals the capital Y_k before that game; this reflects the fairness of the game.

From (8.2) we get the unconditional expectation

$$E(Y_{k+1}) = E(Y_k), \tag{8.3}$$

that is, the average of Y_{k+1} will equal that of Y_k when no reference is made to what has gone before.

Much of the theory goes through if (8.2) is weakened to inequality. We say that $(Y_k, \mathcal{F}_k)$ is a *supermartingale* if for $k = 0, 1, 2, \ldots$ we have $E(|Y_k|) < \infty$ and

$$E(Y_{k+1} | \mathcal{F}_k) \leq Y_k \tag{8.4}$$

or a *submartingale* if

$$E(Y_{k+1} | \mathcal{F}_k) \geq Y_k. \tag{8.5}$$

(Thus, in the supermartingale case, the game favours the casino, and in the submartingale case, the gambler has an advantage.)

Martingales may be thought of in an analytic rather than a probabilistic way, as a sequence of functions defined on a hierarchy of sets. A simple case may be visualised graphically, as in Figure 8.1. Let E be a set, let μ be a finite Borel measure on E, and let $C_0, C_1, \ldots$ be finite collections of disjoint Borel subsets of E of positive measure such that $E = \cup \{A \in C_k\}$ for each k, with every set of C_k a disjoint union of sets in C_{k+1}. We let $\mathcal{F}_k$ be the σ-field generated by C_k, which in this case is the finite family of sets formed by unions of sets in C_k. For $k = 0, 1, 2, \ldots$ let $g_k : E \to [0, \infty)$ be such that for

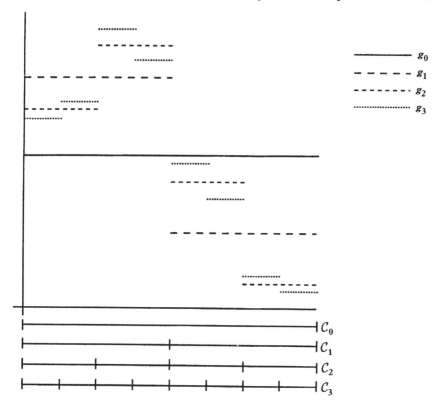

Figure 8.1 The sequence of funtions $\{g_k\}$ is a martingale: g_k is constant on each interval of C_k, and the average of g_{k+1} over an interval A of C_k equals $g_k(x)$ for all $x \in A$

each $A \in C_k$

$$g_k(x) \text{ is constant for } x \in A, \tag{8.6}$$

with

$$g_k(x) = \mu(A)^{-1} \int_A g_{k+1}(y)d\mu(y) \tag{8.7}$$

for all $x \in A$, that is the value of g_k at any point of a set A of C_k equals the average of g_{k+1} over A.

Here we have interpreted the probabilistic notions analytically, with random variables replaced by functions, and expectations by integrals. Condition (8.6) says that g_k is $\mathcal{F}_k$-measurable, and (8.7) is the martingale condition (8.2), so that $(g_k, \mathcal{F}_k)$ may be considered a martingale.

The main reason why martingales and supermartingales are important is that, under very weak assumptions, they converge with probability 1 (or, in the analytic formulation, for almost all x). The standard proof of the martingale convergence theorem depends on the notion of upcrossings. We fix numbers $a < b$ and consider the number of upcrossings of the interval $[a,b]$, that is the number of times Y_k changes from less than a to more than b. More formally, we define the number U_n of *upcrossings* of $[a,b]$ made by Y_k up until time n to be the largest integer M such that there are integers R_i, S_i with

$$0 \leq R_1 < S_1 < R_2 < S_2 < \ldots < R_M < S_M \leq n \qquad (8.8)$$

and

$$Y_{R_i} < a \quad \text{and} \quad Y_{S_i} > b \quad \text{for all} \quad 1 \leq i \leq m.$$

We assume that the random integers S_i and R_i are chosen so that S_i is the least integer greater than R_i, and R_i is the least integer greater than S_{i-1}, for which (8.8) is true.

The key to the proof of the martingale convergence theorem is the following bound on the expected number of upcrossings.

Proposition 8.1 (upcrossing lemma)

Let Y_k be a supermartingale and let U_n be the number of upcrossings of $[a,b]$ up until time n. Then

$$\mathsf{E}(U_n) \leq \frac{\mathsf{E}(|Y_n|) + |a|}{b - a}. \qquad (8.9)$$

Proof   We define a new process Z_k that 'shadows' Y_k. In the gambling context we think of Z_k as the capital (which is allowed to be negative) after the k-th game of a second gambler B whose stake in each game depends on that of gambler A (the gambler with capital Y_k described above) as follows. Gambler B has initial capital $Z_0 = 0$ and does not bet until the first time R_1 that gambler A's capital is less than a. Gambler B then stakes equal amounts to A in each game until the time S_1 is reached when A's capital exceeds b (if this ever happens) and then stops betting. When A's capital next falls below a, gambler B recommences betting, staking equal amounts to A until A's capital again exceeds b, and so on, see Figure 8.2. There cannot normally be too many upcrossings, otherwise gambler B's system of shadowing A would lead to a profit above the odds. Mathematically, with the upcrossings as in (8.8), we define

$$Z_{k+1} = \begin{cases} Z_k & (\text{if } 0 \leq k < R_1 \text{ or } S_i \leq k < R_{i+1}) \\ Z_k + Y_{k+1} - Y_k & (\text{if } R_i \leq k < S_i). \end{cases} \qquad (8.10)$$

(The first case is when gambler B 'sticks', and in the second case B's winnings

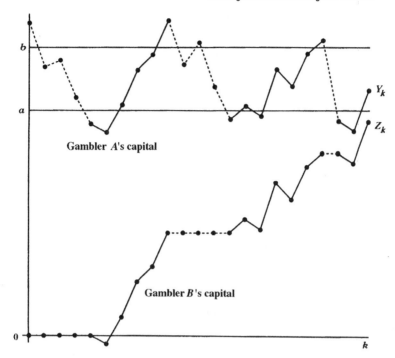

Figure 8.2 Proof of the upcrossing lemma. Gambler B (with capital Z_k after the k-th game) bets during the 'upcrossings' of gambler A (with capital Y_k) indicated by a solid line. A sequence of upcrossings leads to a profit for B. If upcrossings are too likely, this would mean that B had a 'winning system', which is not possible in the martingale or supermartingale situation

are the same as A's.) Since Y_k is a supermartingale, (8.10) and (8.4) give

$$E(Z_{k+1}|\mathcal{F}_k) = \begin{cases} Z_k & (\text{if } 0 \le k < R_1 \text{ or } S_i \le k < R_{i+1}) \\ Z_k + E(Y_{k+1}|\mathcal{F}_k) - Y_k & (\text{if } R_i \le k < S_i) \end{cases}$$
$$\le Z_k.$$

(This says that whatever the outcome of the previous games, gambler B's expected capital after each game is at most his capital before that game. The system that B has been playing, even though it depends on A's system, still cannot yield an expected profit.) On taking unconditional expectations $E(Z_{k+1}) \le E(Z_k)$ so that $E(Z_n) \le E(Z_0) = 0$.

Since Z_k increases by at least $(b-a)$ for each upcrossing of $[a,b]$, we have

$$Z_n \ge (b-a)U_n + \min\{0, Y_n - a\} \tag{8.11}$$

(the second term allows for the possibility that Y_n is less than a at time n).

Taking the unconditional expectation of (8.11),

$$0 \geq E(Z_n) \geq (b-a)E(U_n) + E(\min\{0, Y_n - a\})$$
$$\geq (b-a)E(U_n) + E(-|Y_n| - |a|),$$

giving (8.9). □

The convergence theorem for supermartingales now follows easily.

Theorem 8.2 (supermartingale convergence theorem)

Let Y_k be a martingale or supermartingale with

$$\sup_k E(|Y_k|) < \infty. \tag{8.12}$$

Then there is a random variable (i.e. a P-measurable function) Y such that $Y_k \to Y$ almost surely. Moreover, $E(|Y|) \leq \liminf_{k \to \infty} E(|Y_k|)$, so $|Y| < \infty$ almost surely.

Proof Proposition 8.1 together with (8.12) implies that for every pair of rational numbers $a < b$, the random variable Y_k makes finitely many upcrossings of the interval $[a, b]$ with probability one. Thus, since there are countably many rational pairs, there is probability one that Y_k makes finitely many upcrossings of *every* rational interval $[a, b]$. However, if a real sequence $(y_k)_{k=0}^{\infty}$ fails to be convergent, we may find rationals a, b such that $\liminf_{k \to \infty} y_k < a < b < \limsup_{k \to \infty} y_k$, so that in particular $(y_k)_{k=0}^{\infty}$ makes infinitely many upcrossings of $[a, b]$. We conclude that Y_k converges with probability one. Define $Y \equiv \liminf_{k \to \infty} Y_k$, so that $Y_k \to Y$ almost surely. By Fatou's lemma

$$E(|Y|) = E(\liminf_{k \to \infty}|Y_k|) \leq \liminf_{k \to \infty} E(|Y_k|) \leq \sup_k E(|Y_k|) < \infty,$$

so $|Y| < \infty$ almost surely. □

Our applications will concern non-negative (super-)martingales (that is with $Y_k \geq 0$ for all k) to which the following corollary applies.

Corollary 8.3

Given a non-negative supermartingale Y_k, there exists a non-negative random variable Y, such that Y_k converges to Y almost surely. Moreover, $0 \leq E(Y) \leq \inf_k E(Y_k)$.

Proof Since $0 \leq E(|Y_k|) = E(Y_k) \leq E(Y_0) < \infty$ for all k, the result follows immediately from Theorem 8.2. □

A disadvantage of the martingale convergence theorem in the form of Theorem 8.2 or Corollary 8.3 is that it is possible to have a non-negative martingale Y_k with $E(Y_k) = E(Y_0) > 0$ for all k, but with limit $Y = 0$ almost everywhere. For many applications we need to be able to conclude that $Y > 0$, at least with positive probability. We now add conditions to ensure that this is so.

A martingale Y_k is called an L^2-*bounded* martingale if

$$\sup_{0 \leq k < \infty} E(Y_k^2) < \infty. \tag{8.13}$$

Corollary 8.4

Let Y_k be an L^2-bounded martingale. Then there is a random variable Y such that $Y_k \to Y$ almost surely, with

$$E(|Y - Y_k|) \leq E((Y - Y_k)^2)^{1/2} \to 0.$$

as $k \to \infty$. In particular $E(Y) = E(Y_k)$ for all k.

Proof   Observe that if $j > k$ then $E(Y_j Y_k | \mathcal{F}_k) = E(Y_j | \mathcal{F}_k) Y_k = Y_k^2$, so taking unconditional expectations $E(Y_j Y_k) = E(Y_k^2)$. Thus for $m \geq 1$

$$\sum_{j=1}^{m} E((Y_j - Y_{j-1})^2) = \sum_{j=1}^{m} (E(Y_j^2) - 2E(Y_j Y_{j-1}) + E(Y_{j-1}^2))$$

$$= \sum_{j=1}^{m} (E(Y_j^2) - E(Y_{j-1}^2))$$

$$= E(Y_m^2) - E(Y_0^2),$$

so $\sum_{j=1}^{\infty} E((Y_j - Y_{j-1})^2) < \infty$, using (8.13). In the same way, for $m > k$

$$E((Y_m - Y_k)^2) = E(Y_m^2) - E(Y_k^2) = \sum_{j=k+1}^{m} E((Y_j - Y_{j-1})^2),$$

so letting $m \to \infty$ and using Fatou's lemma,

$$E((Y - Y_k)^2) \leq \sum_{j=k+1}^{\infty} E((Y_j - Y_{j-1})^2) \to 0$$

as $k \to \infty$. Using the Cauchy–Schwartz inequality,

$$|E(|Y|) - E(|Y_k|)| \leq E(|Y - Y_k|) \leq E((Y - Y_k)^2)^{1/2} \to 0.$$

As Y_k is a martingale, $E(Y_k)$ is constant by (8.3), so the final conclusion follows.   □

The martingale convergence theorems may be interpreted analytically to give conditions for a sequence of functions to converge almost everywhere.

Corollary 8.5

For $k = 0, 1, 2, \ldots$ let C_k be finite collections of disjoint Borel sets of positive measure, as above, and let $g_k : E \to [0, \infty)$ satisfy (8.6) and (8.7). Then there is a (Borel measurable) $g : E \to [0, \infty)$ such that $g_k(x) \to g(x)$ for μ-almost all $x \in E$.

Proof This is just Corollary 8.3 expressed analytically. Condition (8.6) implies that g_k is $\mathcal{F}_k$-measurable, where $\mathcal{F}_k$ is the σ-algebra generated by C_k, and (8.7) says that the average of g_{k+1} over A is $g_k(x)$, which is (8.2). Thus Corollary 8.3 tells us that $g_k(x)$ is convergent μ-almost everywhere. $\square$

8.2 A random cut-out set

A variety of random fractal constructions have been proposed, for one class of examples see FG, Section 15.1. Random fractals may often be analysed using martingale techniques: the usual procedure is to define a random measure on the random set as the limit of a sequence of measures associated with the construction, and to use martingales to deduce properties of the limiting measure. Here we find the dimension of a random fractal obtained by removing a sequence of random intervals of decreasing length from the unit interval. (Note that this construction differs from that of Section 3.2 in that the intervals removed may overlap.)

Let $(a_k)_{k=1}^{\infty}$ be a given decreasing sequence of numbers convergent to 0, with $0 < a_k < \frac{1}{2}$. Let $A_1, A_2, \ldots$ be a sequence of random open subintervals of $[0,1]$ such that A_k has length a_k with the midpoints of the intervals independently uniformly distributed on $[0,1]$. It is convenient to identify the ends 0 and 1 of the interval $[0,1]$, so if A_k has centre x and $0 \leq x < \frac{1}{2}a_k$ (respectively $1 - x < \frac{1}{2}a_k \leq 1$) then A_k is taken to consist of the two end intervals $[0, x + \frac{1}{2}a_k) \cup (1 - \frac{1}{2}a_k + x, 1]$ (respectively $[0, x + \frac{1}{2}a_k - 1) \cup (x - \frac{1}{2}a_k, 1]$). We define a sequence of random closed sets E_k by 'cutting out' the intervals A_k, so that $E_0 = [0, 1]$ and $E_k = E_{k-1} \backslash A_k$ for $k = 1, 2, \ldots$. Then the cut-out set

$$E = \bigcap_{k=0}^{\infty} E_k = [0,1] \backslash \bigcup_{k=1}^{\infty} A_k \tag{8.14}$$

is a random closed set, see Figure 8.3. It is easy to see that if $\sum_{1}^{\infty} a_k < \infty$ then there is positive probability that $\mathcal{L}(E) > 0$ (where $\mathcal{L}$ is Lebesgue measure or 'length') and if $\sum_{1}^{\infty} a_k = \infty$ then $\mathcal{L}(E) = 0$ almost surely. It turns out that the

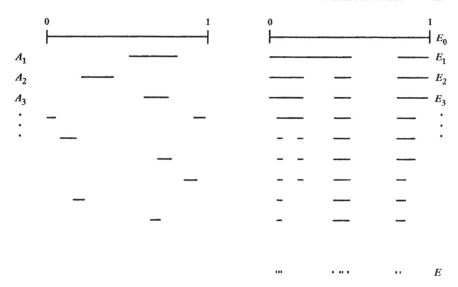

Figure 8.3   Construction of the random cut-out set E by repeated removal of randomly placed intervals. The sequence of intervals shown on the left are cut out to form the sets on the right

critical case is when

$$a_k \sim \frac{t}{k} \tag{8.15}$$

as $k \to \infty$ for some constant t with $0 < t < 1$. We show that in this case E is generally a fractal, provided it is non-empty, and we show that with probability one its Hausdorff dimension is at most $1 - t$ and frequently equals $1 - t$.

We first estimate the probabilities that a given point, and a given pair of points, are in E_k. Clearly for all $x \in [0, 1]$ we have $\mathsf{P}(x \in A_i) = a_i$. We define

$$p_k \equiv \prod_{i=1}^{k}(1 - a_i) \quad (k = 1, 2, \ldots). \tag{8.16}$$

Our calculations depend on estimates of this product. Since $\sum_{i=1}^{\infty} a_i = \infty$ we have that

$$\log p_k = \sum_{i=1}^{k} \log(1 - a_i) \sim -\sum_{i=1}^{k} a_i \sim -\sum_{i=1}^{k} t/i \sim -t \log k. \tag{8.17}$$

Thus by (8.15)

$$\log p_k \sim -t \log k \sim t \log a_k. \tag{8.18}$$

For $x, y \in [0, 1]$ write $d(x, y) = \min\{|x - y|, \ 1 - |x - y|\}$, that is the distance between x and y with 0 and 1 identified.

Lemma 8.6

(a) *For all $x \in [0,1]$ and $k = 1, 2, \ldots$*

$$\mathsf{P}(x \in E_k) = \prod_{i=1}^{k}(1 - a_i) = p_k. \tag{8.19}$$

(b) *Given $\epsilon > 0$ there exists $c > 0$ such that*

$$\frac{\mathsf{P}(x \in E_k \text{ and } y \in E_k)}{p_k^2} \leq cd(x,y)^{-t(1+\epsilon)} \tag{8.20}$$

for all $x, y \in [0,1]$ and $k = 1, 2, \ldots$.

Proof We have that $x \in E_k$ if and only if $x \notin A_i$ for all $i = 1, \ldots, k$. But the events $(x \notin A_i)_{i=1}^{k}$ are independent, so

$$\mathsf{P}(x \in E_k) = \prod_{i=1}^{k} \mathsf{P}(x \notin A_i) = \prod_{i=1}^{k}(1 - a_i) = p_k.$$

Given $\epsilon > 0$ we have from (8.18) that there exists $c_1 > 0$ such that

$$\prod_{j=1}^{k}(1 - a_j) = p_k \geq c_1 a_k^{t(1+\epsilon)} \tag{8.21}$$

for all $k = 1, 2, \ldots$. By considering the positions of A_j for which A_j excludes both x and y, and using the uniform distribution of the centre of A_j, we have

$$\mathsf{P}(x \notin A_j \text{ and } y \notin A_j) = \begin{cases} 1 - a_j - d(x,y) \leq 1 - a_j & \text{if } a_j \geq d(x,y) \\ 1 - 2a_j \qquad\qquad \leq (1 - a_j)^2 & \text{if } a_j < d(x,y). \end{cases}$$

Thus

$$\frac{\mathsf{P}(x \notin A_j \text{ and } y \notin A_j)}{(1 - a_j)^2} \leq \begin{cases} (1 - a_j)^{-1} & \text{if } a_j \geq d(x,y) \\ 1 & \text{if } a_j < d(x,y), \end{cases}$$

so by independence of the cut-outs,

$$\frac{\mathsf{P}(x \in E_k \text{ and } y \in E_k)}{p_k^2} = \prod_{j=1}^{k} \frac{\mathsf{P}(x \notin A_j \text{ and } y \notin A_j)}{(1 - a_j)^2}$$

$$\leq \prod_{j: a_j \geq d(x,y)} (1 - a_j)^{-1}$$

$$= \left(p_{j(d(x,y))} \right)^{-1}$$

$$\leq c_1^{-1} a_{j(d(x,y))}^{-t(1+\epsilon)}$$

for a suitable c_1 by (8.21), where $j(d(x,y))$ is the largest integer j such that $a_j \geq d(x,y)$. From (8.15) $a_{k+1}/a_k \to 1$, so $a_{j(d(x,y))} \sim d(x,y)$, giving (8.20) for a suitable c. $\square$

With $1_{E_k \times E_k}$ as the indicator function of $E_k \times E_k$, inequality (8.20) becomes

$$p_k^{-2} \, \mathsf{E}(1_{E_k \times E_k}(x,y)) = p_k^{-2} \, \mathsf{P}(x \in E_k \text{ and } y \in E_k)$$
$$\leq cd(x,y)^{-t(1+\epsilon)}. \tag{8.22}$$

We introduce a sequence of random measures by setting $\mu_k = p_k^{-1} \mathcal{L}|_{E_k}$, that is the restriction of Lebesgue measure to E_k scaled by a factor p_k^{-1}, so that for a set A

$$\mu_k(A) = p_k^{-1} \mathcal{L}(A \cap E_k).$$

We use the martingale convergence theorem to show that μ_k converges to a random measure μ with probability one. Recall that a *binary interval* is an interval of the form $[p2^{-m}, (p+1)2^{-m})$ for integers m and p.

Lemma 8.7

With probability one there exists a Borel measure μ supported by E with $0 \leq \mu(E) < \infty$ such that $\mu_k(A) \to \mu(A)$ for every set A that is a finite union of binary intervals. Moreover, $\mu(E) > 0$ with positive probability.

Proof Write $\mathcal{F}_k$ for the σ-field underlying the random positions of the centres of $A_1, \ldots, A_k$. (Formally $\mathcal{F}_k$ is the σ-field generated by a k-fold product of Borel subsets of $[0,1]$.) For each binary interval A we have by independence that

$$\mathsf{E}(\mu_{k+1}(A)|\mathcal{F}_k) = \mathsf{E}(p_{k+1}^{-1} \mathcal{L}(A \cap E_k \cap ([0,1] \setminus A_{k+1}))|\mathcal{F}_k)$$
$$= p_{k+1}^{-1} \mathcal{L}([0,1] \setminus A_{k+1}) \mathcal{L}(A \cap E_k)$$
$$= p_{k+1}^{-1}(1 - a_{k+1}) p_k \mu_k(A)$$
$$= \mu_k(A).$$

Thus $\mu_k(A)$ is a non-negative martingale for each binary interval A, so by Corollary 8.3 there are random variables $\mu(A)$ such that, with probability one, $\mu_k(A) \to \mu(A)$ for each of the countably many binary intervals. Then μ extends to a Borel measure supported by E in the usual way, and by additivity, $\mu_k(A) \to \mu(A)$ whenever A is a finite union of binary intervals. (In fact, almost surely, this convergence occurs for all Borel sets.)

Now consider $(\mu_k[0,1])^2$. We have

$$
\begin{aligned}
\mathsf{E}((\mu_k[0,1])^2) &= p_k^{-2}\mathsf{E}(\mathcal{L}(E_k)^2) \\
&= p_k^{-2}\mathsf{E}((\mathcal{L} \times \mathcal{L})(x,y) : x \in E_k \text{ and } y \in E_k)) \\
&= p_k^{-2}\mathsf{E}\left(\int \int 1_{E_k \times E_k}(x,y)\mathrm{d}x\,\mathrm{d}y \right) \\
&\leq c \int_0^1 \int_0^1 d(x,y)^{-t(1+\epsilon)}\mathrm{d}x\,\mathrm{d}y < \infty
\end{aligned}
$$

using (8.22), where ϵ is chosen so that $t(1+\epsilon) < 1$. (Note that $d(x,y)^{-1}$ has a singularity like $|x-y|^{-1}$ for x near y.) Thus $\mu_k[0,1]$ is an L^2-bounded martingale, so $\mathsf{E}\left(\mu(E)\right) = \mathsf{E}(\mu[0,1]) = \mathsf{E}(\mu_0[0,1]) = 1$ by Corollary 8.4, giving that $\mathsf{P}\left(\mu(E) > 0\right) > 0$. $\quad\square$

Proposition 8.8

With probability one the random cut-out set E has $\dim_{\mathrm{H}} E \leq \overline{\dim_{\mathrm{B}}} E \leq 1 - t$, and with positive probability $1 - t \leq \dim_{\mathrm{H}} E$.

Proof Let δ be given, and let $k(\delta)$ be the greatest integer k such that $a_k = |A_k| > 2\delta$. We note that if $x \in E_\delta$, where E_δ is the δ-neighbourhood of E, and $j \leq k(\delta)$, then $x \notin A_j^-$, where A_j^- is the open interval with the same midpoint as A_j and length $a_j - 2\delta$. By the independence of the cut-outs we have that for all $x \in [0,1]$

$$
\begin{aligned}
\mathsf{P}(x \in E_\delta) &\leq \mathsf{P}(x \notin A_j^- \text{ for all } j = 1, \dots, k(\delta)) \\
&\leq \prod_{j=1}^{k(\delta)} \mathsf{P}(x \notin A_j^-) \\
&\leq \prod_{j=1}^{k(\delta)} (1 - (a_j - 2\delta)). \qquad (8.23)
\end{aligned}
$$

But

$$
\begin{aligned}
\log \prod_{j=1}^{k(\delta)} (1 - (a_j - 2\delta)) &\leq -\sum_{j=1}^{k(\delta)} (a_j - 2\delta) \\
&= -\sum_{j=1}^{k(\delta)} a_j + 2\delta k(\delta) \\
&\sim -t \log k(\delta) + 2\delta k(\delta) \\
&\sim t \log \delta + t \qquad (8.24)
\end{aligned}
$$

as $\delta \to 0$, using (8.17) and that $t/k(\delta) \sim a_{k(\delta)} \sim 2\delta$ (a consequence of (8.15) and that $a_{k+1}/a_k \to 0$). It follows from (8.23) that, given $\epsilon > 0$, there exists c such that

$$E(\mathcal{L}(E_\delta)) = P(x \in E_\delta) \leq c\delta^{t-\epsilon}$$

for all $\delta \leq 1$. Thus

$$E\left(\sum_{\delta=2^{-k}:\, k=1,2,\dots} \mathcal{L}(E_\delta)\delta^{-t+2\epsilon}\right) \leq c \sum_{\delta=2^{-k}\,:\, k=1,2,\dots} \delta^\epsilon < \infty.$$

We conclude that with probability one, $\sum_{\delta=2^{-k}:k=1,2,\dots} \mathcal{L}(E_\delta)\delta^{-t+2\epsilon} < \infty$, so that $\mathcal{L}(E_\delta)\delta^{-t+2\epsilon}$ is bounded above for δ of the form 2^{-k}, and thus (since $\mathcal{L}(E_\delta)$ increases with δ) for all $0 < \delta < 1$. It follows from (2.5) that $\overline{\dim}_B E \leq 1 - t + 2\epsilon$. Since $\epsilon > 0$ is arbitrary, we conclude that $\overline{\dim}_B E \leq 1 - t$ with probability one.

For the lower bound we use a potential theoretic calculation. Let $\epsilon > 0$ and let μ_k and μ be the random measures on E_k and E introduced above. Since, almost surely, $\mu_k(A) \to \mu(A)$ on binary intervals, we get, using Fatou's lemma,

$$E\left(\int\int |x - y|^{-d} d\mu(x) d\mu(y)\right)$$

$$\leq E\left(\lim_{k\to\infty} \int\int |x - y|^{-d} d\mu_k(x) d\mu_k(y)\right)$$

$$\leq \liminf_{k\to\infty} E\left(\int\int |x - y|^{-d} d\mu_k(x) d\mu_k(y)\right)$$

$$= \liminf_{k\to\infty} p_k^{-2} E\left(\int\int |x - y|^{-d} 1_{E_k \times E_k}(x, y) dx\, dy\right)$$

$$\leq c \int_0^1 \int_0^1 d(x, y)^{-d} d(x, y)^{-t(1+\epsilon)} dx\, dy$$

$$< \infty$$

provided that $d < 1 - t(1 + \epsilon)$, using (8.22). It follows that, for all $d < 1 - t$, we have $\int\int |x - y|^{-d} d\mu(x) d\mu(y) < \infty$ with probability one, where μ is a random measure supported by E. By Proposition 2.5, $\dim_H E \geq 1 - t$ provided that $\mu(E) > 0$, an occurrence of positive probability. $\square$

There are many natural variations and extensions of this random cut-out construction. For example, in two dimensions, we can remove a sequence of discs (or indeed any convex sets), with radii $a_1 \geq a_2 \geq \dots$ and independent uniformly distributed centres, from the unit square (with opposite sides identified) to get a fractal subset E of the square, see

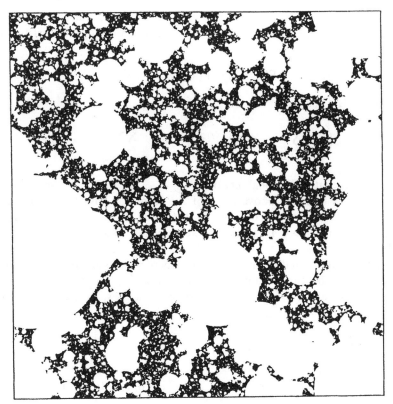

Figure 8.4 A random cut-out set obtained by cutting out discs of decreasing radius and random centres. Here the k-th disc has radius $\frac{1}{5}k^{-1/2}$ and the cut-out set has dimension 1.87

Figure 8.4. A similar analysis enables the dimensions of the cut-out set to be found, see Exercise 8.4.

A variety of other random sets may be studied by using martingales to define random measures on the sets. For example, instead of cutting out intervals, one can consider the points that are covered by infinitely many intervals. With suitable conditions on the interval lengths, it is possible to calculate the almost sure dimension of the set of points that belong to infinitely many random intervals.

Again, martingale techniques may be used to investigate the natural random generalisations of self-similar sets. Such *statistically self-similar sets* are discussed in FG, Section 15.1, where a random measure is defined on the set using an L^2-bounded martingale.

8.3 Bi-Lipschitz equivalence of fractals

In this section we apply martingales in a very different way, to study the existence of bi-Lipschitz mappings between two (non-random) self-similar sets.

The sets E and F are called *bi-Lipschitz equivalent* if there exists a *bi-Lipschitz* mapping $f : E \to F$, that is, a bijection satisfying

$$c_1|x - y| \le |f(x) - f(y)| \le c_2|x - y| \quad (x, y \in E), \qquad (8.25)$$

where $0 < c_1 \le c_2$. Bi-Lipschitz mappings preserve many 'fractal properties' of sets; in particular $\dim E = \dim F$ where 'dim' denotes Hausdorff dimension, lower or upper box dimension, or indeed any other reasonable definition of dimension, see (2.12). Just as a major aim in topology is to classify sets to within homeomorphism (regarding two sets as equivalent if there is a continuous bijection with continuous inverse between them), one approach to fractal geometry is to classify sets to within bi-Lipschitz equivalence. In topology one seeks homeomorphism invariants for sets, that is, quantities associated with sets (such as the Euler–Poincaré characteristic) that are equal for homeomorphic sets. In the same way, a dimension may be thought of as a bi-Lipschitz invariant. For two sets to be bi-Lipschitz equivalent they must certainly be homeomorphic and have the same dimensions. However, in general this is far from enough to guarantee equivalence.

We use martingales to show that certain self-similar sets of equal dimension are *not* bi-Lipschitz equivalent. For ease of exposition, we give the proof for a particular pair of sets, but the method works for much more general self-similar sets.

Let E be the middle-third Cantor set and let F be the self-similar set obtained from the unit interval by repeated replacement of intervals by three equally spaced subintervals of lengths $\beta = 3^{-\log 3 / \log 2}$ times that of the parent interval, see Figure 8.5. Then E and F are homeomorphic, see Exercise 8.6, and β has been chosen so $\dim_H E = \dim_H F = \log 2 / \log 3$, so that bi-Lipschitz equivalence cannot be ruled out on topological or dimensional grounds.

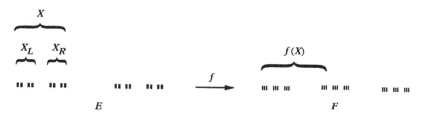

Figure 8.5 The mapping f between two self-similar sets. For X a k-th level subset of E, the image $f(X)$ is a complete union of $m(k)$-th level subsets of F

Proposition 8.9

The sets E and F described above are not bi-Lipschitz equivalent.

Proof We assume that there is a bi-Lipschitz mapping $f: E \to F$ satisfying (8.25), and derive a contradiction. By a *k-th level subset* of E we mean $E \cap X$, where X is one of the intervals of length 3^{-k} that occur in the usual construction of the Cantor set; we let C_k denote the family of k-th level subsets of E. Similarly, a *k-th level subset* of F is one of the form $F \cap Y$ where Y is one of the 3^k intervals of length β^k in the construction of F.

Let c_1, c_2 be as in (8.25). For $k = 0, 1, 2, \ldots$ define $m(k)$ to be the least integer such that

$$\beta^{m(k)} < c_1 3^{-k}. \tag{8.26}$$

A consequence of f being bi-Lipschitz is that, if X is a k-th level subset of E, then $f(X)$ is a complete union of some n of the $m(k)$-th level subsets of F where n is an integer satisfying

$$3^{m(k)} 2^{-k} c_1 \leq n \leq 3^{m(k)} 2^{-k} c_2. \tag{8.27}$$

To see this, note that if $f(X)$ contains any point of an $m(k)$-th level subset Y of F, then $f(X) \supset Y$; otherwise there would be points $x \in X$ and $w \in E \setminus X$ such that $f(x), f(w) \in Y$, giving

$$\beta^{m(k)} = |Y| \geq |f(x) - f(w)| \geq c_1 |x - w| \geq c_1 3^{-k},$$

contradicting (8.26). The bounds on n in (8.27) follow since

$$c_1 \leq \frac{\mathcal{H}^s(f(X))}{\mathcal{H}^s(X)} = \frac{n 3^{-m(k)}}{2^{-k}} \leq c_2, \tag{8.28}$$

using (2.12) and that $\mathcal{H}^s(E) = \mathcal{H}^s(F) = 1$.

For $k = 0, 1, 2, \ldots$ we define $g_k : E \to \mathbb{R}$ by

$$g_k(x) \equiv \mathcal{H}^s(f(X)) / \mathcal{H}^s(X) \tag{8.29}$$

where X is the k-th level subset of E with $x \in X$. Thus, g_k is constant on each k-th level subset of E. Moreover, if $x \in X$, and X_L and X_R are the $(k+1)$-th level subsets of the k-th level set X,

$$\begin{aligned}
g_k(x) &= 2^k \mathcal{H}^s(f(X)) \\
&= \tfrac{1}{2} [2^{k+1} \mathcal{H}^s(f(X_L)) + 2^{k+1} \mathcal{H}^s(f(X_R))] \\
&= \tfrac{1}{2} [g_{k+1}(x_L) + g_{k+1}(x_R)] \\
&= \mathcal{H}^s(X)^{-1} \int_X g_{k+1}(x) d\mathcal{H}^s(x), \tag{8.30}
\end{aligned}$$

where x_L and x_R are any points of X_L and X_R respectively. This is just the martingale condition (8.7), and we conclude from Corollary 8.5 that $g_k(x)$ converges for almost all $x \in E$.

Choose x to be any such point of convergence, with $g_k(x) \to c$, say. By (8.28) and (8.29) $c_1 \le c \le c_2$.

From (8.28) and (8.29),

$$g_k(x) = 2^k 3^{-m(k)} n \tag{8.31}$$

where n satisfies (8.27). Thus

$$\frac{g_{k+1}(x)}{g_k(x)} = \frac{2^{k+1} 3^{-m(k+1)} n'}{2^k 3^{-m(k)} n} = \frac{2n'}{3^{m(k+1)-m(k)} n}$$

where $c_1^{1-\log 2/\log 3} \le n, n' \le 3b\, c_1^{-\log 2/\log 3}$ using (8.26) and (8.27), and $m(k+1) - m(k)$ is bounded for all k, by virtue of the definition (8.26). Thus $g_{k+1}(x)/g_k(x)$ can only take finitely many distinct values; since $g_k(x)$ converges to a non-zero limit, this requires that $g_k(x)$ is eventually constant, that is $g_k(x) = c$ for all sufficiently large k. For all such k, there are, by (8.31), integers n_k with

$$c = g_k(x) = 2^k 3^{-m(k)} n_k, \tag{8.32}$$

so c is rational, say $c = p/q$ where p and q are co-prime positive integers. Hence $2^k 3^{-m(k)} n_k = p/q$ for all sufficiently large k, that is

$$2^k n_k q = 3^{m(k)} p, \tag{8.33}$$

so p is divisible by 2^k for arbitrarily large integers k, which is absurd. We conclude that our hypothesis that f is bi-Lipschitz is false. $\square$

In the above proof, we used martingales to deduce that f was 'differentiable with respect to the measures' in a certain sense, and the geometry of the sets E and F to deduce that f was 'locally linear'.

In a very similar way, we can show that if E, respectively F, are self-similar sets constructed by replacing intervals by m, respectively n, equal and equally spaced subintervals, then for E and F to be bi-Lipschitz equivalent, it is necessary (and, indeed, sufficient) that $\dim_H E = \dim_H F$ and either $m = n^q$ or $n = m^q$ for some positive integer q.

More generally, a similar method gives necessary conditions for the equivalence of self-similar sets constructed using unequal similarity ratios. We write $E(r_1, \ldots, r_m)$ for the attractor of some IFS on $\mathbb{R}^n$ consisting of similarities of ratios $r_1, \ldots, r_m$ and satisfying the strong separation condition, so that $E(r_1, \ldots, r_m)$ is totally disconnected.

Proposition 8.10

The following conditions are necessary for the sets $E(r_1, \ldots, r_m)$ and $E(t_1, \ldots, t_q)$ to be bi-Lipschitz equivalent.

(a) $\dim_H E(r_1, \ldots, r_m) = \dim_H E(t_1, \ldots, t_q) \equiv s$, *say;*

(b) $\mathbb{Q}(r_1^s, \ldots, r_m^s) = \mathbb{Q}(t_1^s, \ldots, t_q^s)$, *where* $\mathbb{Q}(a_1, \ldots, a_m)$ *denotes the sub-field of* $(\mathbb{R}, +, \times)$ *consisting of the rational functions of* $a_1, \ldots, a_m$;

(c) *there exist positive integers* p *and* p' *such that*

$$\text{sgp}(r_1^p, \ldots, r_m^p) \subset \text{sgp}(t_1, \ldots, t_q),$$
$$\text{sgp}(t_1^{p'}, \ldots, t_q^{p'}) \subset \text{sgp}(r_1, \ldots, r_m),$$

where $\text{sgp}(a_1, \ldots, a_m)$ *denotes the sub-semigroup of* $(\mathbb{R}^+, \times)$ *generated by* $a_1, \ldots, a_m$, *that is, the set of products of the form* $a_1^{\alpha_1} \cdots a_m^{\alpha_m}$ *with the* α_i *non-negative integers.*

Proof The idea of the proof is similar to that of Proposition 8.9; we omit the details. □

The general problem of determining whether two sets are bi-Lipschitz equivalent is complicated. The approach above applies to certain self-similar sets that are homeomorphic to Cantor sets and in particular are totally disconnected. At the other extreme, equivalence amongst a large class of quasi-self-similar fractals that are homeomorphic to the circle is completely determined by Hausdorff dimension; this is discussed in FG, Section 14.4.

8.4 Notes and references

The book by Williams (1991) gives a detailed treatment of martingales with some nice applications. Many texts on probability, such as Grimmett and Stirziker (1992), include introductions to martingale theory.

Mandelbrot (1972, 1982) introduced the random cut-out model, or random trema model, as he termed it. His dimension calculations were based on a birth and death process; the approach here is due to Zähle (1984), who gives many generalisations. For other constructions based on random translates of intervals, etc, see Kahane (1985) and the references therein. The natural random analogues of the Cantor set and other self-similar constructions are described in FG, Chapter 15.

The martingale approach to bi-Lipschitz equivalence of self-similar sets is that of Falconer and Marsh (1992); see Cooper and Pignataro (1988) for a different approach.

Exercises

8.1 Check that if Y_k is a martingale then $E(Y_k) = E(Y_0)$ for all k. Verify that if Y_k is a martingale, then $E(Y_{m+k}|\mathcal{F}_k) = Y_k$ for all $m, k \geq 0$.

8.2 Let Y_k be a supermartingale with respect to $\mathcal{F}_k$. Verify that Y_k^2 is also a supermartingale, provided that $\sup_k E(Y_k^2) < \infty$.

8.3 In the random cut-out model, show that if $\sum_1^\infty a_k < \infty$ then $\mathcal{L}(E) > 0$ with positive probability, and if $\sum_1^\infty a_k = \infty$ then $\mathcal{L}(E) = 0$ almost surely.

8.4 Let a_k be a decreasing sequence with $a_k \sim tk^{-1/2}$ where $0 < t < (2/\pi)^{1/2}$. Let E be the random cut-out set obtained by removing from the unit square (with opposite sides identified) a sequence of discs of radii a_k and independent uniformly distributed random centres. Modify the arguments of Section 8.2 to show that $\dim_H E \le \overline{\dim}_B E \le 2 - \pi t^2$, with a positive probability of equality.

8.5 Let E and F be self-similar subsets of $\mathbb{R}$ constructed by replacing intervals by m, respectively n, equal and equally spaced subintervals. Show that there exists a bi-Lipschitz map between E and F if and only if $\dim_H E = \dim_H F$ and either $m = n^q$ or $n = m^q$ for some positive integer q. (Hint: mimic the proof of Proposition 8.9; for a simpler problem show that no such map exists in the case when m has a prime factor that does not divide n.)

8.6 Verify that the two sets considered in Proposition 8.9 are homeomorphic. (Hint: map the 'gaps' of the sets to each other in a systematic way.)

Chapter 9   Tangent measures

Tangent measures provide a means of studying infinitesimal features of sets and measures. In particular, many of the classical results in geometric measure theory concerning densities, rectifiability and projections of sets may be proved in a natural way using tangent measures. Some very powerful and technical results have been obtained using these methods; here we are content to describe the basic properties of tangent measures and give a few simple but elegant applications.

9.1 Definitions and basic properties

Tangent measures describe the structure of a set in the neighbourhood of a point that becomes apparent when viewed through a microscope with ever-increasing magnification. Thus we look at the limiting sets or measures that can be realised by a sequence of enlargements about a point. These limits or 'tangent measures' are in some way like derivatives: they carry a great deal of information about the local form of a set or a measure, but certain regularity properties of tangent measures mean that they are often easier to work with than the original sets or measures.

Throughout this chapter, μ will be a finite (or locally finite) Borel measure on $\mathbb{R}^n$; by far the most important instance is where μ is the restriction of s-dimensional Hausdorff measure to an s-set E (that is a Borel set $E \subset \mathbb{R}^n$ with $0 < \mathcal{H}^s(E) < \infty$), so that $\mu = \mathcal{H}^s|_E$. We also assume throughout this chapter that there is a number s such that the s-dimensional densities of μ are bounded away from 0 and ∞ at μ-almost all x, that is

$$0 < \underline{D}^s(\mu, x) \equiv \liminf_{r \to 0} \frac{\mu(B(x,r))}{(2r)^s} \leq \limsup_{r \to 0} \frac{\mu(B(x,r))}{(2r)^s} = \overline{D}^s(\mu, x) < \infty. \quad (9.1)$$

(In the case when μ is the restriction of $\mathcal{H}^s$ to an s-set then $\overline{D}^s(\mu, x) < \infty$ necessarily holds at μ-almost all x.)

For $x \in \mathbb{R}^n$ and $r > 0$ we define the similarity mapping $F_{x,r} : \mathbb{R}^n \to \mathbb{R}^n$ by

$$F_{x,r}(y) = (y - x)/r; \quad (9.2)$$

thus $F_{x,r}$ translates x to the origin and scales by a factor $1/r$, so the ball $B(x,r)$ is mapped onto the unit ball $B(0,1)$. We are particularly interested in the way these similarities transform measures, and we define the induced similarity mapping between measures by

$$(F_{x,r}[\mu])(A) = \mu(x + rA) \tag{9.3}$$

where μ is a measure on $\mathbb{R}^n$ and $x + rA = \{x + ra : a \in A\}$ for all sets $A \subset \mathbb{R}^n$. Thus $F_{x,r}[\mu]$ is thought of as the enlargement of μ about x by a factor $1/r$.

To define tangent measures we examine possible limits of $F_{x,r}[\mu]$ as $r \searrow 0$. To obtain limiting measures that are positive and locally finite, the magnitudes of these measures need to be scaled appropriately. Thus we define ν to be a *tangent measure* of μ at a point $x \in \mathbb{R}^n$ if there is a sequence $r_k \searrow 0$ such that

$$\nu = \lim_{k \to \infty} r_k^{-s} F_{x,r_k}[\mu], \tag{9.4}$$

where the limit is the weak limit of the sequence of measures, see Section 1.4. The set of all tangent measures of μ at x is termed the *tangent space* of μ at x and denoted by $\mathrm{Tan}(\mu, x)$. [Note that what we term tangent measures are referred to as 'standardised tangent measures' in some accounts.] Of course, $\mathrm{Tan}(\mu, x)$ depends on the value of s, but for every μ there is at most one s for which (9.1) is satisfied.

Using (9.3) and the definition (1.24) of a weak limit, (9.4) says that for every continuous $g : \mathbb{R}^n \to \mathbb{R}$ with bounded support

$$\int g(y)\mathrm{d}\nu(y) = \lim_{k \to \infty} r_k^{-s} \int g((y - x)/r_k)\mathrm{d}\mu(y). \tag{9.5}$$

Tangent measures are locally finite, but in general are infinite measures of unbounded support, even though this need not be so for μ.

Some examples should help in understanding this concept. Fix E as a bounded Borel subset of $\mathbb{R}^n$ with $0 < \mathcal{L}^n(E) < \infty$, where $\mathcal{L}^n$ is n-dimensional Lebesgue measure, and let $\mu = \mathcal{L}^n|_E$. The Lebesgue density theorem (2.20) states that $\mathcal{L}^n(B(x,r) \cap E)/\mathcal{L}^n(B(x,r)) \to 1$ as $r \to 0$ for $\mathcal{L}^n$-almost all $x \in E$. Intuitively, this means that at almost all $x \in E$, small balls centred at x are nearly filled by E, so that enlargements of μ about such x approach Lebesgue measure, the unique tangent measure. To see this formally, if g is continuous with support in $B(0, R)$, then

$$r^{-n} \int g \mathrm{d}F_{x,r}[\mu] = r^{-n} \int g((y - x)/r)\mathrm{d}\mu(y)$$

$$= r^{-n} \int g((y - x)/r)1_E(y)\mathrm{d}y$$

$$= \int g(z)1_E(x + rz)\mathrm{d}z$$

$$\to \int g(z)\mathrm{d}z$$

as $r \to 0$. (For the last step we use that by the Lebesgue density theorem

$\int_{B(0,R)} |1_E(x+rz) - 1| dz \to 0$, for almost all $x \in E$, where R is chosen so $\mathrm{spt}\, g \subset B(0, R)$.) Thus $\lim_{r \to 0} r^{-n} F_{x,r}[\mu] = \mathcal{L}^n$ for almost all $x \in E$, so $\mathcal{L}^n$ is the unique tangent measure of μ at x.

For a second example let μ be 'length measure' on the unit circle C in $\mathbb{R}^2$, that is $\mu(A) = \mathcal{H}^1|_C$ where $\mathcal{H}^1$ is 1-dimensional Hausdorff measure. For $x \in C$, the portion of C near x approaches a line segment under increasing magnification, so the unique tangent measure might be expected to be one-dimensional measure restricted to the line through the origin parallel to the tangent to C at x. Again, this may be verified formally using (9.5). More generally, this is true if C is any smooth curve.

A more interesting situation arises when μ is a fractal measure. Let $\mu = \mathcal{H}^s|_E$, where $s = \log 2 / \log 3$ and E is the middle-third Cantor set. In this case, the tangent space has a very rich structure. For almost all x, $\mathrm{Tan}(\mu, x)$ contains infinitely many measures, each the restriction of $\mathcal{H}^s$ to an unbounded extension of the Cantor set E that looks locally like E itself. Writing $E = \{0 \cdot b_1 b_2 \dots \text{ in base 3, where } b_i = 0 \text{ or } 2 \text{ for all } i\}$, almost every $x \in E$ has a base 3 expansion which contains every finite sequence of 0s and 2s infinitely often, and thus at such x the tangent space $\mathrm{Tan}(\mu, x)$ contains measures supported by enlargements of E about all of its points. In particular, $\mathrm{Tan}(\mu, x)$ contains the restriction of the measure $\mathcal{H}^s$ to the extended Cantor set $E' = \{a_n a_{n-1} \dots a_1 \cdot b_1 b_2 \dots \text{ in base 3, where } a_i, b_i = 0 \text{ or } 2 \text{ for all } i\}$, together with its restriction to all similar copies of E' that contain the origin. For a further example, see Figure 9.1.

Observe in all these examples that, whenever $\nu \in \mathrm{Tan}(\mu, x)$ and $z \in \mathrm{spt}\nu$ (where $\mathrm{spt}\nu$ is the support of ν), the translate $F_{z,1}[\nu]$ of ν that brings z to the origin is also in $\mathrm{Tan}(\mu, x)$. (In the first two examples this is because the unique measure in $\mathrm{Tan}(\mu, x)$ is unchanged by such translation, in the third example it depends on the wealth of measures in $\mathrm{Tan}(\mu, x)$.) As we shall see in Proposition 9.3, this 'shift invariance' holds for all tangent spaces, and this is one of the properties that makes tangent measures such a useful tool.

We now derive some general properties of tangent measures. Taking A as a ball in (9.3) we have that for $r, R > 0$,

$$\begin{aligned} r^{-s}(F_{x,r}[\mu])(B(0,R)) &= r^{-s}\mu(x + rB(0,R)) \\ &= r^{-s}\mu(B(x,rR)) \\ &= R^s (rR)^{-s} \mu(B(x,rR)). \end{aligned} \tag{9.6}$$

This observation leads to several basic properties of tangent measures, including their relationship with densities.

Lemma 9.1

Let μ be a measure on $\mathbb{R}^n$. Then for all x satisfying

$$0 < \underline{D}^s(\mu, x) \le \overline{D}^s(\mu, x) < \infty \tag{9.7}$$

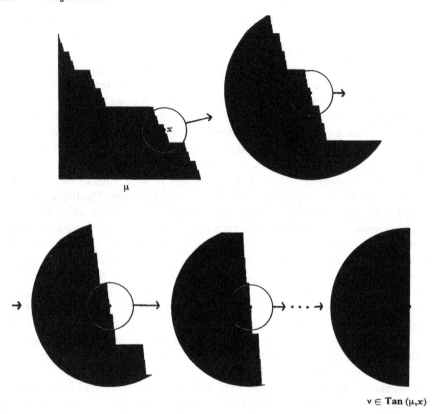

$\nu \in \mathbf{Tan}\,(\mu,x)$

Figure 9.1 A tangent measure $\nu \in \mathrm{Tan}(\mu, x)$ is obtained as the weak limit of a sequence of enlargements about x of (scalar multiples of) the measure μ. Here μ is Lebesgue measure restricted to the region under a 'devil's staircase' and the tangent measure ν is Lebesgue measure restricted to a half-plane

we have:

(a) $\mathrm{Tan}(\mu, x)$ *is non-empty,*
(b) $\underline{D}^s(\mu, x) \le (2R)^{-s}\nu(B(0, R)) \le \overline{D}^s(\mu, x)$ *for all* $\nu \in \mathrm{Tan}(\mu, x)$ *and* $R > 0$,
(c) $0 \in \mathrm{spt}\nu$ *for all* $\nu \in \mathrm{Tan}(\mu, x)$.

Proof

(a) If (9.7) holds at x, then by (9.6) there is a number $d > 0$ such that for each R, if r is sufficiently small,

$$r^{-s}(F_{x,r}[\mu])(B(0, R)) \le R^s d.$$

By weak compactness (Proposition 1.9) there is a sequence $r_k \searrow 0$ and a measure ν such that $r_k^{-s} F_{x,r_k}[\mu] \to \nu$, so $\nu \in \operatorname{Tan}(\mu, x)$.

(b) Let ν be given by (9.4) for some $r_k \searrow 0$. Then for all $R > 0$

$$(2R)^s \underline{D}^s(\mu, x) = R^s \liminf_{r \to 0} (rR)^{-s} \mu(B(x, rR))$$

$$\leq \limsup_{k \to \infty} r_k^{-s}(F_{x,r_k}[\mu])(B(0, R))$$

$$\leq \nu(B(0, R))$$

using (9.1), (9.6) and (1.25). In the same way, but using (1.26), we see that $\nu(B^0(0, R)) \leq (2R)^s \overline{D}^s(\mu, x)$ for the open ball $B^0(0, R)$. Since $\nu(B(0, R)) \leq \nu(B^0(0, R_1))$ for all $R_1 > R$, the upper bound follows on taking R_1 arbitrarily close to R.

(c) Since $0 < \underline{D}^s(\mu, x)$, part (b) gives that $0 < \nu(B(0, R))$ for all $R > 0$, so $0 \in \operatorname{spt}\nu$. $\square$

Our remaining aim in this section is to prove the shift invariance of tangent spaces. Recall that, for a measure μ and measurable set E in $\mathbb{R}^n$, a point $x \in \operatorname{spt}\mu$ is a *density point* of E if

$$\lim_{r \to 0} \frac{\mu(E \cap B(x, r))}{\mu(B(x, r))} = 1. \tag{9.8}$$

In particular, according to Proposition 1.7, μ-almost all $x \in E$ are density points. It is easy to see that if $\nu \in \operatorname{Tan}(\mu, x)$ and $z \in \operatorname{spt}\nu$ then there exists a sequence $x_k \in \operatorname{spt}\mu$ with $F_{x,r_k}(x_k) \to z$. The following lemma extends this to allow the x_k to be chosen from a prescribed set E.

Lemma 9.2

Let E be a μ-measurable set and let $x \in \operatorname{spt}\mu$ be a density point of E. Let $\nu = \lim_{k \to \infty} r_k^{-s} F_{x,r_k}[\mu] \in \operatorname{Tan}(\mu, x)$. Then for all $z \in \operatorname{spt}\nu$ there exist points $x_k \in E$ such that

$$F_{x,r_k}(x_k) \equiv \frac{x_k - x}{r_k} \to z \tag{9.9}$$

as $k \to \infty$.

Proof For (9.9) to hold we must arrange for x_k to be 'near' $x + r_k z$. Thus for all k, choose $x_k \in E$ such that

$$|x + r_k z - x_k| < \operatorname{dist}(x + r_k z, E) + k^{-1} r_k \tag{9.10}$$

If $z = 0$, then $\operatorname{dist}(x + r_k z, E) = 0$, and (9.9) is immediate. Suppose that (9.9) is false and $z \neq 0$. Then there exists δ with $0 < \delta < |z|$ such that

$$|x + r_k z - x_k| \geq 2\delta r_k \tag{9.11}$$

for infinitely many k. For such k with $k > \delta^{-1}$, (9.10) and (9.11) give

$$r_k \delta \leq r_k(2\delta - k^{-1}) < \text{dist}(x + r_k z, E),$$

so in particular

$$B(x + r_k z, r_k \delta) \subset B(x, r_k|z| + r_k \delta) \setminus E \subset B(x, 2r_k|z|) \setminus E.$$

Since x is a density point of E,

$$
\begin{aligned}
1 &= \lim_{k \to \infty} \frac{\mu(E \cap B(x, 2r_k|z|))}{\mu(B(x, 2r_k|z|))} \\
&= \lim_{k \to \infty} \frac{\mu(B(x, 2r_k|z|)) - \mu(B(x, 2r_k|z|) \setminus E)}{\mu(B(x, 2r_k|z|))} \\
&\leq 1 - \liminf_{k \to \infty} \frac{\mu(B(x + r_k z, r_k \delta))}{\mu(B(x, 2r_k|z|))} \\
&= 1 - \liminf_{k \to \infty} \frac{(F_{x,r_k}[\mu])(B(z, \delta))}{(F_{x,r_k}[\mu])(B(0, 2|z|))} \\
&\leq 1 - \frac{\nu(B^0(z, \delta))}{\nu(B(0, 2|z|))} < 1,
\end{aligned}
$$

using (1.25), (1.26) and that $\nu(B^0(z, \delta)) > 0$ since $z \in \text{spt}\nu$. This contradiction means that (9.11) cannot hold for infinitely many k, so (9.9) follows. $\square$

The following proof of shift invariance uses properties of weak convergence of measures in a delicate way.

Proposition 9.3

For μ-almost all x, the tangent space $\text{Tan}(\mu, x)$ is shift invariant, that is $F_{z,1}[\nu] \in \text{Tan}(\mu, x)$ for all $\nu \in \text{Tan}(\mu, x)$ and $z \in \text{spt}\nu$.

Proof Let R, ϵ, δ be positive numbers and let

$$E(R, \epsilon, \delta) = \{x : \text{there exists } \nu_x \in \text{Tan}(\mu, x) \text{ and } z(x) \in \text{spt}\nu_x \text{ with}$$

$$d_R(F_{z(x),1}[\nu_x], r^{-s} F_{x,r}[\mu]) > \epsilon \text{ for all } r < \delta\}. \tag{9.12}$$

(See (1.28) for the definition of the pseudo-metric d_R on measures.) The result will follow easily once we have shown that $\mu(E(R, \epsilon, \delta)) = 0$ for all such R, ϵ and δ.

Suppose to the contrary, that $\mu(E(R, \epsilon, \delta)) > 0$ for some R, ϵ and δ. For all $x \in E(R, \epsilon, \delta)$ choose $\nu_x \in \text{Tan}(\mu, x)$ and $z(x) \in \text{spt}\nu_x$ such that the condition in (9.12) holds. We may find a Borel set $E \subset E(R, \epsilon, \delta)$ with $\mu(E) > 0$ such that

$$d_R(F_{z(x),1}[\nu_x], F_{z(y),1}[\nu_y]) < \tfrac{1}{2}\epsilon \tag{9.13}$$

for all $x, y \in E$. This follows from the separability by Lemma 1.11: if (μ_k) is a countable dense sequences of measures then

$$E \equiv \{x \in E(R, \epsilon, \delta) : d_R(F_{z(x),1}[\nu_x], \mu_k) < \tfrac{1}{4}\epsilon\}$$

has positive μ-measure for some k.

Choose x to be any density point of E, with $\nu_x = \lim_{k \to \infty} r_k^{-s} F_{x,r_k}[\mu]$. By Lemma 9.2 we may find $x_k \in E$ such that $(x_k - x)/r_k \to z(x)$. Since $F_{x_k,r_k} = F_{(x_k-x)/r_k,1} \circ F_{x,r_k}$, it follows that

$$r_k^{-s} F_{x_k,r_k}[\mu] = F_{(x_k-x)/r_k,1}[r_k^{-s} F_{x,r_k}[\mu]]$$
$$\to F_{z(x),1}[\nu_x] \tag{9.14}$$

as $k \to \infty$. Since $x_k \in E(R, \epsilon, \delta)$, if k is sufficiently large

$$\epsilon < d_R(F_{z(x_k),1}[\nu_{x_k}], r_k^{-s} F_{x_k,r_k}[\mu])$$
$$\leq d_R(F_{z(x_k),1}[\nu_{x_k}], F_{z(x),1}[\nu_x]) + d_R(F_{z(x),1}[\nu_x], r_k^{-s} F_{x_k,r_k}[\mu])$$
$$< \tfrac{1}{2}\epsilon + \tfrac{1}{2}\epsilon,$$

using (9.13), Lemma 1.10 and (9.14), which is the contradiction sought.

We conclude that $\mu(E(R, \epsilon, \delta)) = 0$ for all $R, \epsilon, \delta > 0$. By Lemma 1.10, the set

$$\{x : \text{ there exists a non-shift-invariant measure } \nu_x \in \text{Tan}(\mu, x)\}$$
$$= \bigcup_{m=1}^{\infty} \bigcup_{n=1}^{\infty} E(m, 1/m, 1/n)$$

and this has μ-measure 0, as required.   □

9.2 Tangent measures and densities

Tangent measures are a natural tool for studying local properties of sets and measures, such as their densities and average densities. The idea is to convert a problem involving a set or measure into a more tractable one involving tangent measures. We give some typical examples of this procedure, starting with the classical result that the density of an s-set cannot exist on a set of positive measure, unless s is an integer. (Recall that the non-existence of density is a manifestation of fractality.) We work with a general measure μ, but the main example that we have in mind is the restriction of $\mathcal{H}^s$ to an s-set E in $\mathbb{R}^n$, so that $\mu = \mathcal{H}^s|_E$ for a Borel set E with $0 < \mathcal{H}^s(E) < \infty$.

To illustrate the use of tangent measures in a particularly simple case, we first prove that the density fails to exist if $0 < s < 1$. After that, we develop the more complicated proof for general non-integral s.

Proposition 9.4

Let $0 < s < 1$ and let μ be a measure on $\mathbb{R}^n$ with $0 < \underline{D}^s(\mu, x) \leq \overline{D}^s(\mu, x) < \infty$ for μ-almost all x. Then $\underline{D}^s(\mu, x) < \overline{D}^s(\mu, x)$ for μ-almost all x.

Proof Suppose that $0 < \underline{D}^s(\mu, x) = \bar{D}^s(\mu, x)$ for a set of x of positive measure. By Proposition 9.3 we may select such an x such that $\mathrm{Tan}(\mu, x)$ is shift invariant; let $d \equiv \underline{D}^s(\mu, x) = \bar{D}^s(\mu, x) > 0$ be the density at this particular x. By Lemma 9.1(b), $d = (2R)^{-s}\nu(B(0, R))$ for all $\nu \in \mathrm{Tan}(\mu, x)$ and all $R > 0$. Fixing any $\nu \in \mathrm{Tan}(\mu, x)$, there exists $z \in \mathrm{spt}\,\nu$ with $|z| = 1$ (otherwise $\nu(B(0, R))$ would be constant for R near 1). By Proposition 9.3, $F_{z,1}[\nu] \in \mathrm{Tan}(\mu, x)$, so $d = (2R)^{-s}(F_{z,1}[\nu])(B(0, R)) = (2R)^{-s}\nu(B(z, R))$ for all $R > 0$. For $0 < R < \frac{1}{2}$ we have $B(z, R) \subset B(0, 1 + 2R) \setminus B(0, 1 - 2R)$, so

$$(2R)^s d = \nu(B(z, R))$$
$$\leq \nu(B(0, 1 + 2R)) - \nu(B^0(0, 1 - 2R))$$
$$= 2^s d[(1 + 2R)^s - (1 - 2R)^s]$$
$$= O(R)$$

for small R, a contradiction since $0 < s < 1$. We conclude that $\underline{D}^s(\mu, x) < \bar{D}^s(\mu, x)$ for μ-almost all x. $\quad\square$

Thus in particular, the density $D^s(E, x)$ of an s-set E fails to exist almost everywhere if $0 < s < 1$.

To extend this result to general non-integral s, we introduce 's-uniform' measures. We show that such measures can only exist if s is an integer, and then show that if the density $D^s(\mu, x)$ exists on a set of positive measure, then μ has an s-uniform tangent measure, so that s must be an integer.

We define a measure ν on $\mathbb{R}^n$ to be *s-uniform* for $s \geq 0$ if there exists $c > 0$ such that

$$\nu(B(x, r)) = cr^s \text{ for all } x \in \mathrm{spt}\nu \text{ and } r > 0. \tag{9.15}$$

Lemma 9.5

Let ν be an *s-uniform measure* on $\mathbb{R}^n$. If $x \in \mathrm{spt}\,\nu$ and $\nu_1 \in \mathrm{Tan}(\nu, x)$ then ν_1 is *s-uniform* on $\mathbb{R}^n$.

Proof Suppose $\nu_1 = \lim_{k\to\infty} r_k^{-s} F_{x,r_k}[\nu]$ and $z \in \mathrm{spt}\nu_1$. We may find points $z_k \in \mathrm{spt}\, F_{x,r_k}[\nu]$ with $z_k \to z$, so there exist $x_k \in \mathrm{spt}\nu$ with $(x_k - x)/r_k = F_{x,r_k}(x_k) = z_k \to z$. Then for all $R > 0$

$$(F_{x,r_k}[\nu])(B(z, R)) = \nu(B(x + r_k z, r_k R))$$
$$= \nu(B(x_k + r_k(z - z_k), r_k R))$$
$$\geq \nu(B(x_k, r_k(R - |z - z_k|)))$$
$$= cr_k^s(R - |z - z_k|)^s$$

by (9.15). Hence by (1.25)

$$\nu_1(B(z, R)) \geq \limsup_{k\to\infty} r_k^{-s} F_{x,r_k}[\nu](B(z, R)) \geq cR^s.$$

That $\nu_1(B^0(z,R)) \leq cR^s$ for all $R > 0$ follows in a similar way using (1.26), and these inequalities combine to give $\nu_1(B(z,R)) = cR^s$ for all R, so that ν_1 is s-uniform. $\square$

In fact s-uniform measures are very regular indeed: they can only exist if s is an integer.

Proposition 9.6

Let ν be an s-uniform measure on $\mathbb{R}^n$, so that for some $s \geq 0$ and $c > 0$ we have

$$\nu(B(x,r)) = cr^s \text{ for all } x \in \text{spt}\nu \text{ and } r > 0. \tag{9.16}$$

Then s is an integer with $0 \leq s \leq n$.

*Proof   Let $g(r) \in \mathbb{R}^n$ be the centre of mass of the restriction of an s-uniform measure ν to $B(0,r)$, so that for all $z \in \mathbb{R}^n$

$$z \cdot g(r) = \int_{B(0,r)} (z \cdot y) d\nu(y), \tag{9.17}$$

where '$\cdot$' denotes the usual scalar product on $\mathbb{R}^n$. The proof depends on the estimate that if ν is any s-uniform measure and $0 \in \text{spt}\nu$ then

$$|z \cdot g(r)|/|z| \to 0 \tag{9.18}$$

as $z \to 0$ through points z in the support of ν.

To derive (9.18) we consider the 'potential integral'

$$Q_r(x) = \int_{B(x,r)} (r^2 - |x - y|^2) d\nu(y). \tag{9.19}$$

Differentiating under the integral sign at $x = 0$, we have

$$\nabla Q_r(0) = -2 \int_{B(0,r)} y d\nu(y) \tag{9.20}$$

where '∇' is the usual 'grad' operator. (Note that the contribution to the derivative from the boundary vanishes, since

$$\left| \int_{B(x,r)} (r^2 - |y|^2) d\nu(y) - \int_{B(0,r)} (r^2 - |y|^2) d\nu(y) \right|$$

$$\leq \int_{B(0,r+|x|) \setminus B(0,r-|x|)} (r^2 - (r - |x|)^2) d\nu(y)$$

$$\leq 2r|x|(\nu(B(0,r+|x|)) - \nu(B(0,r-|x|)))$$

$$= 2r|x|c((r + |x|)^s - (r - |x|)^s)$$

$$= O(|x|^2)$$

as $|x| \to 0$, using (9.16).) Thus, from (9.17) and (9.20) and the directional derivative formula,

$$-2z \cdot g(r) = z \cdot \nabla Q_r(0)$$
$$= Q_r(z) - Q_r(0) + o(z)$$
$$= o(z),$$

since if $0, z \in \mathrm{spt}\nu$ we have $Q_r(0) = Q_r(z)$ $(= 2cr^{2+s}/(2+s)$ by direct calculation from (9.19) using (9.16)). Thus (9.18) follows.

We now assume that the stated proposition is false, so that for some non-integral s and some integer n with $0 < s < n$ there exists an s-uniform measure ν_0 on $\mathbb{R}^n$. Let n be the least integer for which this is possible; we obtain a contradiction by exhibiting an s-uniform measure in $\mathbb{R}^{n-1}$ (where $\mathbb{R}^0 = \{0\}$).

We must have that $\mathrm{spt}\nu_0$ is a proper subset of $\mathbb{R}^n$; otherwise packing the unit ball $B(0,1)$ with small balls and using (9.16) with $s < n$ would imply that $\nu_0(B(0,1)) = \infty$. (Note that this is where we use that s is non-integral.) Choose $y \in \mathbb{R}^n \setminus \mathrm{spt}\nu_0$, let $r > 0$ be the least number such that $B(y,r) \cap \mathrm{spt}\nu_0 \neq 0$, and take $x \in B(y,r) \cap \mathrm{spt}\nu_0$. Then choosing $\nu \in \mathrm{Tan}(\nu_0, x)$ it is easy to see that $0 \in \mathrm{spt}\nu \subset H$, where H is the half-space $\{z \in \mathbb{R}^n : z \cdot e \geq 0\}$ with e as the outward unit normal to $B(y,r)$ at x. (Under magnification about x, the ball $B(y,r)$, which contains no points of $\mathrm{spt}\nu$ in its interior, enlarges to the half-space $\mathbb{R}^n \backslash H$.) Using Lemma 9.5 it follows that ν is s-uniform, that is $\nu(B(z,r)) = c_0 r^s$ for all $z \in \mathrm{spt}\nu$ and $r > 0$. By Lemma 9.1(c), $0 \in \mathrm{spt}\nu$. Defining $g(r)$ to be the centre of mass of this measure ν restricted to $B(0,r)$, (9.18) holds.

If $g(r) = 0$ for all $r > 0$, then since $\mathrm{spt}\nu$ lies in the half-space H and the centre of mass of ν restricted to $B(0,r)$ is in the bounding hyperplane ∂H, we must have $\mathrm{spt}\nu \cap B(0,r) \subset \partial H$ for all $r > 0$. Thus $\mathrm{spt}\nu \subset \partial H$, so by identifying ∂H with $\mathbb{R}^{n-1}$ we have that ν is an s-uniform measure on $\mathbb{R}^{n-1}$.

On the other hand, if $g(r) \neq 0$ for some $r > 0$, we may take $\nu_1 = \lim_{k\to\infty} r_k^{-s} F_{0,r_k}[\nu] \in \mathrm{Tan}(\nu, 0)$, so that ν_1 is s-uniform by Lemma 9.5. Given $\eta > 0$ and $R > 0$ define $A = \{z \in B^0(0,R) : |z \cdot g(r)| > \eta|z|\}$. Then by the weak convergence property (1.26)

$$\nu_1(A) \leq \liminf_{k\to\infty} r_k^{-s} F_{0,r_k}[\nu](A) = \liminf_{k\to\infty} r_k^{-s}\nu(r_k A)$$
$$= \liminf_{k\to\infty} r_k^{-s}\nu\{z \in B(0, r_k R) : |z \cdot g(r)| > \eta|z|\}$$
$$= 0,$$

since if $z \in B(0, rR) \cap \mathrm{spt}\nu$ and r is sufficiently small then $|z \cdot g(r)| \leq \eta|z|$ by (9.18). Thus $\nu_1(A) = 0$ for all $\eta > 0$ and $R > 0$, so $\mathrm{spt}\nu_1 \subset \{z : z \cdot g(r) = 0\}$, and again by identifying this hyperplane with $\mathbb{R}^{n-1}$ we get an s-uniform measure on $\mathbb{R}^{n-1}$. $\square$

Given this property of s-uniform measures, it is straightforward to deduce the density result.

Theorem 9.7

Let μ be a finite measure on $\mathbb{R}^n$ and let $s \geq 0$. Suppose that

$$0 < \underline{D}^s(\mu, x) = \overline{D}^s(\mu, x) < \infty \qquad (9.21)$$

(that is the density exists and is positive and finite) for a set of x of positive μ-measure. Then s is an integer with $0 \leq s \leq n$.

Proof By Lemma 9.1(*a*) μ has a tangent measure at all x for which (9.21) holds, so by Proposition 9.3 we may choose such an x for which there is a shift invariant $\nu \in \mathrm{Tan}(\mu, x)$. Thus for all $z \in \mathrm{spt}\nu$ we have $F_{z,1}[\nu] \in \mathrm{Tan}(\mu, x)$, and so by Lemma 9.1(*b*)

$$\underline{D}^s(\mu, x) \leq (2R)^{-s}(F_{z,1}[\nu])(B(0, R)) = (2R)^{-s}\nu(B(z, R)) \leq \overline{D}^s(\mu, x)$$

for all $R > 0$. It follows from (9.21) that $\nu(B(z, R)) = 2^s D^s(\mu, x)R^s$ for all $z \in \mathrm{spt}\nu$ and all $R > 0$, so s is an integer by Proposition 9.6. $\square$

The corresponding result for densities of s-sets follows easily.

Corollary 9.8

Let E be an s-set in $\mathbb{R}^n$ where s is not an integer. Then for $\mathcal{H}^s$-almost all $x \in E$ we have $\underline{D}^s(E, x) < \overline{D}^s(E, x)$.

Proof Let $\mu = \mathcal{H}^s|_E$. Then $0 < \overline{D}^s(\mu, x) = \overline{D}^s(E, x) < \infty$ for μ-almost all x (see Exercise 9.1). Thus either $0 \leq \underline{D}^s(\mu, x) < \overline{D}^s(\mu, x)$ for μ-almost all x, as required (since $\underline{D}^s(\mu, x) = \underline{D}^s(E, x)$ and $\overline{D}^s(\mu, x) = \overline{D}^s(E, x)$), or $0 < \underline{D}^s(\mu, x) = \overline{D}^s(\mu, x)$ for a set of positive measure, in which case s would be an integer by Theorem 9.7. $\square$

The existence of the density $D^s(\mu, x)$ almost everywhere implies not only that s is an integer, but also that μ is an s-rectifiable measure. (A measure μ is *s-rectifiable* if it is absolutely continuous with respect to $\mathcal{H}^s$, so that $\mu(A) = 0$ whenever $\mathcal{H}^s(A) = 0$ and there exists an s-rectifiable set E with $\mu(\mathbb{R}^n \setminus E) = 0$, see Section 2.1.) Let s be an integer and μ a measure satisfying $0 < \underline{D}^s(\mu, x) \leq \overline{D}^s(\mu, x) < \infty$ for almost all x. It may be shown that μ is s-rectifiable if and only if for almost all x the tangent space $\mathrm{Tan}(\mu, x)$ contains a single measure ν that is the restriction of a scalar multiple of $\mathcal{H}^s$ to an s-dimensional subspace of $\mathbb{R}^n$. It is perhaps not surprising that developing further properties of s-uniform measures and applying them to tangent measures leads to the following difficult theorem and its corollary.

Theorem 9.9

Let μ be a measure on $\mathbb{R}^n$ such that for μ-almost all x the density $D^s(\mu, x)$ exists with $0 < D^s(\mu, x) < \infty$. Then s is an integer and μ is s-rectifiable.

Proof Omitted. $\square$

Corollary 9.10

Let E be an s-set in $\mathbb{R}^n$. Then E is s-rectifiable if and only if $D^s(E, x)$ exists for $\mathcal{H}^s$-almost all $x \in E$.

Proof Taking $\mu = \mathcal{H}^s|_E$ in Theorem 9.9 shows that this condition is sufficient, since $\overline{D}^s(E, x) > 0$ almost everywhere. $\square$

The interplay between the 'metric' concept of density and the 'geometric' concept of rectifiability is at the heart of geometric measure theory, and tangent measures are a major tool for relating these concepts.

Next we use tangent measures to compare densities and average densities for non-integral s. For this we study the 'density function' $g : \mathbb{R} \to \mathbb{R}$ of μ at x given by

$$g(t) = \frac{\mu(B(x, e^{-t}))}{(2e^{-t})^s} \tag{9.22}$$

By definition the (s-dimensional) lower density of μ at x is given by

$$\underline{D}^s(\mu, x) = \liminf_{t \to \infty} g(t) \tag{9.23}$$

and the lower average density by

$$\underline{A}^s(\mu, x) = \liminf_{T \to \infty} \frac{1}{T} \int_0^T g(t) \mathrm{d}t, \tag{9.24}$$

with similar expressions for upper densities, see (6.21).

We first show that if a density and the corresponding average density are equal, then $g(t)$ is 'nearly constant' over long intervals.

Lemma 9.11

Let μ be a measure on $\mathbb{R}^n$, let $s > 0$, and let x be such that

$$0 < \underline{D}^s(\mu, x) \le \overline{D}^s(\mu, x) < \infty. \tag{9.25}$$

Suppose that for some $d > 0$ either

$$\underline{A}^s(\mu, x) = \underline{D}^s(\mu, x) = d \tag{9.26}$$

or

$$\overline{A}^s(\mu, x) = \overline{D}^s(\mu, x) = d. \tag{9.27}$$

Then given $\epsilon > 0$ and $\lambda_0 > 0$ there exist arbitrarily large T such that for all $t \in [T, T + \lambda_0]$

$$d - \epsilon < g(t) < d + \epsilon. \tag{9.28}$$

Proof We give the proof under hypothesis (9.27); the case (9.26) is similar. We have control over the rate of increase of f: for $t > u$

$$\begin{aligned} g(t) &= 2^{-s} e^{s(t-u)} e^{su} \mu(B(x, e^{-t})) \\ &\le e^{s(t-u)} 2^{-s} e^{su} \mu(B(x, e^{-u})) \\ &\le e^{s(t-u)} g(u). \end{aligned} \tag{9.29}$$

Clearly the right-hand inequality of (9.28) holds for all sufficiently large t by (9.27). In particular, given $0 < \delta < 1$ we may find T_0 such that if $t > T_0$

$$g(t) \le d + \tfrac{1}{2} \eta (\lambda_0 + 1)^{-1} \tag{9.30}$$

where $\eta = d(\delta - (1 - e^{-s\delta})/s) > 0$. Suppose that $\tau \ge T_0$ and there exists $u \in [\tau, \tau + \lambda_0]$ with $g(u) \le de^{-s\delta}$. By (9.29) $g(t) \le de^{s(t-u-\delta)} \le d$ for $u \le t \le u + \delta$, so

$$\int_u^{u+\delta} (g(t) - d)dt \le \int_u^{u+\delta} d(e^{s(t-u-\delta)} - 1)dt = d((1 - e^{-s\delta})/s - \delta) = -\eta.$$

Writing $\lambda = \lambda_0 + 1$,

$$\int_\tau^{\tau+\lambda} (g(t) - d)dt \le \int_u^{u+\delta} (g(t) - d)dt + \int_{[\tau, \tau+\lambda] \setminus [u, u+\delta]} (g(t) - d)dt$$

$$\le -\eta + \lambda \times \tfrac{1}{2} \eta \lambda^{-1} = -\tfrac{1}{2} \eta \tag{9.31}$$

using (9.30). Suppose that for some $T_1 \ge T_0$ it is the case that for all $m = 0, 1, 2, \ldots$ there exists $u \in [T_1 + m\lambda, T_1 + m\lambda + \lambda_0]$ with $g(u) \le de^{-s\delta}$. Then, if $T \ge T_1$ and M is the greatest integer such that $M\lambda \le T - T_1$,

$$\frac{1}{T} \int_0^T (g(t) - d)dt = \frac{1}{T} \int_0^{T_1} (g(t) - d)dt + \frac{1}{T} \sum_{m=0}^{M-1} \int_{T_1+m\lambda}^{T_1+(m+1)\lambda} (g(t) - d)dt$$

$$+ \frac{1}{T} \int_{T_1+M\lambda}^T (g(t) - d)dt.$$

As $T \to \infty$, the first and third terms in this sum tend to 0, so by (9.31)

$$\overline{A}^s(\mu, x) - d = \limsup_{T \to \infty} \frac{1}{T} \sum_{m=0}^{M-1} \int_{T_1+m\lambda}^{T_1+(m+1)\lambda} (g(t) - d)\,dt$$

$$\leq \limsup \frac{-M\eta}{2T} = \frac{-\eta}{2\lambda} < 0$$

which contradicts (9.27). We conclude that there exist arbitrarily large T such that $g(u) > de^{-s\delta}$ for all $u \in [T, T + \lambda_0]$, which is the left-hand inequality of (9.28) if $\delta > 0$ is chosen small enough so that $d - \epsilon < de^{-s\delta}$. $\square$

We may interpret Lemma 9.11 in terms of tangent measures.

Lemma 9.12

Let μ, x and s be as in Lemma 9.11, satisfying (9.25) and either (9.26) or (9.27). Then there exists $\nu \in \mathrm{Tan}(\mu, x)$ such that $\nu(B(0,R)) = (2R)^s d$ for all $R > 0$.

Proof By Lemma 9.11 we may find a sequence $t_k \nearrow \infty$ such that

$$d - 1/k < g(t) < d + 1/k$$

for all $t \in [t_k - \log k, t_k + \log k]$, so letting $r_k = e^{-t_k}$ and using the definition (9.22) of g

$$d - 1/k < \mu(B(x,r))/(2r)^s < d + 1/k$$

for all $r \in [r_k/k, kr_k]$. For all $R > 0$ there exists k_0 such that for all $k \geq k_0$ we have $r_k/k \leq r_k R \leq kr_k$ giving

$$r_k^{-s} F_{x,r_k}[\mu](B(0, R)) = r_k^{-s} \mu(B(x, r_k R))$$
$$= (2R)^s (2r_k R)^{-s} \mu(B(x, r_k R)) \to 2^s R^s d$$

as $k \to \infty$. The same is true for the open ball $B^0(0, R)$, so taking $\nu \in \mathrm{Tan}(\mu, x)$ to be the weak limit of a subsequence of $r_k^{-s} F_{x,r_k}[\mu]$ we get that $\nu(B(0, R)) = (2R)^s d$, using (1.25) and (1.26). $\square$

With Lemma 9.12 at our disposal, the proof of Proposition 9.4 is easily adapted to yield a stronger result involving average densities.

Theorem 9.13

Let μ be a measure on $\mathbb{R}^n$ with $0 < \underline{D}^s(\mu, x) \leq \overline{D}^s(\mu, x) < \infty$ for μ-almost all x. If $0 < s < 1$ then

$$\underline{D}^s(\mu, x) < \underline{A}^s(\mu, x) \leq \overline{A}^s(\mu, x) < \overline{D}^s(\mu, x) \tag{9.32}$$

for μ-almost all x.

Proof   Suppose that $0 < \underline{D}^s(\mu, x) = \overline{A}^s(\mu, x)$ for a set of x of positive measure. By Proposition 9.3 we may select such an x such that every $\nu \in \text{Tan}(\mu, x)$ is shift invariant; let $d \equiv \underline{D}^s(\mu, x) = \overline{A}^s(\mu, x) > 0$ for this particular x. By Lemma 9.12 we may choose some $\nu \in \text{Tan}(\mu, x)$ such that $\nu(B(0, R)) = d(2R)^s$. By shift invariance, for all $z \in \text{spt}\,\nu$ we have $F_{z,1}[\nu] \in \text{Tan}(\mu, x)$, so by Lemma 9.1(*b*) $d \le (2R)^{-s}(F_{z,1}[\nu])(B(0, R)) = (2R)^{-s}\nu(B(z, R))$ for all $R > 0$. We may choose $z \in \text{spt}\,\nu$ with $|z| = 1$, otherwise $\nu(B(0, R))$ would be constant over an interval for R close to 1. For $0 < R < 1$, we have $B(z, R) \subset B(0, 1 + R) \setminus B^0(0, 1 - R)$ so

$$
\begin{aligned}
d(2R)^s &\le \nu(B(z, R)) \\
&\le \nu(B(0, 1 + R)) - \nu(B(0, 1 - R)) \\
&= 2^s d[(1 + R)^s - (1 - R)^s] \\
&= O(R)
\end{aligned}
$$

for small R, a contradiction, since $0 < s < 1$. We conclude that the left-hand inequality of (9.32) holds; the proof of the right-hand inequality is similar. $\square$

Just as with Corollary 9.8 we can specialise this result to *s*-sets.

Corollary 9.14

Let E be an s-set in $\mathbb{R}^n$ with $0 < s < 1$. Then $\overline{A}^s(E, x) < \overline{D}^s(E, x)$ for $\mathcal{H}^s$-almost all $x \in E$.

Proof   With $\mu = \mathcal{H}^s|_E$ we have $0 < \overline{D}^s(\mu, x) = \overline{D}^s(E, x) < \infty$ for μ-almost all x (see Exercise 9.1). Thus either $0 \le \overline{A}^s(\mu, x) < \overline{D}^s(\mu, x)$ for μ-almost all x as required, or $0 < \overline{A}^s(\mu, x) = \overline{D}^s(\mu, x)$ on a set of positive measure, which cannot happen by Theorem 9.13. $\square$

In fact it has been shown that the conclusion of Theorem 9.13 holds for all non-integral *s*.

9.3 Singular integrals

We give a very brief introduction to the use of tangent measures in another area where they have proved a powerful tool, namely the theory of singular integrals.

One of the most familiar singular intergrals is the *Hilbert transform,* defined by the real integral

$$
(Hf)(x) = \int \frac{f(y)}{y - x} \, dy \tag{9.33}
$$

for $f \in L^1(\mathbb{R})$. This integral must be interpreted in terms of its principal value, that is,

$$(Hf)(x) = \lim_{\epsilon \to 0} \int_{|y-x| \ge \epsilon} \frac{f(y)}{y - x} \, dy. \tag{9.34}$$

The fundamental property of the Hilbert transform is that the integral (9.33) does indeed exist and is finite for almost all $x \in R$, provided that $f \in L^1(\mathbb{R})$. In other words, at a typical point x the singular contributions to the integral on either side of x more or less cancel.

It is natural to consider analogues of the Hilbert transform on fractal domains. For a simple example, let E be the middle-third Cantor set, of dimension $s = \log 2/\log 3$, and let $\mu = \mathcal{H}^s|_E$ be the 'Cantor measure'. For $f \in L^1(\mu)$ define

$$(Hf)(x) = \int \frac{(y - x)f(y)}{|y - x|^{s+1}} \, d\mu(y) \equiv \lim_{\epsilon \to 0} \int_{|y-x| \ge \epsilon} \frac{(y - x)f(y)}{|y - x|^{s+1}} \, d\mu(y). \tag{9.35}$$

[Since

$$c_1 r^s \le \mu(B(x, r)) \le c_2 r^s \tag{9.36}$$

for $x \in E$ and $r < 1$ where $0 < c_1 \le c_2 < \infty$, see Exercise 2.11, it is natural to consider a kernel with exponent $-s$; use of $(y - x)/|y - x|^{s+1}$ is merely a device to express the sign change across the singularity. Recall that we considered the singular behaviour of the absolute integral of (9.35) in (6.36).] We use tangent measures to show that (9.35) fails to exist for almost all x.

Proposition 9.15

Let $\mu = \mathcal{H}^s|_E$ be the Cantor measure and let $f \in L^1(\mu)$. Then for μ-almost all x such that $f(x) \ne 0$

$$\limsup_{\epsilon \to 0} \left| \int_{|y-x| \ge \epsilon} \frac{(y - x)f(y)}{|y - x|^{s+1}} \, d\mu(y) \right| = \infty \tag{9.37}$$

and so $(Hf)(x)$ does not exist.

Proof To avoid unnecessarily awkward notation, we make the convention of writing $(y - x)^{-s}$ for $(y - x)|y - x|^{-s-1}$, so that $(y - x)^s$ is understood to be real with the sign of $y - x$. We give the proof in the case when $f(x) = 1$ for all x; little modification is needed for general $f \in L^1(\mu)$, see Exercise 9.8.

Suppose that there exists a set $X_0 \subset E$ with $\mu(X_0) > 0$ such that for all $x \in X_0$ the integral $\int_{|y-x| \ge \epsilon} (y - x)^{-s} d\mu(y)$ is bounded for $0 < \epsilon \le 1$, that is there are numbers $c(x)$ such that for all $0 < \delta \leq \epsilon \leq 1$

$$\left| \int_{\delta \leq |y - x| \leq \epsilon} (y - x)^{-s} d\mu(y) \right| \leq c(x) \tag{9.38}$$

for all $x \in X_0$. Then we may find a number $c < \infty$ and a compact set $X \subset X_0$ with $\mu(X) > 0$ such that $c(x) \leq c$ for all $x \in X$. Fix $x \in \mathrm{spt}\mu$ as any density point of X (see (9.8)); we examine the tangent measures of μ at x. Choose $\nu = \lim_{k \to \infty} r_k^{-s} F_{x, r_k}[\mu] \in \mathrm{Tan}\,(\mu, x)$ and let $z \in \mathrm{spt}\nu$. By Lemma 9.2 there is a sequence $x_k \in X$ such that $z_k \equiv (x_k - x)/r_k \to z$. By Lemma 9.1(b) and (9.36) ν has no atoms (that is, $\nu(\{y\}) = 0$ for all y), so

$$\int_{r \leq |y - z| \leq R} (y - z)^{-s} d\nu(y)$$

$$= \lim_{k \to \infty} \int_{r \leq |y - z_k| \leq R} (y - z_k)^{-s} d\nu(y)$$

$$= \lim_{k \to \infty} r_k^{-s} \int_{r \leq |y - z_k| \leq R} (y - z_k)^{-s} d(F_{x, r_k}[\mu])(y)$$

$$= \lim_{k \to \infty} r_k^{-s} \int_{r \leq |(y - x)/r_k - z_k| \leq R} ((y - x)/r_k - z_k)^{-s} d\mu(y)$$

$$= \lim_{k \to \infty} \int_{r r_k \leq |y - x_k| \leq R r_k} (y - x_k)^{-s} d\mu(y) \tag{9.39}$$

using an extension of (9.5) and the definition of z_k. Since $x_k \in X$ it follows from (9.38) that

$$\left| \int_{r \leq |y - z| \leq R} (y - z)^{-s} d\nu(y) \right| \leq \limsup_{k \to \infty} c(x_k) \leq c \tag{9.40}$$

for all $\nu \in \mathrm{Tan}\,(\mu, x)$, all $z \in \mathrm{spt}\nu$ and $0 < r \leq R$.

But for μ-almost all $x \in E$ the measure ν given by the restriction of $\mathcal{H}^s$ to the extended Cantor set

$$E' = \{a_m \ldots a_1 \cdot b_1 b_2 \ldots \text{ in base 3, where } a_i, b_i = 0 \text{ or } 2 \text{ for all } i\}$$

is in $\mathrm{Tan}(\mu, x)$. Taking such an $x \in X$ and $0 \in \mathrm{spt}\nu$, we have on integrating by parts and using that $c_1 r^s \leq \nu(B(0, r)) \leq c_2 r^s$ for some $c_1, c_2 > 0$ by Lemma 9.1(b),

$$\int_{r \leq y \leq R} (y - 0)^{-s} d\nu(y) = [y^{-s} \nu(B(0, y))]_r^R + s \int_r^R y^{-s-1} \nu(B(0, y)) dy$$

$$\geq -c_2 + c_1 s \int_r^R y^{-1} dy$$

which may be made arbitrarily large by taking r small enough. This contradicts (9.40) and the result follows. $\quad\square$

The proof of Proposition 9.15 used (9.38), the boundedness of the integral with respect to μ, to imply (9.40), the boundedness of the integral with respect to the tangent measures. For a suitable choice of tangent measure, the latter property is more readily shown to be false. Proposition 9.15 and its proof hold for a very much wider class of measure μ, for example the restriction of $\mathcal{H}^s$ to any self-similar s-set in $\mathbb{R}^n$ with the strong separation condition. In fact the following even more general result concerning vector integrals is true. Note that (9.42) is a very strong statement to the effect that the centre of mass of ν restricted to each ball centred in $\mathrm{spt}\nu$ is at the centre of the ball.

Proposition 9.16

Let $s > 0$ and let μ be a finite measure on $\mathbb{R}^n$ such that for almost all x we have $0 < \underline{D}^s(\mu, x) \leq \overline{D}^s(\mu, x) < \infty$, with the limit

$$\lim_{\varepsilon \to 0} \int_{|x-y| \geq \varepsilon} \frac{(y - x)}{|y - x|^{s+1}} \, d\mu(y) \tag{9.41}$$

existing and finite. Then for almost all x

$$\int_{B(z,r)} y \, d\nu(y) = z\nu(B(z,r)) \tag{9.42}$$

for all $\nu \in \mathrm{Tan}(\mu, x)$ and for all $z \in \mathrm{spt}\nu$ and $r > 0$. In particular, this implies that s is an integer and that μ is an s-rectifiable measure.

Sketch of proof   The existence of the limit (9.41) implies that

$$\lim_{0 < \varepsilon < \delta \to 0} \int_{\varepsilon \leq |y-x| \leq \delta} (y - x)/|y - x|^{s+1} d\mu(y) = 0, \tag{9.43}$$

and using Egoroff's theorem we may find a set X with $\mu(X) > 0$ on which this convergence is uniform. Proceeding just as in the proof of Proposition 9.15 we again get (9.39), so on passing to tangent measures, (9.43) gives

$$\int_{r \leq |y-z| \leq R} (y - z)/|y - z|^{s+1} d\nu(y) = 0 \tag{9.44}$$

for all $\nu \in \mathrm{Tan}(\mu, x)$, all $z \in \mathrm{spt}\nu$ and all $0 < r \leq R$. To deduce (9.42) consider the integral

$$I_\phi = \int (y - z)\phi(|y - z|) d\nu(y)$$

where $\phi : [0, \infty) \to \mathbb{R}$. By (9.44) $I_\phi = 0$ if $\phi(t) = t^{-s-1} 1_{[r, R]}(t)$, where 1 is the indicator function, and approximating $1_{[0, R]}(t)$ by a linear combination of such functions gives $\int_{0 \leq |y-z| \leq R} (y - z) d\nu(y) = 0$, which is (9.42).

Now suppose that ν is a measure satisfying (9.42) with $c_1 r^s \leq \nu(B(z, r)) \leq c_2 r^s$ for $z \in \text{spt}\nu$ (as before, this latter condition follows for $\nu \in \text{Tan}(\mu, x)$ from Lemma 9.1(b) and the density bounds on μ). It may be shown, using an argument in the spirit of Proposition 9.6, that s is an integer and that ν is the restriction of $\mathcal{H}^s$ to an s-dimensional subspace of $\mathbb{R}^n$. Pulling this tangent measure property back to the original measure μ gives that $\text{spt}\mu$ is rectifiable, with μ absolutely continuous with respect to $\mathcal{H}^s$ restricted to $\text{spt}\mu$. We omit the details. $\square$

Given the existence almost everywhere of the integral (9.41) when $s = n$ and μ is the restriction of n-dimensional Lebesgue measure to a bounded Borel set, it may be verified that the principal value integral (9.41) exists when μ is the restriction of $\mathcal{H}^s$ to a rectifiable s-set in $\mathbb{R}^n$ where s is an integer, and further for any μ that is absolutely continuous with respect to such measures. (Remember that a rectifiable set is made up of pieces that are subsets of s-dimensional C^1 sets (see Section 2.1). Thus Proposition 9.16 lends to a pleasing characterisation of those measures on $\mathbb{R}^n$ for which singular integrals (9.35) exist. Typically, singular integrals do not exist for fractal measures.

9.4 Notes and references

Tangent measures were introduced in the fundamental paper of Preiss (1987) which unified and extended the theory of densities, tangents and rectifiability that had developed over the previous 60 years. For a comprehensive account of this work and many further applications of tangent measures, see the book of Mattila (1995a) which also contains a very substantial bibliography for this area of geometric measure theory. Preiss (1987) and Mattila (1995a) derive strong properties of s-uniform measures, and use these to relate densities to rectifiability. The original proof of Proposition 9.4 and Theorem 9.7 on density and integral dimension were due to Marstrand (1954) with extensions to average density by Falconer and Springer (1995), and for general s by Marstrand (1996). The use of tangent measures to study singular integrals was pioneered by Mattila (1995b) and Mattila and Preiss (1995) and again a full account of this area may be found in Mattila (1995a). A related topological theory for the local form of sets is given by Wicks (1991).

Exercises

9.1 Verify that if E is an s-set then $0 < \overline{D}^s(E, x) < \infty$ for $\mathcal{H}^s$-almost all $x \in E$.

9.2 Find $\text{Tan}(\mu, x)$ for all $x \in \mathbb{R}^2$ where μ is the restriction of plane Lebesgue measure to the unit disc $B(0, 1) \subset \mathbb{R}^2$.

9.3 Let μ be the middle-third Cantor set, let $s = \log 2/\log 3$ and let μ be the restriction of $\mathcal{H}^s$ to E. Describe $\text{Tan}(\mu, 0)$.

9.4 Let μ be a measure on $\mathbb{R}^n$, let $f: \mathbb{R}^n \to [0, \infty)$ be continuous, and define a measure μ_f on $\mathbb{R}^n$ by $\mu_f(A) = \int_A f \, d\mu$ for Borel sets A. Show that if $f(x) > 0$ then $\nu \in \mathrm{Tan}(\mu, x)$ if and only if $f(x)\nu \in \mathrm{Tan}(\mu_f, x)$. Show that if $f: \mathbb{R}^n \to [0, \infty)$ is merely locally integrable, this remains true for μ-almost all x such that $f(x) > 0$. (Hint: for the final part use Lusin's theorem, that for $\varepsilon > 0$ there is a set A with $\mu(\mathbb{R}^n \setminus A) < \varepsilon$, such that the restriction of f to A is continuous.)

9.5 Let E be the middle-third Cantor set, let $s = \log 2 / \log 3$ and let $\mu = \mathcal{H}^s|_E$. Let $f: E \to E$ be given by $f(x) = 3x \pmod 1$. Show that $\mathrm{Tan}(\mu, x) = \mathrm{Tan}(\mu, f(x))$ for all $x \in E$. (This observation leads to various ergodic properties of $\mathrm{Tan}(\mu, x)$.)

9.6 Extend Proposition 9.3 to show that for μ-almost all x and every $\nu \in \mathrm{Tan}(\mu, x)$ we have that $F_{z,r}[\nu] \in \mathrm{Tan}(\mu, x)$ for all $z \in \mathrm{spt}\, \nu$ and $r > 0$.

9.7 Tangent measures may be used to study the angular distribution of sets. Let E be an s-set in $\mathbb{R}^n$ with $0 < s < n$. Let $\eta > 0$ and for each $x \in \mathbb{R}^n$ and unit vector θ define the cone $C(x, \theta, \eta) = \{y \in \mathbb{R}^n : (y - x) \cdot \theta > \eta |y - x|\}$. Show that for $\mathcal{H}^s$-almost all $x \in E$ there exists a unit vector θ such that

$$\liminf_{r \to 0} r^{-s} \mathcal{H}^s(E \cap B(x, r) \cap C(x, \theta, \eta)) = 0.$$

(Hint: As in the proof of Proposition 9.6 there must exist a tangent measure of μ contained in a half-space. This is not possible if the result stated here is false.)

9.8 Prove Proposition 9.15 for all $f \in L^1(\mu)$. (Use Lusin's theorem given in Exercise 9.4.)

9.9 Show that if $f: \mathbb{R} \to \mathbb{R}$ is a Lipschitz function of compact support then the Hilbert transform (9.34) exists at all $x \in \mathbb{R}$. (Hint: show that $\int_{|y-x| \geq \varepsilon} f(y)(y - x)^{-1} dy$ satisfies the Cauchy criterion for convergence as $\varepsilon \to 0$.)

Chapter 10   Dimensions of measures

Measures have been a fundamental tool in the study of the sets now termed 'fractals' ever since such irregular sets first attracted the attention of mathematicians early in the 20th century. We have already seen many instances where a set is analysed by studying properties of measures supported on the set. Often however, fractal structures are in essence already measures. For example, if the attractor of a dynamical system is displayed on a computer screen by plotting a sequence of iterates of a point, what is actually observed is a measure rather than an attracting set: the measure of a region is given by the proportion of the iterates lying in that region.

In the next two chapters we study measures as fractal entities in their own right, relating their properties to those of associated sets. We develop the idea of the dimension and local dimension of measures. We examine sets such as E_s at which the local dimension of a given measure μ is s; these sets may be 'large' for a range of s. We measure such sets by μ itself in this chapter, and by dimension in Chapter 11 leading to the multifractal spectrum.

10.1 Local dimensions and dimensions of measures

Much of the theory of Hausdorff and packing dimensions of sets depends on local properties of suitably defined measures. We now study such properties in their own right. Throughout this chapter μ will be a finite Borel regular measure on $\mathbb{R}^n$, so that in particular $0 < \mu(\mathbb{R}^n) < \infty$.

Recall from (2.15) and (2.16) that the *lower* and *upper local* or *pointwise dimensions* of μ at $x \in \mathbb{R}^n$ are given by

$$\underline{\dim}_{\mathrm{loc}} \mu(x) = \liminf_{r \to 0} \frac{\log \mu(B(x,r))}{\log r} \tag{10.1}$$

and

$$\overline{\dim}_{\mathrm{loc}} \mu(x) = \limsup_{r \to 0} \frac{\log \mu(B(x,r))}{\log r}, \tag{10.2}$$

and we say that the *local dimension* exists at x if these are equal, writing $\dim_{\mathrm{loc}} \mu(x)$ for the common value. Thus, the local dimensions describe the power law behaviour of $\mu(B(x,r))$ for small r, with $\dim_{\mathrm{loc}} \mu(x)$ small if μ is 'highly concentrated' near x. Note that $\dim_{\mathrm{loc}} \mu(x) = \infty$ if x is outside the support of μ and $\dim_{\text{loc}}\mu(x) = 0$ if x is an atom of μ. For technical completeness, we remark that routine methods show that the mapping $x \mapsto \underline{\dim}_{\text{loc}}\mu(x)$ is Borel measurable, so that sets such as $\{x : \underline{\dim}_{\text{loc}}\mu(x) < c\}$ are Borel sets for all c, with similar properties for upper local dimensions.

Writing $\mu|_E$ for the restriction of μ to the Borel set E (so that $\mu|_E(A) = \mu(E \cap A)$), we note that for μ-almost all $x \in E$

$$\underline{\dim}_{\text{loc}}\mu|_E(x) = \underline{\dim}_{\text{loc}}\mu(x) \quad \text{and} \quad \overline{\dim}_{\text{loc}}\mu|_E(x) = \overline{\dim}_{\text{loc}}\mu(x). \quad (10.3)$$

This follows easily from Proposition 1.7, that almost all points of a set are density points, see Exercise 10.2.

The fundamental relationships between Hausdorff and packing measures of sets and local properties of measures supported on the sets were stated in Propositions 2.3 and 2.4. We rewrite these relationships to give explicit expressions for the dimensions of sets.

Proposition 10.1

Let $E \subset \mathbb{R}^n$ be a non-empty Borel set. Then

$$\dim_{\text{H}} E = \sup\{s : \text{there exists } \mu \text{ with } 0 < \mu(E) < \infty$$

$$\text{and } \underline{\dim}_{\text{loc}}\mu(x) \geq s \text{ for } \mu\text{-almost all } x \in E\} \quad (10.4)$$

$$= \inf\{s : \text{ there exists } \mu \text{ with } 0 < \mu(\bar{E}) < \infty$$

$$\text{and } \underline{\dim}_{\text{loc}}\mu(x) \leq s \text{ for all } x \in E\} \quad (10.5)$$

and

$$\dim_{\text{P}} E = \sup\{s : \text{there exists } \mu \text{ with } 0 < \mu(E) < \infty$$

$$\text{and } \overline{\dim}_{\text{loc}}\mu(x) \geq s \text{ for } \mu\text{-almost all } x \in E\} \quad (10.6)$$

$$= \inf\{s : \text{ there exists } \mu \text{ with } 0 < \mu(\bar{E}) < \infty$$

$$\text{and } \overline{\dim}_{\text{loc}}\mu(x) \leq s \text{ for all } x \in E\}. \quad (10.7)$$

Proof These expressions follow directly from Propositions 2.3 and 2.4. □

Note that it is the lower local dimension that relates to the Hausdorff measure of a set and the upper local dimension to the packing measure. Moreover, lower bounds for the local dimensions of measures lead to lower bounds for the dimensions of the sets, and similarily for upper bounds.

These relationships suggest that it might be appropriate to use local dimensions to assign dimensions to the measures themselves. Thus we define the *Hausdorff* and *packing dimensions* of a finite Borel measure μ by

$$\dim_{\text{H}} \mu = \sup\{s : \underline{\dim}_{\text{loc}}\mu(x) \geq s \text{ for } \mu\text{-almost all } x\} \quad (10.8)$$

and

$$\dim_P \mu = \sup\{s : \overline{\dim}_{\text{loc}} \mu(x) \geq s \text{ for } \mu\text{-almost all } x\}. \tag{10.9}$$

It is hardly surprising that these dimensions of measures may be expressed in terms of the dimensions of associated sets.

Proposition 10.2

For a finite Borel measure μ

$$\dim_H \mu = \inf\{\dim_H E : E \text{ is a Borel set with } \mu(E) > 0\} \tag{10.10}$$

and

$$\dim_P \mu = \inf\{\dim_P E : E \text{ is a Borel set with } \mu(E) > 0\}. \tag{10.11}$$

Proof We apply Proposition 10.1 to definitions (10.8) and (10.9). First take $s < \dim_H \mu$, so that $\underline{\dim}_{\text{loc}} \mu(x) \geq s$ for all $x \in E_0$ for some Borel set E_0 with $\mu(\mathbb{R}^n \backslash E_0) = 0$. Given a Borel set E with $\mu(E) > 0$, it follows that $\underline{\dim}_{\text{loc}} \mu(x) \geq s$ for all $x \in E \cap E_0$ where $\mu(E \cap E_0) = \mu(E) > 0$, so by Proposition 2.3(a) or (10.4) $s \leq \dim_H(E \cap E_0) \leq \dim_H E$ for all such E, so $\dim_H \mu$ is no greater than the right-hand side of (10.10).

For the opposite inequality, let $s > \dim_H \mu$, so by (10.8) $\underline{\dim}_{\text{loc}} \mu(x) < s$ for all x in some Borel set E with $\mu(E) > 0$. By Proposition 2.3(b) or (10.5) $\dim_H E \leq s$, as required.

The proof of (10.11) is similar, using Proposition 2.3(c), (d) or (10.6) and (10.7). $\square$

It is occasionally useful to work with the *upper* Hausdorff and packing dimensions of measures defined by

$$\dim_H^* \mu = \inf\{s : \underline{\dim}_{\text{loc}} \mu(x) \leq s \text{ for } \mu\text{-almost all } x\} \tag{10.12}$$

and

$$\dim_P^* \mu = \inf\{s : \overline{\dim}_{\text{loc}} \mu(x) \leq s \text{ for } \mu\text{-almost all } x\}. \tag{10.13}$$

Clearly

$$\dim_H \mu \leq \dim_H^* \mu \quad \text{and} \quad \dim_P \mu \leq \dim_H^* \mu.$$

These dimensions may also be expressed in terms of dimensions of sets.

Proposition 10.3

For a finite Borel measure μ

$$\dim_H^* \mu = \inf\{\dim_H E : E \text{ is a Borel set with } \mu(\mathbb{R}^n \backslash E) = 0\} \tag{10.14}$$

and

$$\dim_{\mathrm{P}}^{*} \mu = \inf\{\dim_{\mathrm{P}} E : E \text{ is a Borel set with } \mu(\mathbb{R}^{n}\backslash E) = 0\}. \qquad (10.15)$$

Proof This is rather similar to the proof of Proposition 10.2, see Exercise 10.4. □

To illustrate these notions we calculate the dimensions and local dimensions of a family of self-similar measures on $[0,1]$. For $0 \le p \le \frac{1}{2}$ we define a self-similar probability measure μ_p on $[0,1]$ by repeated subdivision of measure between binary intervals in the ratio $p : (1-p)$, see Figure 10.1. Writing $X_{i_1,\ldots,i_k}$ for the closed interval of numbers with binary expansion beginning $0 \cdot i_1 \ldots i_k$, where $i_j = 0$ or 1 for $j = 1,\ldots,k$ we have

$$\mu_p(X_{i_1,\ldots,i_k}) = p^{n_0}(1-p)^{n_1} \qquad (10.16)$$

where n_0 and n_1 denote the number of times that the integers 0 and 1, respectively, occur in the sequence $(i_1,\ldots,i_k)$. Thus the self-similar measure μ_p is that defined by (2.44)–(2.45) where $F_1(x) = \frac{1}{2}x$, $F_2(x) = \frac{1}{2}(x+1)$, $p_1 = p$ and $p_2 = 1 - p$.

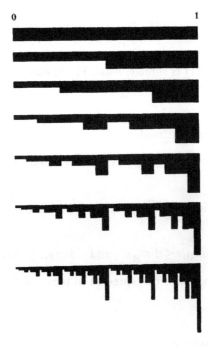

Figure 10.1 Construction of the self-similar measure of Proposition 10.4 by repeated subdivision of measure in the ratio $\frac{1}{3} : \frac{2}{3}$, that is with $p = \frac{1}{3}$

The following argument to calculate the dimensions of the measure μ_p is similar to that in FG, Proposition 10.1. The final conclusion of this proposition will be needed in Section 10.2.

Proposition 10.4

For $0 \le p \le \frac{1}{2}$ let μ_p be the probability measure defined above and write

$$s(p) = -(p \log p + (1-p) \log (1-p))/\log 2. \tag{10.17}$$

Then

$$\dim_H \mu_p = \dim_H^* \mu_p = s(p) = \dim_P \mu_p = \dim_P^* \mu_p \tag{10.18}$$

and

$$\underline{\dim}_{\text{loc}} \mu_p(x) = s(p) = \overline{\dim}_{\text{loc}} \mu_p(x) \tag{10.19}$$

for μ-almost all x. Moreover, there is a family of Borel sets F_p that increases with p such that $\dim_H F_p = s(p)$ and $\mu_p(F_p) = 1$.

Proof   First we note that μ_0 is the unit point mass located at 1 and $s(0) = 0$, so setting $F_0 = \{1\}$ gives the result when $p = 0$.

Now fix $0 < p \le \frac{1}{2}$. We may think of μ_p as the probability measure on $[0, 1]$ that gives rise to a random number x such that the k-th binary digit of x equals 0 with probability p and equals 1 with probability $1 - p$, independently for all digits. For $i = 0, 1$ we let $n_i(x|_k)$ denote the number of occurences of the digit i in the first k digits of the binary expansion of $x \in [0, 1]$. The strong law of large numbers tells us that 'with probability 1', that is for μ_p-almost all x, we have that

$$n_0(x|_k)/k \to p \quad \text{and} \quad n_1(x|_k)/k \to (1-p)$$

as $k \to \infty$. Defining Borel sets by $K_0 = F_0 = 1$ and

$$K_p = \{x \in [0, 1] : \lim_{k \to \infty} n_0(x|_k)/k = p\}, \tag{10.20}$$

for $0 < p \le \frac{1}{2}$ it follows that $\mu_p(K_p) = 1$.

For each $x \in [0, 1]$ write $X_k(x)$ for the binary interval $X_{i_1,\dots,i_k}$ of length 2^{-k} that contains x. From (10.16)

$$\log \mu_p(X_k(x)) = n_0(x|_k) \log p + n_1(x|_k) \log (1-p).$$

If $x \in K_p$ then for $t \ge 0$

$$\frac{1}{k} \log \frac{\mu_p(X_k(x))}{2^{-kt}} = \frac{1}{k} n_0(x|_k) \log p + \frac{1}{k} n_1(x|_k) \log(1-p) + t \log 2 \tag{10.21}$$

$$\to p \log p + (1-p) \log (1-p) + t \log 2$$

as $k \to \infty$. Thus if $t < s(p)$ (see (10.17)) then $\lim_{k \to \infty} \mu_p(X_k(x))/|X_k(x)|^t = 0$, so by a straightforward variant of Proposition 2.3(a) with the balls centred at x

replaced by the binary intervals containing x, we get that $\dim_H E \geq s(p)$ for every Borel set E with $\mu_p(E) > 0$, so in particular $\dim_H K_p \geq s(p)$, and also $\dim_H \mu_p \geq s(p)$ by (10.10).

On the other hand, if $0 < q \leq p$ and $x \in K_q$ then (10.21) gives

$$\frac{1}{k} \log \frac{\mu_p(X_k(x))}{2^{-kt}} \to q \log p + (1-q) \log(1-p) + t \log 2$$
$$\geq p \log p + (1-p) \log(1-p) + t \log 2,$$

since $q(\log p - \log(1-p)) = q \log(p/(1-p))$ decreases as q increases if $p \leq \frac{1}{2}$. Thus if $t > s(p)$ then $\lim \mu_p(X_k(x))/|X_k(x)|^t = \infty$ for all $x \in F_p \equiv \cup_{0 \leq q \leq p} K_q$. Again, the analogue of Proposition 2.3(b) for binary intervals gives that $\dim_H F_p \leq s(p)$ and equality follows since $s(p) \leq \dim_H K_p \leq \dim_H F_p$ from above. Noting that $\mu_p(F_p) \geq \mu_p(K_p) = 1$ it follows from (10.14) that $\dim_H^* \mu_p \leq s(p)$, so recalling that $s(p) \leq \dim_H \mu_p \leq \dim_H^* \mu_p$ we get the first two equalities of (10.18). That $\underline{\dim}_{loc} \mu(x) = s(p)$ for μ-almost all x is now immediate using (10.8) and (10.12).

The right-hand two equalities of (10.18) may be obtained by a parallel argument (which we leave to the reader) using the parts of Proposition 2.3 pertaining to packing measures, and the corresponding definitions and properties of $\dim_P \mu_p$ and $\dim_P^* \mu_p$. It then follows from (10.9) and (10.13) that $\overline{\dim}_{loc} \mu(x) = s(p)$ for μ-almost all x. $\square$

In the above examples, $\underline{\dim}_{loc} \mu(x)$ and $\overline{\dim}_{loc} \mu(x)$ are both constant for μ-almost all x. Measures with this property occur frequently in practice and are termed 'exact-dimensional' or 'unidimensional': we say that a measure μ has *exact lower dimension* s if $\underline{\dim}_{loc} \mu(x) = s$ for μ-almost all x, and *exact upper dimension* s if $\overline{\dim}_{loc} \mu(x) = s$ for μ-almost all x. Clearly from (10.8), (10.12), (10.9) and (10.13) μ has exact lower dimension s if and only if

$$\dim_H \mu = \dim_H^* \mu = s$$

and μ has exact upper dimension s if and only if

$$\dim_P \mu = \dim_P^* \mu = s.$$

As might be expected, exactness may be expressed in terms of dimensions of sets.

Proposition 10.5

The measure μ is of exact lower dimension s if and only if there exists a Borel set E_0 with $\mu(\mathbb{R}^n \setminus E_0) = 0$ and $\dim_H E = s$ for every $E \subset E_0$ with $\mu(E) > 0$. (We may take $E_0 = \{x : \underline{\dim}_{loc} \mu(x) = s\}$). Similarly, μ is of exact upper dimension s if and only if there exists a Borel set E_0 with $\mu(\mathbb{R}^n \setminus E_0) = 0$ and $\dim_P E = s$ for every $E \subset E_0$ with $\mu(E) > 0$.

Proof If μ is of exact lower dimension s then, by definition, there is a Borel set E_0 with $\mu(\mathbb{R}^n \setminus E_0) = 0$ such that $\underline{\dim}_{loc}\mu(x) = s$ for all $x \in E_0$. Thus if $E \subset E_0$ and $\mu(E) > 0$ then $\dim_H E = s$ by (10.4) and (10.5).

Conversely, if there exists E_0 with $\mu(\mathbb{R}^n \setminus E_0) = 0$ and $\dim_H E = s$ for all $E \subset E_0$ with $\mu(E) > 0$, then $\dim_H \mu = s$ by (10.10), so $\underline{\dim}_{loc}\mu(x) \geq s$ for μ-almost all x by (10.8). On the other hand, $\dim_H^* \mu \leq s$ by (10.14) so $\underline{\dim}_{loc}\mu(x) \leq s$ for μ-almost all x by (10.12), and thus μ is of exact lower dimension s.

The proof for exact upper dimension is parallel, using the corresponding properties of upper local dimension and packing dimension.   □

In the example analysed in Proposition 10.4 the measures μ_p are of exact lower and upper dimensions $s(p)$. It may be shown in a similar way that the more general self-similar measures defined by (2.43)–(2.45) with the strong separation condition holding are exact-dimensional.

Another method for demonstrating exact dimensionality, which also may be applied to self-similar measures, uses ergodicity. Recall from Section 6.1 that a measure μ is *invariant* under $f : X \rightarrow X$ if whenever $A \subset X$ is μ-measurable then so is $f^{-1}(A)$ with $\mu(f^{-1}(A)) = \mu(A)$, and that μ is *ergodic* if whenever $f^{-1}(A) = A$ for a measurable set A then either $\mu(A) = 0$ or $\mu(X \setminus A) = 0$.

Proposition 10.6

Let X be a closed subset of $\mathbb{R}^n$, let $f : X \rightarrow X$ be a Lipschitz function, and let μ be a finite measure on X that is invariant and ergodic under f. Then μ has exact lower dimension and exact upper dimension.

Proof The proof utilises the ergodic theorem. Let a be the Lipschitz constant of f so that $|f(z) - f(w)| \leq a|z - w|$ for all $z, w \in X$. Thus for $x, y \in X$ and $r > 0$ we have that $|f(x) - f^{j+1}(y)| \leq a|x - f^j(y)|$ (where f^j is the j-th iterate of f) so if $f^j(y) \in B(x, r)$ then $f^{j+1}(y) \in B(f(x), ar)$. In terms of indicator functions this is

$$1_{B(x,r)}(f^j(y)) \leq 1_{B(f(x),ar)}(f^{j+1}(y)). \tag{10.22}$$

Applying the ergodic theorem, Theorem 6.1, to the indicator functions $1_{B(x,r)}$ and $1_{B(f(x),ar)}$ (for fixed x and r) it follows from (6.4) and (6.5) that for μ-almost all y

$$\mu(B(x,r)) = \int 1_{B(x,r)} d\mu = \mu(x) \lim_{k \to \infty} \frac{1}{k} \sum_{j=0}^{k-1} 1_{B(x,r)}(f^j(y))$$

and

$$\mu(B(f(x), ar)) = \int 1_{B(f(x),ar)} d\mu = \mu(x) \lim_{k \to \infty} \frac{1}{k} \sum_{j=0}^{k-1} 1_{B(f(x),ar)}(f^j f(y))$$

(in the second instance we have applied the ergodic theorem at $f(y)$, noting that 'almost all $f(y)$' corresponds to 'almost all y' using the invariance of μ). By (10.22) it follows that $\mu(B(x,r)) \leq \mu(B(f(x),ar))$ for all $x \in X$ and $r > 0$. Then

$$\underline{\dim}_{\text{loc}} \mu(f(x)) = \liminf_{r \to 0} \log \mu(B(f(x),r))/\log r$$

$$= \liminf_{r \to 0} \log \mu(B(f(x),ar))/\log ar$$

$$\leq \liminf_{r \to 0} \log \mu(B(x,r))/(\log r + \log a)$$

$$= \underline{\dim}_{\text{loc}} \mu(x).$$

Since μ is invariant under f,

$$\int \underline{\dim}_{\text{loc}} \mu(f(x)) \, d\mu(x) = \int \underline{\dim}_{\text{loc}} \mu(x) d\mu(x),$$

so we conclude that $\underline{\dim}_{\text{loc}} \mu(f(x)) = \underline{\dim}_{\text{loc}} \mu(x)$ for μ-almost all x, that is μ has exact lower dimension. Exactness of the upper dimension follows in the same way, using upper limits in the final sequence of inequalities. $\square$

We saw in Lemma 6.4(b) that the self-similar measures defined by (2.43)–(2.45) and satisfying the strong separation condition are invariant and ergodic under the mapping $f: E \to E$ defined by F_i^{-1} on E_i, so by Proposition 10.6 they have exact lower and upper dimensions. Similarly, many invariant ergodic measures may be defined on cookie-cutter sets, and such measures are therefore exact-dimensional. For another ergodicity criterion for exact dimensionality, see Exercise 10.6.

Most approaches to calculating the dimensions of sets involve estimating the dimension of a measure concentrated on the set. However, finding such measures with dimensions equal to or close to that of the set under consideration can be a major problem.

One situation in which fractal properties of measures rather than sets are important is in dynamical systems. In a computer experiment to display the attractor of a dynamical system $x \mapsto f(x)$, one might hope to represent the attractor by plotting a large number of iterates $x, f(x), f^2(x), \ldots, f^k(x)$ of an initial point x. Essentially, what is displayed on the computer screen is the *residence measure* μ defined by $\mu(A) = \lim_{k \to \infty} \frac{1}{k} \#\{j \leq k : f^j x \in A\}$ (assuming that this converges in some sense), which is supported by the attractor of the system. It is this measure that is observed rather than the attracting set itself, parts of which may be very sparsely occupied by the iterates. Methods of finding the dimension of an attracting set often lead to estimates of the dimension of this measure rather than of the set. Certainly, many results that relate the dimension of an attractor to other parameters of a dynamical system concern the dimension of an attracting measure rather than of the set itself.

10.2 Dimension decomposition of measures

If a measure does not happen to be exact-dimensional, it is natural to try to decompose it into measures which are of exact dimension s for a range of s. In this section we describe one approach to this problem. We give the details for lower local dimensions and Hausdorff dimension; there is a parallel theory for upper local dimensions and packing dimension. As before, μ is a finite Borel regular measure on $\mathbb{R}^n$.

For $s \geq 0$ we consider the sets on which the (lower) local dimension is at most s:

$$\underline{E}_{\leq s} = \{x : \underline{\dim}_{\mathrm{loc}} \mu(x) \leq s\}. \tag{10.23}$$

Then $\underline{E}_{\leq s}$ is a Borel set and by Proposition 2.3(b)

$$\dim_H \underline{E}_{\leq s} \leq s. \tag{10.24}$$

Clearly, $\underline{E}_{\leq s}$ increases with s and we have the upper continuity property

$$\underline{E}_{\leq s} = \bigcap_{t > s} \underline{E}_{\leq t}. \tag{10.25}$$

We write $\underline{E}_s$ for the set of points at which the (lower) local dimension is exactly s, so that

$$\underline{E}_s \equiv \{x : \underline{\dim}_{\mathrm{loc}} \mu(x) = s\} \tag{10.26}$$

$$= \underline{E}_{\leq s} \setminus \bigcup_{t < s} \underline{E}_{\leq t}. \tag{10.27}$$

To obtain the desired decomposition of μ we first study the measures μ_s obtained by restricting μ to the sets $\underline{E}_{\leq s}$. Thus $\mu_s = \mu|_{\underline{E}_{\leq s}}$ is defined by

$$\mu_s(A) = \mu(A \cap \underline{E}_{\leq s}). \tag{10.28}$$

for all sets A. Clearly

$$\mu_s(\mathbb{R}^n \setminus \underline{E}_{\leq s}) = 0$$

and from (10.25) and the continuity of measures (1.13),

$$\lim_{t \searrow s} \mu_t(A) = \lim_{t \searrow s} \mu(A \cap \underline{E}_{\leq t}) = \mu(A \cap \underline{E}_{\leq s}) = \mu_s(A). \tag{10.29}$$

We may express μ_s in terms of dimensions of sets.

Proposition 10.7

For $0 \leq s \leq n$ and all Borel sets A

$$\mu_s(A) = \sup\{\mu(A \cap E) : E \text{ is a Borel set with } \dim_H E \leq s\}.$$

Proof   Using (10.24)

$$\mu_s(A) = \mu(A \cap \underline{E}_{\leq s}) \leq \sup\{\mu(A \cap E) : \dim_H E \leq s\}.$$

On the other hand, for μ-almost all $x \notin \underline{E}_{\leq t}$, that is for $\mu|_{(\mathbb{R}^n \setminus \underline{E}_{\leq t})}$-almost all x, we have $\underline{\dim}_{\text{loc}} \mu|_{(\mathbb{R}^n \setminus \underline{E}_{\leq t})}(x) = \underline{\dim}_{\text{loc}} \mu(x) > t$, using (10.3) and (10.23). Thus, by (10.8) $\dim_H \mu|_{(\mathbb{R}^n \setminus \underline{E}_{\leq t})} \geq t$, so by (10.10) if E is any Borel set with $\dim_H E < t$ then $0 = \mu|_{(\mathbb{R}^n \setminus \underline{E}_{\leq t})}(E) = \mu(E \setminus \underline{E}_{\leq t})$. Hence if A is a Borel set, $\mu(A \cap E) = \mu(A \cap E \cap \underline{E}_{\leq t}) \leq \mu_t(A)$, so

$$\sup\{\mu(A \cap E) : \dim_H E < t\} \leq \mu_t(A).$$

Thus for all $t > s$

$$\sup\{\mu(A \cap E) : \dim_H E \leq s\} \leq \mu_t(A),$$

and using (10.29) completes the proof. $\square$

Taking $s = n$ in Proposition 10.7 gives

$$\mu_n(A) = \mu(A) \tag{10.30}$$

for all Borel sets A.

Next we define a finite Borel measure $\hat{\mu}$ on the real interval $[0, n]$ by setting

$$\hat{\mu}([0, s]) = \mu_s(\mathbb{R}^n) \tag{10.31}$$

and extending this to subsets of $[0, n]$ in the usual way. From (10.28)

$$\hat{\mu}([0, s]) = \mu(\underline{E}_{\leq s}). \tag{10.32}$$

The measure $\hat{\mu}$ is sometimes called the *dimension measure* of μ, since $\hat{\mu}(B)$ records the μ-measure of the set of points with (lower) local dimension in the set of real numbers B.

The following 'dimension disintegration formula' expresses the measures μ_s as integrals of certain measures ν_t with respect to $\hat{\mu}$; the measures ν_t are termed the *dimension derivative family* of μ.

Proposition 10.8

There exist real numbers $\nu_t(A)$ defined for each Borel subset A of $\mathbb{R}^n$ and for all $0 \leq t \leq n$, with $0 \leq \nu_t(A) \leq 1$, and such that

(a) ν_t is a Borel probability measure on $\mathbb{R}^n$ for $0 \leq t \leq n$, and
(b) for all $0 \leq s \leq n$ and all Borel sets A

$$\mu_s(A) = \int_{[0,s]} \nu_t(A) d\hat{\mu}(t). \tag{10.33}$$

Proof For each Borel set A we may, analogously to (10.31), define a Borel measure $\hat{\mu}_A$ on $[0,n]$ by $\hat{\mu}_A([0,s]) = \mu_s(A) = \mu(A \cap \underline{E}_{\leq s})$, and extending this to subsets of $[0,n]$. For $0 \leq s < t$ we have

$$0 \leq \hat{\mu}_A(s,t] = \mu_t(A) - \mu_s(A) = \mu(A \cap (\underline{E}_{\leq t} \setminus \underline{E}_{\leq s})) \leq \mu(\underline{E}_{\leq t} \setminus \underline{E}_{\leq s})$$
$$= \mu_t(\mathbb{R}^n) - \mu_s(\mathbb{R}^n) = \hat{\mu}(s,t]$$

so that

$$0 \leq \hat{\mu}_A(B) \leq \hat{\mu}(B) \tag{10.34}$$

for all $B \subset [0,n]$. Thus $\hat{\mu}_A$ is absolutely continuous with respect to $\hat{\mu}$ so that we can represent $\hat{\mu}_A$ in the form

$$\hat{\mu}_A(B) = \int_B \nu_t(A) \mathrm{d}\hat{\mu}(t)$$

where $\nu_t(A)$ is the Radon–Nikodym derivative $\nu_t(A) = \mathrm{d}\hat{\mu}_A(t)/\mathrm{d}\hat{\mu}$. In particular, taking $B = [0,s]$ and using $\hat{\mu}_A([0,s]) = \mu_s(A)$, (10.33) holds for $0 \leq s \leq n$.

There is a technical difficulty in showing that ν_t is a measure on $\mathbb{R}^n$ since $\mathrm{d}\hat{\mu}_A(t)/\mathrm{d}\hat{\mu}$ is defined only to within $\hat{\mu}$-almost all t. We may cope with this as follows. Let

$$\mathcal{A} = \{[k_1 2^{-m}, (k_1 + 1)2^{-m}) \times \cdots \times [k_n 2^{-m}, (k_n + 1)2^{-m})$$
$$: m \in \mathbb{Z}^+, \, k_1, \ldots, k_n \in \mathbb{Z}\}$$

be the collection of binary cubes in $\mathbb{R}^n$. For each $A \in \mathcal{A}$ choose a representation $\nu_t(A) = \mathrm{d}\hat{\mu}_A(t)/\mathrm{d}\hat{\mu}$ of the Radon–Nikodym derivative; by virtue of (10.34) we may assume that $0 \leq \nu_t(A) \leq 1$ for all $A \in \mathcal{A}$ and $t \in [0,n]$. If $A_1, \ldots, A_k$ are disjoint sets with $A = \cup_{i=1}^k A_i$, then

$$\mathrm{d}\hat{\mu}_A(t)/\mathrm{d}\hat{\mu} = \mathrm{d}\hat{\mu}_{A_1}(t)/\mathrm{d}\hat{\mu} + \cdots + \mathrm{d}\hat{\mu}_{A_k}(t)/\mathrm{d}\hat{\mu} \tag{10.35}$$

or

$$\nu_t(A) = \nu_t(A_1) + \cdots + \nu_t(A_k) \tag{10.36}$$

for $\hat{\mu}$-almost all t. Since $\mathcal{A}$ contains countably many binary cubes, each a disjoint union of 2^n binary cubes of half of their side-length, there is a set $W \subset [0,n]$ with $\hat{\mu}([0,n] \setminus W) = 0$ such that for all $t \in W$ (10.36) holds whenever A is a binary cube and $A_1, \ldots, A_k$ are the 2^n binary sub-cubes of half the side-length. Thus for all $t \in W$, we may extend $\nu_t(A)$ in a consistent additive manner to the family of sets A that are finite unions of binary cubes, with (10.33) remaining valid for these sets. We may continue this extension process in the usual way so that for all $t \in W$ the measure ν_t is countably additive on the Borel sets, with $0 \leq \nu_t(\mathbb{R}^n) \leq 1$ and with (10.33) extending to the Borel sets. From (10.30) $\mu(\mathbb{R}^n) = \mu_n(\mathbb{R}^n) = \int_{[0,n]} \nu_t(\mathbb{R}^n) \mathrm{d}\hat{\mu}(t)$, so $\nu_t(\mathbb{R}^n) = 1$ for $\hat{\mu}$-almost all t, since $\hat{\mu}([0,n]) = \mu(\mathbb{R}^n)$. Thus for $\hat{\mu}$-almost all $t \in [0,n]$ the measure ν_t is a probability measure. By redefining ν_t on a set of t of $\hat{\mu}$-measure zero we may take ν_t to be a probability measure for all $t \in [0, n]$, with (10.33) remaining valid. ☐

The next proposition summarises the principal properties of the measures ν_t.

Proposition 10.9

Let ν_t satisfy the conclusions of Proposition 10.8. *Then*

(a) *for every Borel set A we have $\nu_t(A) = 0$ for $\hat{\mu}$-almost all $t \in (\dim_H A, n]$,*
(b) *for all s we have $\nu_t(\underline{E}_{\leq s}) = 1$ for $\hat{\mu}$-almost all $t \in [0, s]$,*
(c) *for $\hat{\mu}$-almost all t, we have $\nu_t(\underline{E}_t) = 1$ and $\nu_t(\mathbb{R}^n \backslash \underline{E}_t) = 0$.*

Proof

(a) Let A be a Borel set and write $s = \dim_H A$. From (10.33)

$$\int_{(s,n]} \nu_t(A) d\hat{\mu}(t) = \mu_n(A) - \mu_s(A)$$

$$\leq \mu(A) - \mu(A \cap A) = 0,$$

using (10.30) and Proposition 10.7. Thus $\nu_t(A) = 0$ for $\hat{\mu}$-almost all $t \in (s, n]$.

(b) Using (10.33), (10.28) and (10.31)

$$\int_{[0,s]} \nu_t(\underline{E}_{\leq s}) d\hat{\mu}(t) = \mu_s(\underline{E}_{\leq s}) = \mu_s(\mathbb{R}^n) = \hat{\mu}([0,s]) = \int_{[0,s]} 1 d\hat{\mu}(t).$$

Since $0 \leq \nu_t(\underline{E}_{\leq s}) \leq 1$ for all t, this implies that $\nu_t(\underline{E}_{\leq s}) = 1$ for $\hat{\mu}$-almost all $t \in [0, s]$.

(c) We note that

$$\{t \in [0, n] : \nu_t(\underline{E}_{\leq q}) < 1 \text{ for some rational } q > t\}$$
$$= \bigcup_{q \in \mathbb{Q}} \{t : \nu_t(\underline{E}_{\leq q}) < 1 \text{ and } t < q\}. \qquad (10.37)$$

From (b) the sets in this union have $\hat{\mu}$-measure 0, so we conclude that for $\hat{\mu}$-almost all t we have $\nu_t(\underline{E}_{\leq q}) = 1$ for all rationals $q > t$. Using (10.25)

$$\nu_t(\underline{E}_{\leq t}) = \lim_{q \searrow t} \nu_t(\underline{E}_{\leq q}) = 1 \qquad (10.38)$$

for $\hat{\mu}$-almost all t.
In the same way,

$$\{t \in [0, n] : \nu_t(\underline{E}_{\leq q}) > 0 \text{ for some rational } q < t\}$$
$$= \bigcup_{q \in \mathbb{Q}} \{t : \nu_t(\underline{E}_{\leq q}) > 0 \text{ and } t > q\},$$

so by (a) this has $\hat{\mu}$-measure 0, since $\dim_H \underline{E}_{\leq q} \leq q$ by (10.24). Thus for $\hat{\mu}$-almost all t we have $\nu_t(\underline{E}_{\leq q}) = 0$ for all rational $q < t$, so

$$\nu_t(\underline{E}_t) = \nu_t(\underline{E}_{\leq t}) - \lim_{q \nearrow t} \nu_t(\underline{E}_{\leq q}) = 1 - 0$$

using (10.27) and (10.38).

Finally we note that

$$\nu_t(\mathbb{R}^n \setminus \underline{E}_t) = \nu_t(\mathbb{R}^n) - \nu_t(\underline{E}_t) = 1 - 1 = 0. \quad \square$$

Proposition 10.9(c) says that (10.33) is a decomposition of the measure μ into components ν_t concentrated on $\underline{E}_t$, the set on which μ has local dimension t. Ideally, one would like ν_t itself to have exact dimension t for $\hat{\mu}$-almost all t. However, since $\nu_t(\underline{E}_{\leq t}) = 1$, (10.10) implies that $\dim_H \nu_t \leq \dim_H \underline{E}_{\leq t}$. Should $\dim_H \underline{E}_{\leq t}$ be strictly less than t, as can happen, then $\dim_H \nu_t < t$, so by (10.8) $\underline{\dim}_{loc} \nu_t(x) < t$ for a set of x of positive ν_t-measure. It would be interesting to have precise conditions that ensure that ν_t is exact for certain t. We shall see that this is the case at least when $\mu(\underline{E}_t) > 0$.

We now obtain an alternative decomposition of μ into components of differing local dimensions, utilising the nature of the measure $\hat{\mu}$ on $[0, n]$. Recall that a number s is an *atom* of $\hat{\mu}$ if $\hat{\mu}(\{s\}) > 0$. By summing these point measures it is clear that the set S of atoms of a finite measure $\hat{\mu}$ is at most countable. The restriction of $\hat{\mu}$ to S is called the *atomic part* of $\hat{\mu}$ and the restriction of $\hat{\mu}$ to $[0, n] \setminus S$ is the *non-atomic part* of $\hat{\mu}$.

We shall see that a finite measure μ has exact-dimensional components corresponding to the atoms s of $\hat{\mu}$. On the other hand, the component μ^D of μ corresponding to the non-atomic part of $\hat{\mu}$ has *diffuse dimension distribution*, that is $\mu^D \{x : \underline{\dim}_{loc} \mu^D(x) = s\} = 0$ for all s.

From (10.27) and (10.33), for a Borel set A,

$$\mu(A \cap \underline{E}_s) = \mu(A \cap \underline{E}_{\leq s}) - \lim_{r \nearrow s} \mu(A \cap \underline{E}_{\leq r})$$

$$= \mu_s(A) - \lim_{r \nearrow s} \mu_r(A)$$

$$= \lim_{r \nearrow s} \int_{(r,s]} \nu_t(A) d\hat{\mu}(t).$$

Taking $A = \mathbb{R}^n$ gives $\mu(\underline{E}_s) = 0$, unless s is an atom of $\hat{\mu}$ in which case

$$\mu(A \cap \underline{E}_s) = \nu_s(A)\hat{\mu}(\{s\}). \tag{10.39}$$

Thus the atoms of $\hat{\mu}$ correspond to those s for which $\mu(\underline{E}_s) > 0$; this leads us to decompose μ into exact-dimensional components by restricting μ to $\underline{E}_s$ for such s.

Proposition 10.10

Let μ be a finite Borel measure on $\mathbb{R}^n$. There exists a finite or countable set $S \subset [0, n]$, a measure μ^s of exact dimension s for all $s \in S$, and a measure μ^D with diffuse dimension distribution, such that

$$\mu = \sum_{s \in S} \mu^s + \mu^D. \tag{10.40}$$

In fact we may take

$$S = \{s \in [0, n] : \mu(\underline{E}_s) > 0\} \tag{10.41}$$

and

$$\mu^s(A) = \mu(A \cap \underline{E}_s) \tag{10.42}$$

and

$$\mu^D(A) = \mu\left(A \setminus \bigcup_{s \in S} \underline{E}_s\right) \tag{10.43}$$

for $A \subset \mathbb{R}^n$, where (as before)

$$\underline{E}_s = \{x : \underline{\dim}_{loc}\mu(x) = s\}. \tag{10.44}$$

Proof The set S in (10.41) must be countable since $\mu(\mathbb{R}^n) < \infty$. For $s \in S$ define Borel measures μ^s by (10.42) and define μ^D by (10.43). By (10.3) and (10.44) we have that $\underline{\dim}_{loc}\mu^s(x) = \underline{\dim}_{loc}\mu(x) = s$ for μ-almost all $x \in \underline{E}_s$. Thus $\underline{\dim}_{loc}\mu^s(x) = s$ for μ^s-almost all x, so μ^s is of exact dimension s.

Using (10.42) and (10.43)

$$\mu(A) = \sum_{s \in S} \mu(A \cap \underline{E}_s) + \mu\left(A \setminus \bigcup_{s \in S} \underline{E}_s\right)$$
$$= \sum_{s \in S} \mu^s(A) + \mu^D(A)$$

for $A \subset \mathbb{R}^n$. To show that μ^D has a diffuse dimension distribution we again use (10.3) to note that $\underline{\dim}_{loc}\mu^D(x) = \underline{\dim}_{loc}\mu(x)$ for μ^D-almost all x. Thus for each $t \in [0, n]$

$$\mu^D(\{x : \underline{\dim}_{loc}\mu^D(x) = t\}) = \mu^D(\underline{E}_t) = \mu\left(\underline{E}_t \setminus \bigcup_{s \in S} \underline{E}_s\right) = 0,$$

since if $\mu(\underline{E}_t) > 0$ then t would be an atom of $\hat{\mu}$ so that $\underline{E}_t = \underline{E}_s$ for some $s \in S$. Thus μ^D has diffuse dimension distribution. $\square$

Measures with exact dimensions are readily found; Propositions 10.4 and 10.6 provide many examples. However, measures with diffuse dimension distribution occur less naturally. We show how such measures may be constructed; in fact we show that there is a measure μ on $\mathbb{R}$ such that the dimension measure $\hat{\mu}$ equals any given probability measure on $[0, 1]$. Essentially, we aggregate measures of the form introduced in Proposition 10.4 to obtain measures with the desired dimension measure.

Lemma 10.11

There exist Borel probability measures ν_s and sets F_s' for all $0 \leq s \leq 1$ such that the F_s' are increasing with s, and such that $\dim_{\mathrm{H}} F_s' = \dim_{\mathrm{H}} \nu_s = s$ and $\nu_s(F_s') = 1$ for all $0 \leq s \leq 1$.

Proof These sets and measures were essentially constructed in Proposition 10.4. We note that for $0 \leq s \leq 1$ we may find a unique $p = p(s)$ such that $0 \leq p \leq \frac{1}{2}$ and $s = -(p \log p + (1 - p) \log (1 - p)) / \log 2$. Setting $F_s' = F_{p(s)}$ and $\nu_s = \mu_{p(s)}$, where $F_{p(s)}$ and $\mu_{p(s)}$ are as in Proposition 10.4, gives sets and measures with the desired properties. $\square$

Proposition 10.12

Let η be a probability measure on $[0, 1]$. Then there exists a Borel measure μ on $\mathbb{R}$ such that $\hat{\mu} = \eta$.

Proof We let ν_s and F_s' be as in Lemma 10.11 and define a Borel measure by

$$\mu(A) = \int_{[0,1]} \nu_t(A) \mathrm{d}\eta(t). \tag{10.45}$$

Since $\nu_t(\mathbb{R}) = 1$ for all $t \in [0, 1]$ we have that $\mu(\mathbb{R}) = 1$. Since $\dim_{\mathrm{H}} \nu_t = t$ it follows from (10.10) that $\nu_t(E) = 0$ for every Borel set E with $\dim_{\mathrm{H}} E < t$. Thus for $0 \leq s \leq 1$, using (10.31), Proposition 10.7, (10.45) and Proposition 10.9(a),

$$\hat{\mu}([0, s]) = \mu_s(\mathbb{R})$$

$$= \sup \left\{ \int_{[0,1]} \nu_t(E) \mathrm{d}\eta(t) : \dim_{\mathrm{H}} E \leq s \right\}$$

$$= \sup \left\{ \int_{[0,s]} \nu_t(E) \mathrm{d}\eta(t) : \dim_{\mathrm{H}} E \leq s \right\}$$

$$= \eta([0, s])$$

since $\nu_t(E) \leq 1$ for all E, and $\nu_t(F_s') = 1$ for $t \leq s$ with $\dim_{\mathrm{H}} F_s' = s$. It follows that $\hat{\mu} = \eta$. $\square$

These arguments may be generalised to obtain a measure μ on $\mathbb{R}^n$ such that $\hat{\mu} = \eta$ for any given probability measure η on $[0, n]$.

10.3 Notes and references

Local dimensions of measures have been around in some form, though not by that name, almost since the birth of geometric measure theory. In particular, Frostman (1935) and Billingsley (1965) use local dimensions of measures to study dimensions of sets. The relationships between Hausdorff and packing measures and local dimensions are discussed in Tricot (1982), Cutler (1986), Haase (1992) and Hu and Taylor (1994). Ergodic criteria for exact dimensionality are given by Cutler (1990) and Fan (1995).

Rogers and Taylor (1959, 1962) obtained a decomposition theorem very like Proposition 10.10 and also constructed measures with diffuse dimension distribution. The approach followed in Section 10.2 is that of Cutler (1986, 1992). Kahane and Katznelson (1990) use Riesz potentials to obtain a similar decomposition.

Exercises

10.1 Let μ be a finite measure on $\mathbb{R}^n$ and let $f : \mathbb{R}^n \to \mathbb{R}^n$ be a Lipschitz mapping. Show that $\underline{\dim}_{loc}\nu(f(x)) \leq \underline{\dim}_{loc}\mu(x)$ for all $x \in \mathbb{R}^n$, where ν is defined by $\nu(A) = \mu(f^{-1}(A))$. Deduce that if f is a similarity or affine transformation then $\underline{\dim}_{loc}\nu(f(x)) = \underline{\dim}_{loc}\mu(x)$ for all $x \in \mathbb{R}^n$.

10.2 Verify (10.3) using the density property, Proposition 1.7.

10.3 Verify that $\{x : \underline{\dim}_{loc}\mu(x) < c\}$ is a Borel set, for μ a finite Borel measure on $\mathbb{R}^n$ and $c \in \mathbb{R}$.

10.4 Prove Proposition 10.3.

10.5 Let $0 < p < 1$, let E be the middle-third Cantor set and let μ be the self-similar measure supported by E obtained by repeated subdivision of the measure on each interval in the usual Cantor set construction in the ratio $p : 1 - p$ between the subintervals at the next stage. Show that μ has exact lower dimension $-(p \log p + (1 - p) \log (1 - p))/\log 3$.

10.6 (Another criterion for exactness.) Let X be a closed subset of $\mathbb{R}^n$ and let μ be an ergodic invariant measure for $f : X \to X$. Suppose that for all Borel sets E we have $\dim_H f^{-1}(E) \leq \dim_H E$. Show that μ is of exact lower dimension. (Hint: if $\mu(E) > 0$ use the definition of ergodicity to show that $\mu(\cup_{j=1}^\infty f^{-j}E) = 1$ and then use (10.10) and (10.14).)

10.7 Let μ_1 and μ_2 be finite measures on $\mathbb{R}^n$ with disjoint support. Show that the dimension measures satisfy $(\mu_1 + \mu_2)\hat{} = \hat{\mu}_1 + \hat{\mu}_2$.

10.8 Show directly that (to within a set of t of $\hat{\mu}$-measure) the measure μ defined by (10.45) has dimension derivative family ν_t.

10.9 With $\underline{E}_{\leq s}$ as in (10.23), show that $\dim_H^* \underline{E}_{\leq s} \leq s$.

Chapter 11   Some multifractal analysis

As we saw in the last chapter, a single measure μ of widely varying intensity may define a whole 'spectrum' of fractal sets, determined by those points at which the local dimension takes particular values. In the last chapter we were concerned with the μ-measure of such sets. Here we undertake a 'finer' analysis, by examining sets which, although they may have μ-measure zero, are nevertheless significant as sets of positive dimension. Multifractal analysis aims to quantify the singularity structure of measures and provide a model for phenomena in which scaling occurs with a range of different power laws.

Recall that for a finite measure μ on $\mathbb{R}^n$ the local dimension (or local Hölder exponent) of μ at x is given by

$$\dim_{\mathrm{loc}}\mu(x) = \lim_{r \to 0} \log \mu(B(x,r))/\log r, \tag{11.1}$$

if this limit exists. For each $\alpha \geq 0$ we consider the set E_α of points x at which $\dim_{\mathrm{loc}}\mu(x)$ exists and equals α. (In multifractal analysis the use of 'α' rather than 's' in this context is almost universal, and we adhere to this convention in this chapter.) For certain measures μ the sets E_α may be non-empty and fractal over a range of α, and when this happens μ is often termed a *multifractal measure*. It is natural to study the *multifractal spectrum* or *singularity spectrum* of μ defined by $f(\alpha) \equiv \dim E_\alpha$ (for some suitable definition of dimension). For example, self-similar measures (see Section 2.2) are generally multifractal measures; their spectra are analysed in detail in Section 11.2.

This idea of getting many fractals for the price of one measure is at first very attractive, but it is beset by technical difficulties when it comes to analysing mathematical properties and when attempting to calculate multifractal spectra in specific cases. For example, it is not always clear when it is more appropriate to work with upper or lower local dimensions, or which definition of dimension should be used in defining $f(\alpha)$. Care is needed in relating the behaviour of $\mu(B(x,r))$ in the limit as $r \to 0$ with its values at small but finite scales, and this leads to fine and coarse multifractal theories. Difficulties also arise from estimates involving $\mu(A)^q$ where q is negative and $\mu(A)$ is small.

Despite this, or perhaps because of it, an enormous amount has been done on the mathematics of multifractals, much of which parallels earlier work on fractal sets. The aim is to give satisfactory definitions and interpretations of multifractal dimension spectra, to study geometrical behaviour (for example by relating the spectra of a measure and its projections onto subspaces), to find general methods of calculating multifractal spectra, and to find the spectra of specific measures.

In this chapter we can do no more than touch on a few mathematical aspects of multifractals. Section 11.1 concerns general theory applicable to general measures and in particular discusses the fine and coarse approaches to multifractals. In Section 11.2 we calculate the spectra of self-similar measures, and in Section 11.3 we use the thermodynamic formalism to extend these calculations to non-linear analogues, that is to Gibbs measures on cookie-cutter sets.

Measures with multifractal features have been observed in many situations. Multifractals have been used to describe residence measures on the attractors of dynamical systems, turbulence in fluids, rainfall distribution, mass distribution in the universe, viscous fingering, neural networks and many other phenomena. Nevertheless, it is not always easy to relate these examples to the mathematical and computational theory.

11.1 Fine and coarse multifractal theories

There are two basic approaches to multifractal analysis: the *fine theory* where we examine the geometry of the sets E_α themselves, and the *coarse theory* where we consider the irregularities of distribution of $\mu(B(x,r))$ for small but positive r. Thus in the fine theory we look at the local limiting behaviour as $r \to 0$ of $\mu(B(x,r))$ and examine global properties of the sets thus defined, whereas in the coarse theory we quantify the global irregularities of $\mu(B(x,r))$ for small r and then take the limit as $r \to 0$. There are many parallels between the fine and the coarse approaches to multifractal analysis, for example both involve Legendre transformation, and both approaches lead to the same multifractal spectra for many basic measures.

The fine theory is perhaps more suited to mathematical analysis, requiring ideas close to those used in studying the Hausdorff dimension of sets. On the other hand, the coarse theory is more convenient when it comes to finding multifractal spectra of physical examples or estimating spectra from computer experiments, and this approach is more reminiscent of box-counting dimension calculations.

We discuss both the fine and coarse approaches and their relationship for general measures. It must be emphasised that many variations are possible in the definitions and conventions adopted, and that similar but different definitions may be encountered elsewhere in the literature.

Let μ be a finite Borel regular measure on $\mathbb{R}^n$. For $\alpha \geq 0$ define

$$E_\alpha = \{x \in \mathbb{R}^n : \dim{}_{\mathrm{loc}}\mu(x) = \alpha\}, \tag{11.2}$$
$$= \{x \in \mathbb{R}^n : \lim_{r \to 0} \log \mu(B(x,r))/\log r = \alpha\}; \tag{11.3}$$

thus E_α is the set of points at which the local dimension exists and equals α. The basic aim of the fine approach to multifractal analysis is to find $\dim E_\alpha$ for $\alpha \geq 0$. (Note that some variants of multifractal theory work with upper or lower local dimension, or replace '$= \alpha$' with '$\geq \alpha$' or '$\leq \alpha$' as in the last chapter.)

In most examples of interest E_α is dense in $\text{spt}\mu$ for values of α for which E_α is non-trivial. Then $\underline{\dim}_B E_\alpha = \underline{\dim}_B \bar{E}_\alpha = \underline{\dim}_B \text{spt}\mu$ (and similarly for upper box dimension), so box-counting dimensions are of little use in discriminating between the sizes of the E_α. Thus it is more natural to work with

$$f_H(\alpha) = \dim_H E_\alpha \quad \text{and} \quad f_P(\alpha) = \dim_P E_\alpha, \tag{11.4}$$

which we term respectively the *Hausdorff* and *packing (fine) multifractal spectra* of μ.

Clearly $0 \leq f_H(\alpha) \leq \dim_H \text{spt}\mu$ for all $\alpha \geq 0$, with similar inequalities for packing dimensions. By Proposition 2.3(*b*) we also have

$$0 \leq f_H(\alpha) \leq \alpha \tag{11.5}$$

for all α.

The definition of the coarse spectrum is along the lines of box-counting dimension. We work with the *r-mesh cubes* in $\mathbb{R}^n$, that is cubes of the form $[m_1 r, (m_1 + 1)r) \times \cdots \times [m_n r, (m_n + 1)r)$ where $m_1, \ldots, m_n$ are integers. For μ a finite measure on $\mathbb{R}^n$ and $\alpha \geq 0$ we write

$$N_r(\alpha) = \#\{r\text{-mesh cubes } A \text{ with } \mu(A) \geq r^\alpha\} \tag{11.6}$$

and define the *coarse multifractal spectrum* of μ as

$$f_C(\alpha) \equiv \lim_{\epsilon \to 0} \lim_{r \to 0} \frac{\log^+(N_r(\alpha + \epsilon) - N_r(\alpha - \epsilon))}{-\log r} \tag{11.7}$$

if the double limit exists. (We write $\log^+ x \equiv \max\{0, \log x\}$; this is merely a device to ensure $f_C(\alpha) \geq 0$.) Definition (11.7) implies that if $\eta > 0$, and $\epsilon > 0$ is small enough, then

$$r^{-f_C(\alpha)+\eta} \leq N_r(\alpha + \epsilon) - N_r(\alpha - \epsilon) \leq r^{-f_C(\alpha)-\eta} \tag{11.8}$$

for all sufficiently small r. Roughly speaking $-f_C(\alpha)$ is the power law exponent for the number of r-mesh cubes A such that $\mu(A) \simeq r^\alpha$. Note that $f_C(\alpha)$ is *not* the box dimension of the set of x such that $\mu(A_r(x)) \simeq r^\alpha$ as $r \to 0$ where $A_r(x)$ is the r-mesh cube containing x; the coarse spectrum provides a global overview of the fluctuations of μ at scale r but gives no information about the limiting behaviour of μ at any point.

In case the limit in (11.7) fails to exist, we define the *lower* and the *upper coarse multifractal spectra* of μ by

$$\underline{f}_C(\alpha) = \lim_{\epsilon \to 0} \liminf_{r \to 0} \frac{\log^+(N_r(\alpha + \epsilon) - N_r(\alpha - \epsilon))}{-\log r} \tag{11.9}$$

and

$$\bar{f}_C(\alpha) = \lim_{\epsilon \to 0} \limsup_{r \to 0} \frac{\log^+(N_r(\alpha + \epsilon) - N_r(\alpha - \epsilon))}{-\log r} \tag{11.10}$$

for $\alpha \geq 0$.

The following lemma gives the basic relationship between the fine and coarse spectra.

Lemma 11.1

Let μ be a finite measure on $\mathbb{R}^n$. Then

$$f_H(\alpha) \leq \underline{f}_C(\alpha) \leq \bar{f}_C(\alpha) \tag{11.11}$$

for all $\alpha \geq 0$.

Proof We need only prove the left-hand inequality in (11.11); the right-hand inequality is obvious. For simplicity we take μ to be a measure on $\mathbb{R}$; the proof is similar in higher-dimensional spaces except that measures of balls and cubes have to be compared instead of measures of intervals.

For fixed $\alpha \geq 0$ write for brevity $f \equiv f_H(\alpha) = \dim_H E_\alpha$; we may assume $f > 0$. Given $0 < \epsilon < f$ then $\mathcal{H}^{f-\epsilon}(E_\alpha) = \infty$. By (11.3) there is a set $E_\alpha^0 \subset E_\alpha$ with $\mathcal{H}^{f-\epsilon}(E_\alpha^0) > 1$ and a number $r_0 > 0$ such that

$$3r^{\alpha+\epsilon} \leq \mu(B(x,r)) < 2^{\epsilon-\alpha} r^{\alpha-\epsilon} \tag{11.12}$$

for all $x \in E_\alpha^0$ and all $0 < r \leq r_0$. We may choose δ with $0 < \delta \leq \frac{1}{2} r_0$ such that $\mathcal{H}_\delta^{f-\epsilon}(E_\alpha^0) \geq 1$.

For each $r \leq \delta$ we consider r-mesh intervals (of the form $[mr, (m+1)r)$ with $m \in \mathbb{Z}$) that intersect E_α^0. Such an interval A contains a point x of E_α^0, with

$$B(x,r) \subset A \cup A_L \cup A_R \subset B(x,2r)$$

where A_L and A_R are the r-mesh intervals immediately on either side of A. By (11.12)

$$3r^{\alpha+\epsilon} \leq \mu(B(x,r)) \leq \mu(A \cup A_L \cup A_R) \leq \mu(B(x,2r)) < r^{\alpha-\epsilon}$$

so that

$$r^{\alpha+\epsilon} \leq \mu(A_0) < r^{\alpha-\epsilon} \tag{11.13}$$

where A_0 is one of A, A_L and A_R. By definition of $\mathcal{H}_\delta^{f-\epsilon}$ there are at least $r^{\epsilon-f} \mathcal{H}_\delta^{f-\epsilon}(E_\alpha^0) \geq r^{\epsilon-f}$ of the r-mesh intervals that intersect E_α^0, so there are at least $\frac{1}{3} r^{\epsilon-f}$ r-mesh intervals A_0 that satisfy (11.13) (note that two intervals A separated by $2r$ or more give rise to different intervals A_0). We conclude that for $r \leq \delta$

$$N_r(\alpha + \epsilon) - N_r(\alpha - \epsilon) \geq \frac{1}{3} r^{\epsilon-f}$$

so from (11.9) $f_C(\alpha) \geq f = f_H(\alpha)$. □

In fact, just as many sets encountered have equal box and Hausdorff dimensions, many common measures have the same coarse and fine spectra. This is so for self-similar measures as we shall see in the next section.

Multifractal spectra are related by Legendre transformation to certain moment sums, and this can provide an alternative way of calculating spectra. Indeed many measures have spectra that are actually equal to the Legendre transforms of a natural auxillary function.

Let $\beta : \mathbb{R} \to \mathbb{R}$ be a convex function. Then there is a range of α, say $\alpha \in [\alpha_{min}, \alpha_{max}]$ for which the graph of β has a line of support L_α of slope $-\alpha$, and for such α this support line is unique. (When $\alpha = \alpha_{min}$ or α_{max} we take L_α to be the asymptote of the graph.) The *Legendre transform* of β is the function $f : [\alpha_{min}, \alpha_{max}] \to \mathbb{R}$ given by the value of the intercept of L_α with the vertical axis. By inspecting Figure 11.1,

$$f(\alpha) = \inf_{-\infty < q < \infty} \{\beta(q) + \alpha q\} \qquad (11.14)$$

and f is continuous in α.

The coarse spectrum is related to the Legendre transform of the power law exponents of moment sums. For $q \in \mathbb{R}$ and $r > 0$ we consider the q-th power moment sums of a measure μ, given by

$$M_r(q) = \sum \mu(A)^q, \qquad (11.15)$$

where the sum is over the r-mesh cubes A for which $\mu(A) > 0$. (There is a problem of stability here for negative q: if a cube A just clips the edge of sptμ,

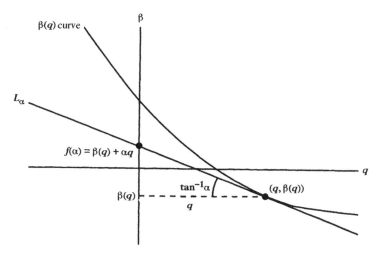

Figure 11.1   The Legendre transform of $\beta(q)$ is $f(\alpha)$, the intersect of the tangent L_α of slope $-\alpha$ with the β axis

then $\mu(A)^q$ can be very large. There are ways around this difficulty, for example by restricting the sums to cubes with a central portion intersecting sptμ, but we do not pursue this here.)

These moment sums are related to the $N_r(\alpha)$: using (11.6) it follows that for all $\alpha \geq 0$

$$M_r(q) = \sum \mu(A)^q \geq r^{q\alpha} N_r(\alpha) \tag{11.16}$$

if $q \geq 0$ and

$$M_r(q) = \sum \mu(A)^q \geq r^{q\alpha} \#\{r\text{-mesh cubes with } 0 < \mu(A) \leq r^\alpha\} \tag{11.17}$$

if $q < 0$. We identify the power law behaviour of $M_r(q)$ by defining

$$\underline{\beta}(q) = \liminf_{r \to 0} \log M_r(q)/ - \log r \tag{11.18}$$

and

$$\bar{\beta}(q) = \limsup_{r \to 0} \log M_r(q)/ - \log r. \tag{11.19}$$

(Note that many texts use $-\tau(q)$ in place of our $\beta(q)$.) The following lemma relates the $\underline{f}_C$ and $\bar{f}_C$ to the Legendre transforms of the $\underline{\beta}$ and $\bar{\beta}$.

Lemma 11.2

Let μ be a finite measure on $\mathbb{R}^n$. Then for all $\alpha \geq 0$

$$\underline{f}_C(\alpha) \leq \underline{f}_L(\alpha) \equiv \inf_{-\infty < q < \infty} \{\underline{\beta}(q) + \alpha q\} \tag{11.20}$$

and

$$\bar{f}_C(\alpha) \leq \bar{f}_L(\alpha) \equiv \inf_{-\infty < q < \infty} \{\bar{\beta}(q) + \alpha q\}. \tag{11.21}$$

Proof First take $q \geq 0$. Then, given $\epsilon > 0$, (11.16) and (11.9) imply that

$$M_r(q) \geq r^{q(\alpha+\epsilon)} N_r(\alpha+\epsilon) \geq r^{q(\alpha+\epsilon)} r^{-\underline{f}_C(\alpha)+\epsilon} \tag{11.22}$$

for all sufficiently small r. It follows from (11.18) that

$$-\underline{\beta}(q) \leq q(\alpha+\epsilon) - \underline{f}_C(\alpha) + \epsilon,$$

so $\underline{f}_C(\alpha) \leq \underline{\beta}(q) + \alpha q$ by taking ϵ arbitrarily small. This inequality also holds when $q < 0$ by a parallel argument, using (11.17) with α replaced by $\alpha - \epsilon$.

The argument for the upper spectra is similar: instead of (11.22) we use that

$$M_r(q) \geq r^{q(\alpha+\epsilon)} N_r(\alpha+\epsilon) \geq r^{q(\alpha+\epsilon)} r^{-\bar{f}_C(\alpha)+\epsilon}$$

for arbitrarily small values of r. $\square$

The Legendre transforms $\underline{f}_L$ and $\overline{f}_L$ defined by (11.20) and (11.21) are sometimes termed the *lower* and *upper Legendre spectra* of μ. There are many measures for which equality occurs in (11.20) and (11.21), indeed the lower and upper values are often all equal, as in the case of self-similar measures. Very often the coarse spectrum is the Legendre transform of a function $\beta(q)$ that can be defined in a more explicit way than as a limit.

The Legendre transform also plays a major rôle in the fine theory of multifractals. Again the aim is to define a suitable β function with a transform that is a good candidate for the multifractal spectrum. A continuous analogue of the mesh cube sum is

$$\beta(q) = \lim_{r \to 0} \log \int \mu(B(x,r))^{q-1} d\mu(x) \Big/ \log r \qquad (11.23)$$

(or an upper or lower limit if this limit fails to exist). For nicely behaved measures the Legendre transform of β gives the multifractal spectrum.

An alternative approach uses measures of Hausdorff type tailored for multifractal purposes. Briefly, given a measure μ on $\mathbb{R}^n$ and $q, \beta \in \mathbb{R}$ we define measures $\mathcal{H}^{q,\beta}$ using the following steps:

$$\mathcal{H}^{q,\beta}_\delta(E) = \inf\left\{ \sum_i \mu(B(x_i, r_i))^q (2r_i)^\beta : E \subset \bigcup_i B(x_i, r_i), \; x_i \in E, r_i \leq \delta \right\},$$

$$\mathcal{H}^{q,\beta}_0(E) = \lim_{\delta \to 0} \mathcal{H}^{q,\beta}_\delta(E),$$

$$\mathcal{H}^{q,\beta}(E) = \sup_{E' \subset E} \mathcal{H}^{q,\beta}_0(E'). \qquad (11.24)$$

(The use of covers by balls centred in E is to avoid difficulties when q is negative. The final step is needed to ensure monotonicity, that is $\mathcal{H}^{q,\beta}(E_1) \leq \mathcal{H}^{q,\beta}(E_2)$ when $E_1 \subset E_2$.) For each q we define $\beta(q)$ analogously to Hausdorff dimension, as the value of β at which $\mathcal{H}^{q,\beta}(\mathbb{R}^n)$ jumps from ∞ to 0, so that $\mathcal{H}^{q,\beta}(\mathbb{R}^n) = \infty$ if $\beta < \beta(q)$ and $\mathcal{H}^{q,\beta}(\mathbb{R}^n) = 0$ if $\beta > \beta(q)$. Then the 'fine' analogue of Lemma 11.2 holds:

$$f_H(\alpha) \leq \inf_{-\infty < q < \infty} \{\beta(q) + q\alpha\}. \qquad (11.25)$$

Again, there is equality for 'nice' measures μ.

This approach to the fine theory using the measures $\mathcal{H}^{q,\beta}$ (and also 'packing' analogues) is mathematically sophisticated, but seems the most appropriate version of multifractal theory for geometrical properties such as the relationship between multifractal features of measures and their projections onto, or intersections with, lower-dimensional subspaces.

In practical situations multifractal spectra are often awkward to estimate and work with. One might hope to compute the coarse spectrum f_C by 'box-counting'. For instance, if μ is a residence measure on the attractor of a dynamical system in the plane, a count of the proportion of the iterates of an initial point that lie in each r-mesh square A might be used to estimate the number of squares for which $\alpha_k \leq \log \mu(A)/\log r < \alpha_{k+1}$, where $0 \leq \alpha_1 < \ldots < \alpha_k$. Examining this 'histogram' for various r enables the power law behaviour of $N_r(\alpha + \epsilon) - N_r(\alpha + \epsilon)$ to be studied and $f(\alpha)$ to be estimated. However, this *histogram method* tends to be computationally slow and awkward.

In general it is more satisfactory to use the *method of moments* for experimental determination of a multifractal spectrum. This uses Legendre transformation: the moment sums (11.15) are estimated for various q and r and the power law behaviour in r examined to find $\beta(q)$ as in (11.18) or (11.19). Legendre transformation of β gives a Legendre spectrum $f_L(\alpha)$ of μ, and there are often good reasons for considering this to be the coarse spectrum. The method of moments is usually more manageable numerically than the histogram method, but even so, practical computation of multifractal spectra is fraught with difficulties.

11.2 Multifractal analysis of self-similar measures

In this section we calculate the multifractal spectra of the self-similar measures introduced in Section 2.2. We do this not only because self-similar measures are important in their own right, but also because the method to be described is a prototype for multifractal calculations for many classes of measures. Self-similar measures are well-behaved, in that the various multifractal spectra introduced in the last section are all equal.

We take μ to be the self-similar measure defined by the probabilistic IFS with similarities $\{F_1, \ldots, F_m\}$ on $\mathbb{R}^n$ with ratios $r_1, \ldots, r_m$ and with associated probabilities $p_1, \ldots, p_m$ (where $p_i > 0$ and $\sum_{i=1}^m p_i = 1$); thus μ satisfies

$$\mu(A) = \sum_{i=1}^m p_i \mu(F_i^{-1}(A)) \tag{11.26}$$

for all sets A, see (2.43)–(2.45). Then $E \equiv \mathrm{spt}\mu$ is the attractor for the IFS $\{F_1, \ldots, F_m\}$, and we assume that the strong separation condition is satisfied, that is $F_i(E) \cap F_j(E) = \emptyset$ for all $i \neq j$, so that E is totally disconnected.

We recall the usual notation, with $I_k = \{(i_1, \ldots, i_k) : 1 \leq i \leq m\}$ and a typical sequence $(i_1, \ldots, i_k)$ abbreviated to i. We take X to be any non-empty compact set with $F_i(X) \subset X$ for all i and $F_i(X) \cap F_j(X) = \emptyset$ if $i \neq j$ (taking $X = E$ would suffice), and write

$$X_i = X_{i_1, \ldots, i_k} = F_{i_1} \circ \cdots \circ F_{i_k}(X). \tag{11.27}$$

For convenience we assume that $|X| = 1$, so for $i = (i_1, i_2, \ldots, i_k)$ we have

$$|X_i| = r_i \equiv r_{i_1} r_{i_2} \cdots r_{i_k} \tag{11.28}$$

and

$$\mu(X_i) = p_i \equiv p_{i_1} p_{i_2} \cdots p_{i_k}. \tag{11.29}$$

We will obtain the multifractal spectrum $f_H(\alpha) = \dim E_\alpha$ as the Legendre transform of an auxilliary function β. Given a real number q, we define $\beta = \beta(q)$ as the positive number satisfying

$$\sum_{i=1}^{m} p_i^q r_i^{\beta(q)} = 1. \tag{11.30}$$

It is easy to see that $\beta : \mathbb{R} \to \mathbb{R}$ is a decreasing real analytic function with

$$\lim_{q \to -\infty} \beta(q) = \infty \quad \text{and} \quad \lim_{q \to \infty} \beta(q) = -\infty. \tag{11.31}$$

Differentiating (11.30) implicitly twice gives

$$0 = \sum_{i=1}^{m} p_i^q r_i^{\beta(q)} \left(\frac{d^2\beta}{dq^2} \log r_i + (\log p_i + \frac{d\beta}{dq} \log r_i)^2 \right),$$

so β is convex in q. Assuming that $\log p_i / \log r_i$ is not the same for all $i = 1, \ldots, m$, then β is strictly convex; we assume that this is the case from now on to avoid degenerate cases. Writing f for the Legendre transform of β, given by

$$f(\alpha) = \inf_{-\infty < q < \infty} \{\beta(q) + \alpha q\}, \tag{11.32}$$

then $f : [\alpha_{min}, \alpha_{max}] \to \mathbb{R}$ where $-\alpha_{min}$ and $-\alpha_{max}$ are the slopes of the asymptotes of the convex function β. Since β is strictly convex, for a given α the infinium in (11.32) is attained at a unique $q = q(\alpha)$; by differentiation this occurs when

$$\alpha = -\frac{d\beta}{dq} \tag{11.33}$$

so that

$$f(\alpha) = \alpha q + \beta(q) = -q \frac{d\beta}{dq} + \beta(q). \tag{11.34}$$

We note that if any one of $q \in \mathbb{R}, \beta \in \mathbb{R}$ and $\alpha \in (\alpha_{min}, \alpha_{max})$ is given, the other two are determined by (11.30) and (11.33). In particular on differentiating (11.30)

$$\alpha = \frac{\sum_{i=1}^{m} p_i^q r_i^\beta \log p_i}{\sum_{i=1}^{m} p_i^q r_i^\beta \log r_i}. \tag{11.35}$$

On inspecting this expression, we see that

$$\alpha_{min} = \min_{1 \le i \le m} \log p_i / \log r_i \quad \text{and} \quad \alpha_{max} = \max_{1 \le i \le m} \log p_i / \log r_i \tag{11.36}$$

corresponding to q approaching ∞ and $-\infty$ respectively.

From the geometry of the Legendre transform, it is easy to see that f is continuous on $[\alpha_{min}, \alpha_{max}]$. Moreover, provided that the numbers $\{\log p_i / \log r_i\}_{i=1}^{m}$ are all different, $f(\alpha_{min}) = f(\alpha_{max}) = 0$, see Exercise 11.2.

Differentiating (11.34) we get, using (11.33),

$$\frac{df}{d\alpha} = \alpha\frac{dq}{d\alpha} + q + \frac{d\beta}{dq}\frac{dq}{d\alpha} = q. \tag{11.37}$$

Since q decreases as α increases, it follows that f is a concave function of α.

There are some values of q that are of special interest. If $q = 0$ then $\beta(q) = \dim_H \mathrm{spt}\mu = \dim_P \mathrm{spt}\mu$, using (11.30) and the formula (2.42) for the dimension of the attractor $\mathrm{spt}\mu$ of the underlying IFS. Moreover, by (11.37) $q = 0$ corresponds to the maximum of $f(\alpha)$; hence $\dim_H \mathrm{spt}\mu = \dim_P \mathrm{spt}\mu = \max_\alpha f(\alpha)$.

When $q = 1$ (11.30) implies $\beta(q) = 0$ so $f(\alpha) = \alpha$ by (11.34). Moreover $\frac{d}{d\alpha}(f(\alpha) - \alpha)) = q - 1 = 0$, so that the $f(\alpha)$ curve lies under the line $f = \alpha$ and touches it just at the point corresponding to $q = 1$. It will follow later that $\alpha(1) = f(\alpha(1)) = \dim_H \mu = \dim_P \mu$ (see (10.8) and (10.9) for the definitions of the dimensions of a measure). The main features of $\beta(q)$ and $f(\alpha)$ for a typical self-similar measure are indicated in Figure 11.2.

Our main aim is to show that the Hausdorff and packing spectra of μ are given by the Legendre transform (11.32) of $\beta(q)$, that is

$$f_H(\alpha) = f_P(\alpha) = f(\alpha) \tag{11.38}$$

for $\alpha \in [\alpha_{\min}, \alpha_{\max}]$, where E_α is the set of points (11.2) of local dimension α, and $f_H(\alpha) = \dim_H E_\alpha$ and $f_P(\alpha) = \dim_P E_\alpha$.

Writing

$$\Phi(q, \beta) = \sum_{i=1}^{m} p_i^q r_i^\beta \tag{11.39}$$

for q and β real, $\beta(q)$ is defined by $\Phi(q, \beta(q)) = 1$, see (11.30). We require the following estimate of Φ near $(q, \beta(q))$.

Lemma 11.3

For all $\epsilon > 0$,

$$\Phi(q + \delta, \beta(q) + (-\alpha + \epsilon)\delta) < 1 \tag{11.40}$$

and

$$\Phi(q - \delta, \beta(q) + (\alpha + \epsilon)\delta) < 1 \tag{11.41}$$

for all sufficiently small $\delta > 0$.

Proof   Recalling that $d\beta/dq = -\alpha$,

$$\beta(q + \delta) = \beta(q) - \alpha\delta + O(\delta^2) < \beta(q) + (-\alpha + \epsilon)\delta$$

if δ is small enough. Since $\Phi(q + \delta, \beta(q + \delta)) = 1$ and Φ is decreasing in its

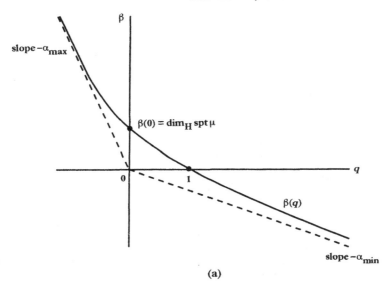

slope $-\alpha_{\max}$

$\beta(0) = \dim_H \operatorname{spt} \mu$

0   1

q

$\beta(q)$

slope $-\alpha_{\min}$

(a)

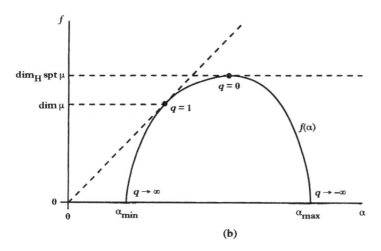

f

$\dim_H \operatorname{spt} \mu$

$\dim \mu$

$q = 0$

$q = 1$

$f(\alpha)$

$q \to \infty$ $q \to -\infty$

0

0        $\alpha_{\min}$ $\alpha_{\max}$ α

(b)

Figure 11.2   Form of the multifractal functions for a typical self-similar measure. (a) The $\beta(q)$ curve, (b) the 'multifractal spectrum' $f(\alpha) = \dim_H E_\alpha$, which is the Legendre transform of $\beta(q)$

second argument, (11.40) follows. Inequality (11.41) is derived in a similar way. $\quad\square$

To prove (11.38) we concentrate a measure ν on E_α and examine the power law behaviour of $\nu(B(x,r))$ as $r \to 0$, so that we can use Proposition 2.3 to find

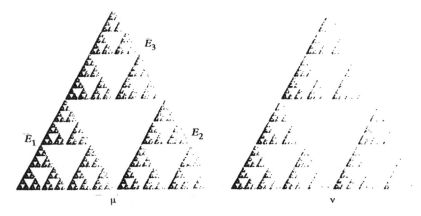

Figure 11.3 This self-similar measure μ supported by the Sierpinski triangle is constructed using probabilities $p_1 = 0.8$, $p_2 = 0.05$, $p_3 = 0.15$. The analysing measure ν, concentrated on the set E_α where μ has local dimension $\alpha = 0.6$ (corresponding to $q = 1.4$), is self-similar with probabilities 0.896, 0.018 and 0.086, giving that $\dim_H E_\alpha = f(\alpha) = 1.138$

the dimensions of E_α. For given $q \in \mathbb{R}$ and $\beta = \beta(q)$ we define a probability measure ν on $\mathrm{spt}\mu$ by

$$\nu(X_{i_1,\ldots,i_k}) = (p_{i_1} p_{i_2} \cdots p_{i_k})^q (r_{i_1} r_{i_2} \cdots r_{i_k})^\beta \tag{11.42}$$

and extend this to a Borel measure in the usual way, see Figure 11.3 for one instance of this. Together with (11.28) and (11.29) this gives three ways of quantifying the X_i for $i \in I$:

$$|X_i| = r_i, \quad \mu(X_i) = p_i, \quad \nu(X_i) = p_i^q r_i^\beta. \tag{11.43}$$

For $x \in \mathrm{spt}\mu$ we write $X_k(x)$ for the k-th level set $X_{i_1,\ldots,i_k}$ that contains x. We shall go back and forth between the set $X_k(x)$ and the ball $B(x,r)$ where $|X_k(x)|$ is comparable with r. In particular, for any α,

$$\lim_{r \to 0} \frac{\log \mu(B(x,r))}{\log r} = \alpha \quad \text{if and only if} \quad \lim_{k \to \infty} \frac{\log \mu(X_k(x))}{\log |X_k(x)|} = \alpha, \tag{11.44}$$

see Exercise 11.3.

Proposition 11.4

With q, β, α and f as above, and with ν defined by (11.42),

(a) $\nu(E_\alpha) = 1$,
(b) *for all $x \in E_\alpha$ we have $\log \nu(B(x,r))/\log r \to f(\alpha)$ as $r \to 0$.*

Proof Let $\epsilon > 0$ be given. Then for all $\delta > 0$

$$\nu\{x : \mu(X_k(x)) \geq |X_k(x)|^{\alpha - \epsilon}\} = \nu\{x : 1 \leq \mu(X_k(x))^{\delta}|X_k(x)|^{(\epsilon - \alpha)\delta}\}$$

$$\leq \int \mu(X_k(x))^{\delta}|X_k(x)|^{(\epsilon - \alpha)\delta}d\nu(x)$$

$$= \sum_{i \in I_k} \mu(X_i)^{\delta}|X_i|^{(\epsilon - \alpha)\delta}\nu(X_i) \qquad (11.45)$$

$$= \sum_{i \in I_k} p_i^{q+\delta} r_i^{\beta + (\epsilon - \alpha)\delta}$$

$$= \left(\sum_{i=1}^{m} p_i^{q+\delta} r_i^{\beta + (\epsilon - \alpha)\delta} \right)^k$$

$$= [\Phi(q + \delta, \beta + (\epsilon - \alpha)\delta)]^k \qquad (11.46)$$

where Φ is given by (11.39), using (11.43) and a multinominial expansion. Choosing δ small enough and using (11.40) gives that

$$\nu\{x : \mu(X_k(x)) \geq |X_k(x)|^{\alpha - \epsilon}\} \leq \gamma^k \qquad (11.47)$$

where $\gamma < 1$ is independent of k. Thus

$$\nu\{x : \mu(X_k(x)) \geq |X_k(x)|^{\alpha - \epsilon} \text{ for some } k \geq k_0\} \leq \sum_{k=k_0}^{\infty} \gamma^k \leq \gamma^{k_0}/(1 - \gamma).$$

It follows that for ν-almost all x we have

$$\liminf_{k \to \infty} \log \mu(X_k(x))/\log |X_k(x)| \geq \alpha - \epsilon.$$

Since this is true for all $\epsilon > 0$, we get the left-hand inequality of

$$\alpha \leq \liminf_{k \to \infty} \log \mu(X_k(x))/\log |X_k(x)|$$

$$\leq \limsup_{k \to \infty} \log \mu(X_k(x))/\log |X_k(x)| \leq \alpha.$$

The right-hand inequality follows in the same way, using (11.41) in estimating $\nu\{x : \mu(X_k(X)) \leq |X_k(x)|^{\alpha + \epsilon}\}$. Using (11.44) we conclude that for ν-almost all x,

$$\lim_{r \to 0} \log \mu(B(x, r))/\log r = \lim_{k \to \infty} \log \mu(X_k(x))/\log |X_k(x)| = \alpha;$$

since ν is a probability measure it follows that $\nu(E_\alpha) = 1$.

For (*b*) note that from (11.43)

$$\frac{\log \nu(X_k(x))}{\log |X_k(x)|} = q \frac{\log \mu(X_k(x))}{\log |X_k(x)|} + \beta \frac{\log |X_k(x)|}{\log |X_k(x)|} \qquad (11.48)$$

so for all $x \in E_\alpha$

$$\frac{\log \nu(X_k(x))}{\log |X_k(x)|} \to q\alpha + \beta = f \tag{11.49}$$

as $k \to \infty$, using (11.34). Part (b) follows, since (11.44) remains true with ν replacing μ. □

Our main result on the multifractal spectrum of self-similar measures follows easily.

Theorem 11.5

Let μ be a self-similar measure as above and let

$$E_\alpha = \{x : \lim_{r \to 0} \log \mu(B(x,r))/\log r = \alpha\}.$$

If $\alpha \notin [\alpha_{\min}, \alpha_{\max}]$ then $E_\alpha = \emptyset$, and if $\alpha \in [\alpha_{\min}, \alpha_{\max}]$ then

$$\dim_H E_\alpha = \dim_P E_\alpha = f(\alpha), \tag{11.50}$$

that is

$$f_H(\alpha) = f_P(\alpha) = f(\alpha).$$

Proof   From (11.43)

$$\log \mu(X_i)/\log |X_i| = \sum_{j=1}^{k} \log p_{i_j} \bigg/ \sum_{j=1}^{k} \log r_{i_j}$$

where $i = (i_1, \ldots, i_k)$, so by (11.36) $\log \mu(X_i)/\log|X_i| \in [\alpha_{\min}, \alpha_{\max}]$ for all i. Thus the only possible limit points of $\log \mu(X_i)/\log |X_i|$, and so (analogously to (11.44) of $\log \mu(B(x,r))/\log r$, are in $[\alpha_{\min}, \alpha_{\max}]$. In particular $E_\alpha = \emptyset$ if $\alpha \notin [\alpha_{\min}, \alpha_{\max}]$.

If $\alpha \in (\alpha_{\min}, \alpha_{\max})$ then by Proposition 11.4 there exists a measure ν concentrated on E_α with $\lim_{r \to 0} \log \nu(B(x,r))/\log r = f(\alpha)$ for all $x \in E_\alpha$, so (11.50) follows from Proposition 2.3. For the cases $\alpha = \alpha_{\min}$ and $\alpha = \alpha_{\max}$, see Exercise 11.5. □

Thus, for a self-similar measure, the dimension of E_α may be calculated by finding $\beta(q)$ and taking its Legendre transform. A specific example of a multifractal spectrum is shown in Figure 11.4.

We remark that instead of E_α we might consider

$$G_\alpha = \{x : \lim_{i \to \infty} \log \mu(B(x,r_i))/\log r_i = \alpha \text{ for some } r_i \searrow 0\}, \tag{11.51}$$

that is the set of x which have α as a limit point of $\log \mu(B(x,r))/\log r$. Clearly $E_\alpha \subset G_\alpha$, so from Theorem 11.5 $f(\alpha)$ is a lower bound for the dimensions of

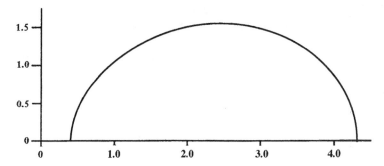

Figure 11.4 Multifractal spectrum of the self-similar measure μ on the Sierpinski triangle shown in Figure 11.3

G_α; in fact little more calculation is required to show that $f(\alpha) = \dim_H G_\alpha = \dim_P G_\alpha$, see Exercise 11.6.

The dimensions of $\mathrm{spt}\mu$ and the measure μ may easily be found from the multifractal spectrum.

Proposition 11.6

Let μ be a self-similar measure as above. Regarding $\alpha = \alpha(q)$ as a function of q,

(a) $f(\alpha)$ takes its maximum when $\alpha = \alpha(0)$, with $f(\alpha(0)) = \dim_H \mathrm{spt}\mu = \dim_P \mathrm{spt}\mu$

(b) $f(\alpha(1)) = \alpha(1) = \dim_H \mu = \dim_P \mu$.

Proof Part (a) and that $f(\alpha(1)) = \alpha(1)$ were noted as a consequence of (11.37). For the dimensions of the measures, if $q = 1$ then $\beta = 0$ from (11.30), so from (11.42) the measure ν is identical to μ. By Proposition 11.4 $\mu(E_{\alpha(1)}) = 1$ and $\dim_{\mathrm{loc}} \mu(x) = \lim_{r \to 0} \log \mu(B(x,r))/\log r = f(\alpha(1))$ for all $x \in E_{\alpha(1)}$, so (b) follows from the definitions of the dimensions of a measure (10.8) and (10.9). $\square$

Next we show that the coarse spectrum of a self-similar measure is also equal to $f(\alpha)$ given by (11.32) if $\alpha \leq \alpha(0)$.

Proposition 11.7

Let μ be a self-similar measure as above. Then

$$f_C(\alpha) \geq f(\alpha) \tag{11.52}$$

for all α, with equality if $\alpha = \alpha(q)$ where $q \geq 0$.

Proof We first note that by Theorem 11.5 and Lemma 11.1 we have $f(\alpha) = f_H(\alpha) \le \underline{f}_C(\alpha) \le \overline{f}_C(\alpha)$, where the coarse spectra are given by (11.9) and (11.10).

To prove the opposite inequality, let d be the minimum separation of X_i and X_j for $i \ne j$, and write $a = 2\sqrt{n}/d$. Given $r < a^{-1}|X|$, let J be the set of all sequences $i = (i_1, \ldots, i_k)$ such that $|X_{i_1,\ldots,i_k}| \le ar$ but $|X_{i_1,\ldots,i_{k-1}}| > ar$. Then

$$ac_{\min}r < |X_i| = r_i \le ar \tag{11.53}$$

if $i \in J$, where $c_{\min} = \min_{1 \le i \le m} r_i$. We note that each point of E lies in exactly one set X_i with $i \in J$, and also that for distinct $i, j \in J$ the sets X_i and X_j have separation at least $ard = 2\sqrt{n}r$.

Suppose $q \ge 0$ and let β, α and f be the corresponding values given by (11.30), (11.33) and (11.34). Then

$$\begin{aligned}
\#\{i \in J : \mu(X_i) \ge a^{-\alpha}|X_i|^\alpha\} &= \#\{i \in J : 1 \le a^{\alpha q} p_i^q r_i^{-\alpha q}\} \\
&\le a^{\alpha q} \sum_{i \in J} p_i^q r_i^{-\alpha q} \\
&= a^{\alpha q} \sum_{i \in J} p_i^q r_i^\beta r_i^{-\beta - \alpha q} \\
&\le a^{\alpha q} (ac_{\min})^{-f} r^{-f} \tag{11.54}
\end{aligned}$$

using (11.53), (11.34) and that $\sum_{i \in J} p_i^q r_i^\beta = 1$, an identity that follows by repeated substitution of $\sum_{i=1}^m p_{i,i}^q r_{i,i}^\beta = p_i^q r_i^\beta$ in itself. Every r-mesh cube is of diameter $\sqrt{n}r$ so intersects at most one of the sets X_i for $i \in J$. With $N_r(\alpha)$ as in (11.6)

$$\begin{aligned}
N_r(\alpha) = \#\{r\text{-mesh cubes } A : \mu(A) \ge r^\alpha\} \\
\le \#\{i \in J : \mu(X_i) \ge a^{-\alpha}|X_i|^\alpha\} \\
\le a^{\alpha q} (ac_{\min})^{-f(\alpha)} r^{-f(\alpha)}.
\end{aligned}$$

It follows that there is a number c such that for sufficiently small ϵ and r

$$N_r(\alpha + \epsilon) - N_r(\alpha - \epsilon) \le N_r(\alpha + \epsilon) \le c r^{-f(\alpha + \epsilon)},$$

so, using (11.10) and that f is continuous, gives $\overline{f}_C(\alpha) \le f(\alpha)$ and thus equality in (11.52). $\quad\square$

The coarse spectra, as we have defined it, is not well-enough behaved for equality to hold in (11.52) for α corresponding to $q < 0$. The problem is that in this case we would need to estimate the number of r-mesh cubes A with $0 < \mu(A) \le r^\alpha$ and this may bear little resemblence to the number of sets X_i of comparable size with $\mu(X_i) \le |X_i|^\alpha$. This difficulty can be overcome by redefining $N_r(\alpha)$ in (11.6) to be the number of r-mesh cubes A with $\mu(A) \ge r^\alpha$ and such that $\mu(A') > 0$ where A' is the mesh cube with the same centre as A and of half the side-length. Then with the coarse spectra defined by (11.9) and

(11.10) using this $N_r(\alpha)$, the argument of Proposition 11.7 may be extended to give equality in (11.52) for all α.

11.3 Multifractal analysis of Gibbs measures on cookie-cutter sets

A rather similar analysis to that of the previous section enables us to calculate the multifractal spectrum of a Gibbs measure supported by a cookie-cutter set of the form introduced in Section 4.1. This is another instance where the thermodynamic formalism enables the theory for self-similar sets to be extended to non-linear analogues.

We assume that $f : X \rightarrow X$ is a cookie-cutter system of the form described in Section 4.1, where X is a closed real interval and $f^{-1} : X \rightarrow X$ has two branches given by F_1 and F_2. We index the iterated sets $F_{i_1} \circ \cdots \circ F_{i_k}(X) = X_{i_1,\ldots,i_k} = X_i$ in the usual way. Theorem 5.3 showed that the dimension $\dim_H E$ of the cookie-cutter attractor E is given in terms of the pressure functional P by the number s satisfying $P(-s \log|f'|) = 0$. Recall that this pressure gives the exponential growth rate in k of $\sum_{i \in I_k} |X_i|^s$. Here we obtain an analogous pressure formula for the multifractal spectrum of a Gibbs measure μ supported by E; in this case we use pressure to estimate the growth rate of $\sum_{i \in I_k} |X_i|^\beta \mu(X_i)^q$.

Let μ be a Gibbs measure on E associated with a C^2 function $\phi : X \rightarrow \mathbb{R}$. We assume that ϕ has pressure 0 so that by (5.6)

$$\mu(X_i) \asymp \exp(S_k \phi(x)) \tag{11.55}$$

for all $i \in I$ and $x \in X_i$, where $S_k \phi(x) = \sum_{i=0}^{k-1} \phi(f^i x)$ (If the pressure of ϕ is non-zero then we may replace ϕ by $\phi - P(\phi)$ which has pressure zero and the same Gibbs measures.) We assume that

$$\phi(x) < 0 \quad \text{for all} \ \ x \in X; \tag{11.56}$$

this ensures that $\mu(X_i)$ satisfying (11.55) remains bounded.

In Section 11.2 we defined $\beta(q)$ for self-similar measures by the requirement that $\sum_{i=1}^m \mu(X_i)^q |X_i|^\beta = 1$, which implies that $\sum_{i \in I_k} \mu(X_i)^q |X_i|^\beta = 1$ for all k. Here we will use pressure to define $\beta = \beta(q)$ to ensure that

$$\sum_{i \in I_k} \mu(X_i)^q |X_i|^\beta \asymp 1; \tag{11.57}$$

for any other choice of β this sum converges at an exponential rate to 0 or ∞. By (4.16)–(4.18)

$$|X_i| \asymp \exp(-(S_k \log|f'|)(x)), \tag{11.58}$$

for all $x \in X_i$, so combining this with (11.55) the condition (11.57) becomes that

$$\sum_{x_i} \exp(S_k(-\beta \log|f'| + q\phi)(x_i)) \asymp 1,$$

where $(x_i) \in X_i$ for each $i \in I_k$. By (5.5) this is achieved by defining $\beta(q)$ by

$$P(-\beta(q)\log|f'| + q\phi) = 0, \tag{11.59}$$

where P is the pressure functional. To see that (11.59) defines a unique β we note that the function

$$(q, \beta) \mapsto P(-\beta \log|f'| + q\phi) \tag{11.60}$$

is strictly decreasing and continuous in both β and q; this may be verified just as in Lemma 5.2, noting that there exist $m_1, m_2 < 0$ such that

$$m_1 \le -\log|f'(x)|, \phi(x) \le m_2$$

for all x in the compact set X. Again as in Lemma 5.2, for all q

$$\lim_{\beta \to \pm\infty} P(-\beta \log|f'| + q\phi) = \mp\infty.$$

We conclude that $q \mapsto \beta(q)$ is continuous and strictly decreasing, and

$$\lim_{q \to -\infty} \beta(q) = \infty \quad \text{and} \quad \lim_{q \to \infty} \beta(q) = -\infty.$$

It may be shown (see Exercise 11.11) that the function (11.60) is convex, that is its graph is a convex surface. Thus $q \mapsto \beta(q)$ is a convex function (with its graph obtained by taking the plane section $P = 0$ of the surface). It may also be shown (though this is rather harder) that β is a differentiable, and indeed a real analytic, function of q. Thus the overall form of β is very much like that indicated in Figure 11.1 for a self-similar measure. To avoid a degenerate analysis, we again assume that β is strictly convex.

To find the dimension of the set

$$E_\alpha = \{x : \lim_{r \to 0} \log \mu(B(x, r))/\log r = \alpha\}$$

we again look to the Legendre transform

$$f(\alpha) = \inf_{-\infty < q < \infty} \{\beta(q) + q\alpha\}.$$

Just as in (11.33)–(11.34) for the self-similar case we have

$$f(\alpha) = \beta(q) + q\alpha \quad \text{where} \quad \alpha = -\frac{d\beta}{dq}, \tag{11.61}$$

and that $f : [\alpha_{\min}, \alpha_{\max}] \to \mathbb{R}^+ \cup \{0\}$ is a continuous concave function. Moreover, the values $q = 0$ and $q = 1$ have the same significance as in the self-similar case.

The proof that $f(\alpha) = \dim_H E_\alpha = \dim_P E_\alpha$ is very similar to that presented for self-similar measures in Section 11.2. Again the crucial step is to introduce a measure ν analogous to (11.42) that is concentrated on E_α; this time, however, the measure we require occurs naturally as a Gibbs measure.

Thus for a given q, we define β, α and f as above, and let ν be a Gibbs measure of the function $-\beta \log|f'| + q\phi$. Since β is defined so that the pressure of this function is zero, we have from (5.6) that

$$\nu(X_i) \asymp \exp(-\beta(S_k \log|f'|)(x) + q(S_k\phi)(x))$$

for all $i \in I_k$ and $k \in \mathbb{Z}^+$ and $x \in X_i$. Using (11.55) and (11.58) it follows that

$$\nu(X_i) \asymp |X_i|^\beta \mu(X_i)^q \qquad (11.62)$$

For $x \in \mathrm{spt}\mu$ we write $X_k(x)$ for the set $X_{i_1,\dots,i_k}$ containing x. As in the case of self-similar measures we may, by virtue of Corollary 4.3, go back and forth between $X_k(x)$ and $B(x,r)$ where $|X_k(x)|$ is comparable with r. Thus for example

$$\lim_{r\to 0}\frac{\log\mu(B(x,r))}{\log r} = \alpha \quad \text{if and only if} \quad \lim_{k\to\infty}\frac{\log\mu(X_k(x))}{\log|X_k(x)|} = \alpha. \qquad (11.63)$$

The following property of pressure is the analogue of Lemma 11.3.

Lemma 11.8

For all $\epsilon > 0$,

$$P(-(\beta(q) + (\epsilon - \alpha)\delta)\log|f'| + (q + \delta)\phi) < 1$$

and

$$P(-(\beta(q) + (\epsilon + \alpha)\delta)\log|f'| + (q - \delta)\phi) < 1$$

for all sufficiently small δ.

Proof   Given that $d\beta/dq = -\alpha$, see (11.61), the proof is almost identical to that of Lemma 11.3.   □

Proposition 11.9

With β, α and f defined in terms of q, and ν as above,
(a) $\nu(E_\alpha) = 1$,
(b) for all $x \in E_\alpha$ we have $\log\nu(B(x,r))/\log r \to f(\alpha)$ as $r \to 0$.

Proof   Let $\epsilon > 0$ be given. Then, just as for (11.45) but using (11.62), for all $\delta > 0$

$$\nu\{x : \mu(X_k(x)) \geq |X_k(x)|^{\alpha-\epsilon}\} \leq \sum_{i\in I_k}\mu(X_i)^\delta|X_i|^{(\epsilon-\alpha)\delta}\nu(X_i)$$

$$\leq c\sum_{i\in I_k}\mu(X_i)^{q+\delta}|X_i|^{\beta+(\epsilon-\alpha)\delta}$$

$$\leq c_1\gamma^k \qquad (11.64)$$

where $\gamma = P(-(\beta + (\epsilon - \alpha)\delta)\log|f'| + (q + \delta)\phi)$ and c and c_1 are independent of k, using (11.62), (11.55), (11.58) and the definition of pressure (5.5). By Lemma 11.8 we have $\gamma < 1$ for δ sufficiently small, so (a) now follows from (11.64) exactly as part (a) of Proposition 11.4 follows from (11.47).

From (11.62)

$$\nu(X_k(x)) \asymp \mu(X_k(x))^q |X_k(x)|^\beta$$

so

$$\lim_{k\to\infty} \frac{\log \nu(X_k(x))}{\log |X_k(x)|} = \lim_{k\to\infty} q\frac{\log \mu(X_k(x))}{\log |X_k(x)|} + \beta$$
$$= q\alpha + \beta$$

for $x \in E_\alpha$. Part (b) follows by using (11.63) first for μ and then for ν. $\square$

Theorem 11.10

Let μ be a Gibbs measure associated with ϕ as above. With

$$E_\alpha = \{x : \lim_{r\to 0} \log \mu(B(x,r))/\log r = \alpha\},$$

we have that

$$\dim_H E_\alpha = \dim_P E_\alpha = f(\alpha)$$

for $\alpha \in (\alpha_{\min}, \alpha_{\max})$.

Proof Just as in Theorem 11.5, this follows from Proposition 2.3 using the measure ν on E_α and (a) and (b) of Proposition 11.9. $\square$

Note that Proposition 11.6 on the significance of $q = 0$ and $q = 1$ also applies in the Gibbs measure situation.

11.4 Notes and references

Much has been written on multifractals, and multifractal spectra have been calculated for many specific measures. We can do no more than mention a selection of references where further details may be found.

The idea of studying measures from a fractal viewpoint is implicit in Mandelbrot's essay (1975, 1982). Legendre transformation was introduced into multifractal analysis in Frisch and Parisi (1985) and Halsey, et al. (1986). Various approaches to multifractals are described at a fairly basic level by Falconer (FG), Feder (1988), Evertsz and Mandelbrot (1992) and Tél (1988). A substantial bibliography on multifractals may be found in Olsen (1994).

Detailed rigorous approaches to multifractal theory, including measures of Hausdorff type (11.24) and the relationships between different types of spectra, are given by Brown, *et al.* (1992) and Olsen (1995). A careful treatment of the coarse theory is given by Riedi (1995).

Self-similar measures are analysed by Cawley and Mauldin (1992), and Edgar and Mauldin (1992) extend this to graph-directed constructions of measures. The multifractal spectra of measures on cookie-cutters is found by Rand (1989), and Lopes (1989) studies invariant measures of rational maps on the complex plane. King (1995) and Olsen (1996) consider self-affine measures, and Mandelbrot and Riedi (1995) consider self-similar measures constructed from IFSs consisting of infinitely many similarity transformations. Falconer and O'Neil (1996) introduce vector-valued multifractal measures.

There are natural random analogues of self-similar measures, when both the similarity ratios and the ratios of division of measure are random variables that are independent and identically distributed at each subdivision. In this case the β function is defined by the expectation equation $E(\sum_{i=1}^{m} P_i^q R_i^{\beta(q)}) = 1$ (where P_i and R_i are the random variables underlying the statistically self-similar measure and the geometry of the construction) and the almost sure multifractal spectrum of the random measure is then given by Legendre transform of β. Such measures are considered by Mandelbrot (1974), Kahane and Peyrière (1976), Olsen (1994) (who also considers a randomised version of graph-directed measures) and Arbeiter and Patzschke (1996) who work with a rather weaker separation condition (which relates to the open set condition in the non-random case).

There are many other ways of studying multifractal aspects of measures. For example, the generalised dimensions introduced by Hentschel and Procaccia (1983) defined by $d_q = \beta(q)/(q-1)$, where β is given by (11.18) or (11.19) in the coarse case or by (11.23) in the fine case, are often studied. (The normalisation by $1/(q-1)$ ensures that $d_q = n$ for all q for measures uniformly distributed over an open region of n-dimensional space.) Multifractal properties of measures are also manifested in the behaviour of their Fourier transforms, see Strichartz (1993), and in the behaviour of their wavelet transforms, see Holschneider (1995).

There are many aspects of multifractal behaviour that are only starting to be understood, such as the interpretation of negative dimensions, see Mandelbrot (1991), and the rigorous geometrical properties of multifractals, such as their behaviour under projection, section or products, see Olsen (1996).

Exercises

11.1 Find the Legendre transform of $\beta(q) = e^{-q}$.

11.2 For the self-similar measures defined by (11.26) or (11.29), show that both asymptotes of the graph of $\beta(q)$ pass through the origin, provided that the numbers $\{\log p_i / \log q_i\}_{i=1}^{m}$ are all different. Deduce that in this case $f(\alpha_{\min}) = f(\alpha_{\max}) = 0$.

11.3 Verify (11.44) directly in the case of a self-similar measure satisfying the strong separation condition. (Note that the more general result for a cookie-cutter measure (11.63) follows easily from Corollary 4.3.)

11.4 Construct a proof of Proposition 11.4 along the lines of that of Proposition 10.4, using the strong law of large numbers.

11.5 Prove (11.50) when $\alpha = \alpha_{\min}$. (Hint: Take α close to $\alpha_{\min}$, and note that Proposition 11.4 remains true with (b) replaced by 'for all x such that $\log \mu(B(x,r))/\log r \leq \alpha$ we have $\lim_{r \to 0} \log \nu(B(x,r))/\log r \leq f(\alpha)$'.)

11.6 With μ as the self-similar measure defined by (11.26) or (11.29) and G_α defined by (11.51), show that $\dim_H G_\alpha = f(\alpha)$. (Hint: show that

$$\left\{ \sum |X_i|^f : i \in I_k \text{ and } \mu(X_i) \geq |X_i|^{\alpha - \epsilon} \right\} \leq \sum_{i \in I_k} |X_i|^\beta \mu(X_i)^{q\alpha/(\alpha - \epsilon)} .$$

11.7 Let μ be a self-similar measure supported by the middle-third Cantor set (so $r_1 = r_2 = \frac{1}{3}$) with the measure divided between the left and right parts in the ratio $p_1 : p_2$, where $p_1 + p_2 = 1$. Find an expression for $\beta(q)$ and hence find α and f in terms of the parameter q.

11.8 Let μ be a self-similar measure constructed by repeated subdivision of the support in the ratios $r_1 = \frac{1}{2}$ and $r_2 = \frac{1}{4}$ and of the measure in ratios p_1 and p_2. Obtain an explicit formula for $\beta(q)$.

11.9 Let μ be a finite measure on $\mathbb{R}^n$ and let $g : \mathbb{R}^n \to \mathbb{R}^n$ be a bi-Lipschitz function. Define the image measure ν on $\mathbb{R}^n$ by $\nu(A) = \mu(g^{-1}(A))$. Show that $f_H(\alpha)$ (given by (11.4)) is the same for both ν and μ. (Hint: see Exercise 10.1.)

11.10 Let μ_1, μ_2 be finite measures on $\mathbb{R}^n$ with disjoint supports and define $\nu = \mu_1 + \mu_2$. Show that $f^\nu_H(\alpha) = \max\{f^1_H(\alpha), f^2_H(\alpha)\}$ where f^ν_H, f^1_H and f^2_H are the Hausdorff spectra of ν, μ_1 and μ_2. Deduce that $f_H(\alpha)$ need not be concave over the range of α for which it does not vanish.

11.11 Show that (11.60) defines a convex function of two variables (q, β). (Hint: use differentiation to show that the sums in the definition of pressure are convex for all k, and take the limit as $k \to \infty$.)

Chapter 12 Fractals and differential equations

Fractal geometry can interact with the theory of differential equations in many ways. For example, solutions of a differential equation can approach a fractal attractor, or we might wish to seek solutions on a domain with fractal boundary, or even a domain that is itself a fractal.

This chapter touches on some of the fascinating interplays between fractal geometry and differential equations. This topic could easily fill a book in its own right and here we merely give a taste of the subject and indicate some basic ideas in a few specific cases. The mathematics in this area is often sophisticated; we make no attempt to give full proofs and we omit many technical details. For instance, solutions to the differential equations ought properly to be considered (and, indeed, shown to exist) in appropriate spaces of functions or distributions.

12.1 The dimension of attractors

In this section we describe a method for estimating the dimension of attractors (which are often fractal) of dynamical systems and differential equations. This method of finding upper bounds for the dimension has been applied to a wide variety of systems, including many of the fundamental partial differential equations of physics, giving an insight into the nature of the attractors. Here we can do little more than state the basic estimate, and illustrate its use in some simple situations.

The underlying idea is simple. We work with an open set X and continuous $f : X \to X$ and study a compact invariant set $E \subset X$; thus $f(E) = E$. The set E is often an attractor of the system in the sense that iterates of points in a neighbourhood U of E approach the set E, so that $E = \cap_{k=0}^{\infty} f^k(U)$. The Hausdorff measures and dimension of E were defined in terms of the sums $\sum_i |U_i|^s$ for covers of E by small sets $\{U_i\}_i$. However, if $\{U_i\}_i$ covers E, then so does the iterated cover $\{f^k(U_i)\}_i$ for each $k = 1, 2, \ldots$, by invariance of E. Dividing the sets $f^k(U_i)$ into appropriate small pieces often gives a cover of E that provides a much better bound for the Hausdorff measures and dimension than the original cover. We estimate the size and shape of $f^k(U_i)$ in terms of the derivative of f, which may be thought of geometrically as an affine transformation that is a local approximation to f.

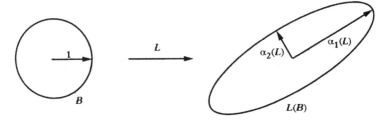

Figure 12.1   Singular values of a linear mapping L

First we consider how a linear mapping (which we will eventually take to be a derivative) transforms a ball into an ellipsoid. Let $L : \mathbb{R}^n \to \mathbb{R}^n$ be linear. We define the *singular values* $\alpha_1(L) \geq \alpha_2(L) \geq \ldots \geq \alpha_n(L) \geq 0$ to be the lengths of the (mutually perpendicular) principal semi-axes of the ellipsoid $L(B)$, where B is the unit ball in $\mathbb{R}^n$, see Figure 12.1. Equivalently, the $\alpha_j(L)$ are the positive square-roots of the eigenvalues of L^*L, where L^* is the adjoint of L. We define the *singular value function* ω_s for $0 \leq s \leq n$ by

$$\omega_s(L) = \alpha_1(L) \cdots \alpha_{m-1}(L)\alpha_m(L)^{s-m+1} \tag{12.1}$$

where m is the integer such that $m - 1 < s \leq m$. Then $\omega_s(L)$ is continuous in s and is increasing at values of s with $m - 1 < s \leq m$ and $\alpha_m(L) > 1$ and decreasing at values of s with $m - 1 < s \leq m$ and $\alpha_m(L) < 1$. Note that if m is an integer then $\omega_m(L)$ is the maximum of $\mathcal{L}^m(L(D))/\mathcal{L}^m(D)$ over all m-dimensional discs D in $\mathbb{R}^n$, where $\mathcal{L}^m$ is the m-dimensional volume of a subset of an m-plane. In particular, $\omega_1(L) = \|L\|$, where $\|\ \|$ is the norm induced by the Euclidean norm on $\mathbb{R}^n$, and $\omega_n(L) = |\det L|$, where det denotes the determinant.

It may be shown (see Exercise 12.1) that ω_s is submultiplicative for each s, that is for all linear L_1, L_2

$$\omega_s(L_1 L_2) \leq \omega_s(L_1)\omega_s(L_2). \tag{12.2}$$

Our dimension calculations depend on covering images of small balls, which are approximate ellipsoids, by small sets.

Lemma 12.1

Let A be an ellipsoid with semi-axes of lengths $\beta_1 \geq \beta_2 \geq \ldots$. For $m = 1, 2, \ldots$ there is a covering of A by at most

$$4^{m-1}\beta_1\beta_2 \cdots \beta_{m-1}\beta_m^{-(m-1)}$$

sets each of diameter at most $(m + 3)^{1/2}\beta_m$.

Proof Note that the j-th principal axis of A has length $2\beta_j$. For each $j = 1, 2, \ldots, m - 1$, slice A by at most $4\beta_j/\beta_m - 1$ parallel hyperplanes distance

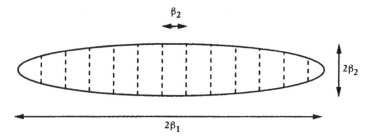

Figure 12.2 Division of an ellipse into slices

β_m apart and perpendicular to the j-th principal axis, see Figure 12.2. These cuts decompose A into at most

$$(4\beta_1/\beta_m)(4\beta_2/\beta_m)\cdots(4\beta_{m-1}/\beta_m)$$

pieces. Each piece projects to the j-th principal axis in a segment of length at most β_m for $j = 1, 2, \ldots, m-1$. Moreover, if V is the subspace perpendicular to these first $m-1$ principal axes, then the projection of each piece onto V is contained in a ball of radius β_m. Thus the diameter of each piece is at most $((m-1)\beta_m^2 + (2\beta_m)^2)^{1/2} = (m+3)^{1/2}\beta_m$, as required. $\square$

Let X be open in $\mathbb{R}^n$ and $f : X \to X$ be continuously differentiable. Let E be a compact invariant subset of X so that $f(E) = E$. We aim to estimate the Hausdorff dimension of E in terms of parameters involving f. We assume that f is *uniformly differentiable* on E, so that the derivative of f at x is a linear mapping $f'(x) : \mathbb{R}^n \to \mathbb{R}^n$ satisfying

$$\lim_{y \to x} |f(y) - f(x) - f'(x)(y - x)|/|y - x| = 0 \qquad (12.3)$$

with convergence uniform over all $x \in E$.

In the following proof we cover E by small balls B_i and note that the sets $f^k(B_i)$ are, roughly, ellipsoids with semi-axes of lengths $\frac{1}{2}|B_i|\alpha_j((f^k)'(x))$ where x is the centre of B_i. By covering each of these ellipsoids according to Lemma 12.1 we get a refined covering of the invariant set $E = f^k(E)$, see Figure 12.3. We define

$$\omega_s \equiv \sup_{x \in E} \omega_s(f'(x)) < \infty \qquad (12.4)$$

where $\omega_s(\cdot)$ is the singular value function (12.1).

Theorem 12.2

Let f be as above, with invariant set E. If $\omega_s < 1$ for some $0 < s \leq n$ then $\dim_H E \leq s$.

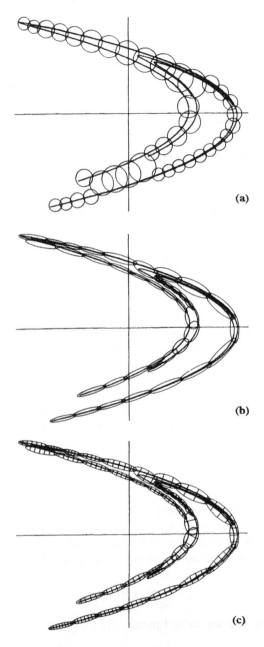

Figure 12.3   Estimation of $\dim_H E$ for a set E that is invariant under a mapping f: (a) E is covered by discs to estimate the dimension. (b) The discs are mapped by f (or more generally by f^k) to approximate ellipsoids. (c) These ellipsoids are sliced into fairly square pieces to get a cover of E giving a better dimension estimate

Proof Let m be the integer such that $m - 1 < s \leq m$, and set $a = 2(m+3)^{1/2}$. Let k be an integer sufficiently large to ensure that

$$w_s^k \leq \min\{(2a)^{-s}, \ 4^{1-m}a^{-s}\}. \tag{12.5}$$

For all $x \in E$,

$$\begin{aligned}
w_s((f^k)'(x)) &= w_s(f'(f^{k-1}x) \circ f'(f^{k-2}x) \circ \cdots \circ f'(x)) \\
&\leq w_s(f'(f^{k-1}x))w_s(f'(f^{k-2}x)) \cdots w_s(f'(x)) \\
&\leq w_s^k, \tag{12.6}
\end{aligned}$$

using the chain rule, (12.2) and (12.4). Note that by (12.1) and (12.5) this implies

$$\alpha_m((f^k)'(x)) \leq w_s((f^k)'(x))^{1/s} \leq (2a)^{-1}. \tag{12.7}$$

Since f is uniformly differentiable on E so is f^k. Thus there exists $r_0 > 0$ such that for all $0 < r \leq r_0$ and $x \in E$ the set $f^k(B(x,r))$ is contained in an ellipsoid with semi-axes of lengths $2r\alpha_1((f^k)'(x)), 2r\alpha_2((f^k)'(x)), \ldots$. We also assume that r_0 is small enough so that $B(x, r_0) \subset X$ for all $x \in E$. By Lemma 12.1 we may cover such an ellipsoid by at most

$$4^{m-1}\alpha_1((f^k)'(x)) \cdots \alpha_{m-1}((f^k)'(x))\alpha_m((f^k)'(x))^{-(m-1)} \tag{12.8}$$

sets of diameters at most $ra\alpha_m((f^k)'(x))$. For $0 < \delta \leq r_0$ suppose $\{U_i\}$ is a δ-cover of E (as in (2.7)); we may assume that each U_i intersects E. Then for each i, $U_i \subset B_i$ for some ball B_i of radius $|U_i|$ and centre $x \in E$. By virtue of (12.8), $f^k(E \cap U_i) \subset f^k(E \cap B_i) \subset \cup_j U_{i,j}$ for sets $\{U_{i,j}\}_j$ of diameter at most $|U_i|a\alpha_m((f^k)'(x)) \leq \frac{1}{2}|U_i| \leq \frac{1}{2}\delta$ (using (12.7)), such that

$$\begin{aligned}
\sum_j |U_{i,j}|^s &\leq 4^{m-1}\alpha_1((f^k)'(x)) \cdots \alpha_{m-1}((f^k)'(x))\alpha_m((f^k)'(x))^{-(m-1)} \\
&\quad \times (|U_i|a\alpha_m((f^k)'(x)))^s \\
&= |U_i|^s 4^{m-1}a^s w_s((f^k)'(x)) \\
&\leq 4^{m-1}a^s|U_i|^s w_s^k \leq |U_i|^s,
\end{aligned}$$

using (12.1) and (12.5). Thus $E = f^k(E) \subset \cup_i f^k(E \cap U_i) \subset \cup_{i,j} U_{i,j}$ where $\sum_{i,j} |U_{i,j}|^s \leq \sum_i |U_i|^s$ and $|U_{i,j}| \leq \frac{1}{2}\delta$ for all i,j. It follows that $\mathcal{H}^s_{\frac{1}{2}\delta}(E) \leq \mathcal{H}^s_\delta(E)$ for all $\delta < r_0$, which implies that $\mathcal{H}^s(E) < \infty$ and $\dim_\mathrm{H} E \leq s$. $\square$

For one way of interpreting this estimate of $\dim_\mathrm{H} E$, write $\alpha_i \equiv w_i/w_{i-1}$, so that α_i reflects the i-th singular values $\alpha_i(f'(x))$, at least in a 'reasonably uniform' situation. Theorem 12.2 gives, on expressing $w_s = 1$ in terms of the α_i,

$$\dim_\mathrm{H} E \leq (m - 1) + \left(\sum_{i=1}^{m-1} \log \alpha_i\right) \Big/ |\log \alpha_{m+1}|, \tag{12.9}$$

where m is the least integer such that $\sum_{i=1}^m \log \alpha_i < 0$.

There are many ways in which Theorem 12.2 can be improved, for example by applying the theorem to f^k rather than to f, noting that $f^k(E) = E$. A similar approach leads to bounds for $\dim_B E$.

Theorem 12.2 and its variants have been used to estimate the dimensions of attractors of a wide variety of dynamical systems and differential equations, though considerable effort is often needed to estimate ω_s. We give three examples of increasing sophistication that show how the method may be used.

The Hénon attractor

Consider the *Hénon mapping* $f : \mathbb{R}^2 \to \mathbb{R}^2$ given by

$$f(x, y) = (y + 1 - ax^2, bx). \tag{12.10}$$

Then f has Jacobian b for all (x, y), so that it contracts area by a constant factor b throughout $\mathbb{R}^2$. The Hénon mapping is, to within a change in coordinates, the most general quadratic mapping with this property, see FG, Section 13.4.

With $a = 1.4$ and $b = 0.3$, the values usually chosen for study, f has an attractor E which looks locally like the product of a Cantor set and a line segment, see Figure 12.4. Taking X to be the quadrilateral region with vertices

$$(1.32, 0.133), (-1.33, 0.42), (-1.06, -0.5) \text{ and } (1.245, -0.14),$$

it may be verified directly that f maps X into itself. The attractor is given by $E = \bigcap_{k=0}^{\infty} f^k(X)$ with the fine structure of E resulting from the repeated stretching and folding effect of iteration by f.

Writing $u = (x, y) \in \mathbb{R}^2$, the derivative $f'(u) : \mathbb{R}^2 \to \mathbb{R}^2$ is given by the matrix

$$f'(u) = \begin{pmatrix} -2ax & 1 \\ b & 0 \end{pmatrix}.$$

In this two-dimensional situation ω_1 and ω_2 are easy to calculate. For $a = 1.4, b = 0.3$ and $u = (x, y) \in X$,

$$\omega_1(f'(u)) = \|f'(u)\| = |4a^2x^2 + 1 + b^2|^{1/2} \le 3.868$$
$$\omega_2(f'(u)) = |\det f'(u)| = |-b| = 0.3.$$

Thus for $1 \le s \le 2$, using (12.1),

$$\omega_s(f'(u)) = \omega_1(f'(u))^{2-s} \omega_2(f'(u))^{s-1} \le 3.868^{2-s} \times 0.3^{s-1}$$

which is less than 1 if $s = 1.53$. Hence $\dim_H E \le 1.53$ by Theorem 12.2.

Numerical estimates suggest that $\dim_H E \simeq 1.26$; our estimate is reasonable given the approximation introduced by taking the worst values of $\omega_s(f'(u))$ over $u \in X$.

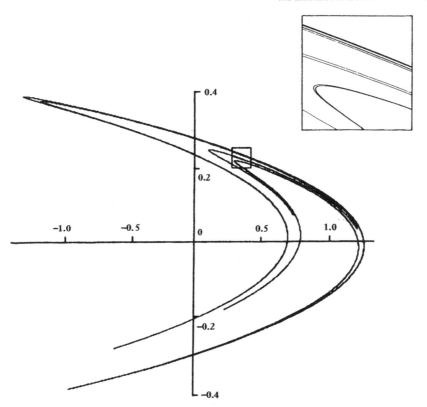

Figure 12.4 The Hénon attractor. The banded fractal structure is becoming apparent in the enlarged square

* The remainder of this section may be omitted on a first reading.

Our next two examples concern attractors of continuous dynamical systems defined by differential equations. We need some more general theory to estimate how infinitesimal line segments, parallelograms and parallelepipeds develop under the system.

Let $g: \mathbb{R}^n \to \mathbb{R}^n$ be sufficiently differentiable, say C^2. We consider the autonomous initial value problem on $\mathbb{R}^n$

$$\frac{\mathrm{d}u}{\mathrm{d}t}(t) = g(u(t)) \quad (t \geq 0) \tag{12.11}$$

$$u(0) = u_0; \tag{12.12}$$

thus $u(t) \in \mathbb{R}^n$ is the position at time t of a particle moving under the system which is at u_0 at time 0. It is useful to write $f_t(u_0) = u(t)$, so that $f_t : \mathbb{R}^n \to \mathbb{R}^n$ specifies where each point of $\mathbb{R}^n$ has moved to at time t. We think of the solutions as defining a 'flow' $u_0 \to f_t(\bar{u}_0)$ on $\mathbb{R}^n$. In our applications, f_t (for suitable t) will play the role of f in the earlier theory. In this notation (12.11) becomes

$$\frac{\mathrm{d}}{\mathrm{d}t} f_t(u_0) = g(f_t(u_0)). \tag{12.13}$$

We examine the evolution of infinitesimal vectors and, more generally, m-dimensional volume elements under this flow. For $t \geq 0, u_0 \in \mathbb{R}^n$ and $\xi \in \mathbb{R}^n$ we define (assuming sufficient differentiability)

$$\xi(t) = f_t'(u_0)\xi, \tag{12.14}$$

where a dash denotes differentiation with respect to the space variable. For small ϵ

$$\epsilon \, \xi(t) = f_t'(u_0)\epsilon \, \xi \simeq f_t(u_0 + \epsilon \xi) - f_t(u_0),$$

so $\xi(t)$ may be thought as the evolution of a vector element being carried by the flow f_t, starting as $\xi = \xi(0)$ and located at u_0 when $t = 0$, see Figure 12.5(a). Assuming that the interchange of order of differentiation is valid and using (12.13),

$$\begin{aligned}
\frac{\mathrm{d}\xi}{\mathrm{d}t}(t) &= \frac{\mathrm{d}}{\mathrm{d}t}(f_t'(u_0)\xi) \\
&= \left(\frac{\mathrm{d}}{\mathrm{d}t} f_t(u_0)\xi\right)' \\
&= (g(f_t(u_0))\xi)' \\
&= g'(f_t(u_0)) \circ f_t'(u_0)\xi \\
&= g'(f_t(u_0))\xi(t).
\end{aligned} \tag{12.15}$$

Equation (12.15) is called the *first variation equation* of the system.

Next we consider the development of an infinitesimal m-dimensional volume element (Figure 12.5(b)). For vectors $\xi_1, \ldots, \xi_m \in \mathbb{R}^n$, we write $|\xi_1 \wedge \xi_2 \wedge \cdots \wedge \xi_m|$ for the m-dimensional volume of the m-dimensional parallelepiped defined by $\xi_1, \ldots, \xi_m$. (This notation is used since this volume is the norm of the exterior product $\xi_1 \wedge \cdots \wedge \xi_n$, though for our brief treatment we avoid any exterior algebra.) With $\xi_1(t), \xi_2(t), \ldots$ as 'infinitesimal vectors' given by (12.14) taking $\xi = \xi_1(0), \xi_2(0), \ldots$, we have that $|\xi_1(t) \wedge \cdots \wedge \xi_m(t)|$ is the m-dimensional volume of the infinitesimal parallelepiped that evolves from that spanned by $\xi_1, \xi_2, \ldots$ and located at u_0 at $t = 0$. We need to extend the first variation equation to a differential equation for $|\xi_1(t) \wedge \cdots \wedge \xi_m(t)|$.

We write $\langle \ , \ \rangle$ for the usual scalar product on $\mathbb{R}^n$, and recall that the *trace* of a linear mapping $L : \mathbb{R}^n \to \mathbb{R}^n$ is defined by $\mathrm{Tr}(L) = \sum_{i=1}^{n}\langle Le_i, e_i\rangle$ where $(e_1, e_2, \ldots)$ is any orthonormal basis of $\mathbb{R}^n$. This definition is independent of the orthonormal basis chosen, and if L is represented by a matrix with respect to an orthonormal basis then $\mathrm{Tr}(L)$ is the sum of the elements on the leading diagonal.

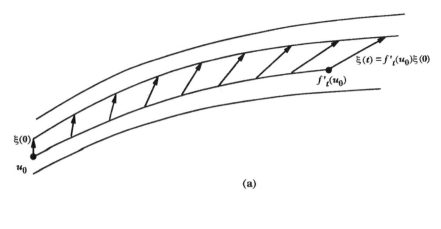

(a)

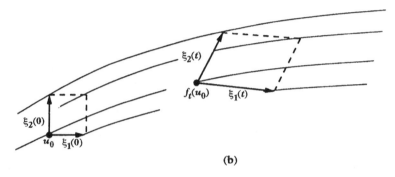

(b)

Figure 12.5 (a) Evolution of an infinitesimal vector $\xi = \xi(0)$ under the flow f_t. (b) Evolution of an infinitesimal rectangle under the flow f_t

Lemma 12.3

The m-dimensional volume of an infinitesimal parallelepiped evolves according to

$$\frac{d}{dt}|\xi_1(t) \wedge \cdots \wedge \xi_m(t)| = |\xi_1(t) \wedge \cdots \wedge \xi_m(t)|\mathrm{Tr}(g'(f_t\,u_0) \circ Q_m(t)) \qquad (12.16)$$

where $Q_m(t) : \mathbb{R}^n \to \mathbb{R}^n$ is orthogonal projection onto $\mathrm{span}\{\xi_1(t), \ldots, \xi_m(t)\}$. Thus

$$|\xi_1(t) \wedge \cdots \wedge \xi_m(t)| = |\xi_1(0) \wedge \cdots \wedge \xi_m(0)| \exp \int_0^t \mathrm{Tr}(g'(f_\tau u_0) \circ Q_m(\tau))d\tau$$

$$(12.17)$$

for $t \geq 0$.

Sketch of proof First assume that $\xi_i \equiv \xi_i(t)$ $(i = 1, \ldots, m)$ are orthonormal (we supress the dependence on t for brevity). Using (12.15), in the time interval $[t, t + \delta t]$ the cube spanned by the infinitesimal vectors $\xi_1, \ldots \xi_m$ is mapped to the parallelepiped spanned by $\xi_1 + g'(f_t u_0)\xi_1 \delta t, \ldots, \xi_m + g'(f_t u_0)\xi_m \delta t$. To the first order, the change in the m-dimensional volume of the cube results from the component of the increase of each ξ_i in the direction of ξ_i. Thus to order δt the volume of the new parallelepiped equals that of the rectangular parallelepiped spanned by

$$\xi_1(1 + \langle g'(f_t u_0)\xi_1 \delta t, \xi_1 \rangle), \ldots, \xi_m(1 + \langle g'(f_t u_0)\xi_m \delta t, \xi_m \rangle).$$

This parallelepiped has volume

$$1 + \delta t \sum_{i=1}^{m} \langle g'(f_t u_0)\xi_i, \xi_i \rangle + O(\delta t^2) = 1 + \delta t \, \text{Tr}(g'(f_t u_0) \circ Q_m) + O(\delta t^2),$$

compared with the original volume $|\xi_1 \wedge \cdots \wedge \xi_m| = 1$. Thus (12.16) holds if $\xi_1, \ldots, \xi_m$ are orthonormal.

If V is an m-dimensional subspace of $\mathbb{R}^n$ and L a linear mapping on $\mathbb{R}^n$, then the m-dimensional volume of $L(A)$ is proportional to that of A for all regions $A \subset V$. Thus the scaling effect of g on the volume of an infinitesimal parallelpiped defined by $\xi_1, \ldots, \xi_m$ depends only on the space spanned by $\xi_1, \ldots, \xi_m$ and not on the choice of spanning vectors. Thus we have lost nothing by assuming $\xi_1, \ldots, \xi_m$ to be orthonormal, and (12.16) holds in the general case.

Equation (12.16) is a first order linear ordinary differential equation which integrates to (12.17). □

To relate this to singular value functions, we note that for a linear mapping $L : \mathbb{R}^n \to \mathbb{R}^n$ and for $m = 1, 2, \ldots$

$$\omega_m(L) = \sup_{\xi_1, \ldots, \xi_m \in B} |L(\xi_1) \wedge \cdots \wedge L(\xi_m)| \tag{12.18}$$

where B is the unit ball in $\mathbb{R}^n$. This is because the m-dimensional volume of the parallelepiped defined by $L(\xi_1), \ldots, L(\xi_m)$ is maximal when these vectors are the m longest principal semi-axes of $L(B)$.

Proposition 12.4

With notation as above

$$\omega_m(f_t'(u_0)) \leq \sup_{\xi_1, \ldots, \xi_m \in B} \exp \int_0^t \text{Tr}(g'(f_\tau u_0) \circ Q_m(\tau))d\tau. \tag{12.19}$$

where $Q_m(\tau)$ is orthogonal projection onto $\text{span}\{\xi_1(\tau), \ldots, \xi_m(\tau)\}$.

Proof From (12.18) and (12.14)

$$\omega_m(f'_t(u_0)) = \sup_{\xi_1,\dots,\xi_m \in B} |f'_t(u_0)(\xi_1) \wedge \cdots \wedge f'_t(u_0)(\xi_m)|$$

$$= \sup_{\xi_1(0),\dots,\xi_m(0) \in B} |\xi_1(t) \wedge \cdots \wedge \xi_m(t)|$$

so (12.19) follows from (12.17) since $\sup_{\xi_1(0),\dots,\xi_m(0)\in B}|\xi_1(0) \wedge \cdots \wedge \xi_m(0)| = 1$.
□

Using Proposition 12.4 to estimate ω_s, Theorem 12.2, gives reasonable upper bounds for the dimension of the attractors of many differential equations.

The Lorenz attractor

The Lorenz equations give an approximate description of the behaviour of fluid convection in cylindrical rolls in a two-dimensional layer heated from below (see FG, Section 13.5). With an appropriate choice of origin, the equations may be written

$$\begin{aligned}
\dot{x} &= \sigma(y - x) \\
\dot{y} &= -\sigma x - y - xz \\
\dot{z} &= xy - bz - b(r + \sigma).
\end{aligned} \tag{12.20}$$

Here x is the rate of rotation of the cylinder, y is the temperature difference between opposite sides of the cylinder, z measures the deviation from a linear vertical temperature gradient, and a 'dot' denotes differentiation with respect to time. The positive constants σ, b and r represent respectively the Prandtl number of the fluid (which depends on the viscosity and thermal conductivity), the width to height ratio of the layer, and the fixed temperature difference between the bottom and top of the system. We assume that $\sigma > b + 1$. Lorenz demonstrated that with

$$\sigma = 10, \; b = \tfrac{8}{3}, \; r = 28 \tag{12.21}$$

there is a chaotic attractor with two 'wings', with trajectories flipping between the wings in a seemingly arbitrary manner (see Figure 12.6).

We regard $u = (x, y, z)$ as a function of time, $u : \mathbb{R} \to \mathbb{R}^3$. From (12.20)

$$\begin{aligned}
\tfrac{1}{2}\frac{\mathrm{d}}{\mathrm{d}t}|u|^2 &= x\dot{x} + y\dot{y} + z\dot{z} \\
&= -\sigma x^2 - y^2 - bz^2 - bz(r + \sigma) \\
&\le -(x^2 + y^2 + z^2) - (b-1)z^2 - bz(r + \sigma) \\
&\le -|u|^2 + \frac{b^2(r + \sigma)^2}{4(b-1)},
\end{aligned}$$

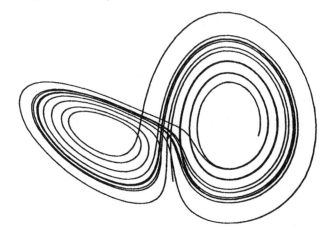

Figure 12.6 A view of the Lorenz attractor for $\sigma = 10$, $b = 8/3$ and $r = 28$. The trajectory spirals round the two 'wings' and 'jumps' from one to the other

using that $\sigma > 1$ and the usual estimate for the maximum of a quadratic expression. Then

$$\frac{\mathrm{d}}{\mathrm{d}t}(|u(t)|^2 e^{2t}) \leq \frac{b^2(r+\sigma)^2}{2(b-1)} e^{2t},$$

and integrating

$$|u(t)|^2 \leq |u(0)|^2 e^{-2t} + \frac{b^2(r+\sigma)^2}{4(b-1)}(1 - e^{-2t}).$$

In particular,

$$\limsup_{t\to\infty} |u(t)| \leq 2\rho_0 \qquad (12.22)$$

say, where $\rho_0 \equiv b(r+\sigma)/4(b-1)^{1/2}$. Thus $u(t)$ is close to, or inside the ball $B(0, 2\rho_0)$ when t is large, and this implies that there is a (maximal) compact set $E \subset B(0, 2\rho_0)$ that is invariant under the solution trajectories, that is with $f_t E = E$ for all $t \geq 0$, where $f_t u_0$ is the solution $u(t)$ such that $u(0) = u_0$. With parameter values (12.21) the set E is the *Lorenz attractor* shown in Figure 12.6.

We apply Theorem 12.2 to estimate the dimension of E, using Proposition 12.4 to estimate $\omega_2(f_t u_0)$ and $\omega_3(f_t u_0)$ for large t. Thinking of (12.20) as $\dot{u} = g(u)$, where $u = (x, y, z)$, the derivative $g'(u) : \mathbb{R}^3 \to \mathbb{R}^3$ has matrix

$$g'(u) = \begin{pmatrix} -\sigma & \sigma & 0 \\ -\sigma - z & -1 & -x \\ y & x & -b \end{pmatrix}.$$

Thus $\mathrm{Tr}(g'(f_t u_0)) = -(\sigma + b + 1)$ for all t and u_0 (representing a constant volume contraction rate), and $Q_3(t)$ is the identity, so by (12.19)

$$\omega_3(f_t'(u_0)) \le \exp(-(\sigma + b + 1)t). \tag{12.23}$$

To estimate ω_2 write

$$g'(u) = L_1 + L_2 + L_3 \tag{12.24}$$

where the linear mappings L_1, L_2 and L_3 are given by

$$L_1 = \begin{pmatrix} -\sigma & 0 & 0 \\ 0 & -1 & 0 \\ 0 & 0 & -b \end{pmatrix} \quad L_2 = \begin{pmatrix} 0 & \sigma & 0 \\ -\sigma & 0 & -x \\ 0 & x & 0 \end{pmatrix}$$

$$L_3 = \begin{pmatrix} 0 & 0 & 0 \\ -z & 0 & 0 \\ y & 0 & 0 \end{pmatrix}.$$

Then if Q_2 denotes projection onto some two-dimensional subspace, $\text{Tr}(L_1 \circ Q_2) \le -b-1$ (since $\sigma > b+1$) and $\text{Tr}(L_2 \circ Q_2) = 0$ (since anti-symmetric mappings have zero trace). For L_3, let $Q_2(t)$ denote orthogonal projection onto $\text{span}\{\xi_1(t), \xi_2(t)\}$ and let v_1, v_2, v_3 be an orthonormal basis of $\mathbb{R}^3$ such that $\text{span}\{v_1, v_2\} = Q_2(t)(\mathbb{R}^3)$. Then writing $v_i = (x_i, y_i, z_i)$

$$\begin{aligned}
\text{Tr}(L_3 \circ Q_2(t)) &= \langle L_3 v_1, v_1 \rangle + \langle L_3 v_2, v_2 \rangle \\
&= -zx_1 y_1 + yx_1 z_1 - zx_2 y_2 + yx_2 z_2 \\
&= zx_3 y_3 - yx_3 z_3,
\end{aligned}$$

since $(x_1, x_2, x_3), (y_1, y_2, y_3)$ and (z_1, z_2, z_3) are also orthonormal. By Cauchy's inequality followed by the arithmetic-geometric mean inequality

$$\begin{aligned}
\text{Tr}(L_3 \circ Q_2(t)) &\le |x_3|(y_3^2 + z_3^2)^{1/2}(z^2 + y^2)^{1/2} \\
&\le \tfrac{1}{2}(x_3^2 + y_3^2 + z_3^2)(z^2 + y^2)^{1/2} \\
&\le \tfrac{1}{2}|u|.
\end{aligned}$$

Given $\delta > 0$, if t is large enough,

$$\text{Tr}(L_3 \circ Q_2(t)) \le \rho_0 + \tfrac{1}{2}\delta$$

using (12.22). Thus for such t the decomposition (12.24) gives

$$\text{Tr}(g'(f_t(u_0)) \circ Q_2(t)) \le -b - 1 + \rho_0 + \tfrac{1}{2}\delta,$$

so from (12.19)

$$\omega_2(f_t'(u_0)) \le \exp((-b - 1 + \rho_0 + \delta)t) \tag{12.25}$$

for sufficiently large t, provided $-b - 1 + \rho_0 > 0$.

By (12.1), (12.23) and (12.25), for all $u_0 \in E$ and sufficiently large t, if $2 \leq s \leq 3$

$$\omega_s(f_t'(u_0)) = \omega_2(f_t'(u_0))^{3-s}\omega_3(f_t'(u_0))^{s-2}$$
$$\leq \exp((3-s)(-b-1+\rho_0+\delta)t - (s-2)(\sigma+b+1)t)$$
$$\leq \exp((s-2)(-\rho_0-\sigma-\delta) + (\rho_0-b-1+\delta))t.$$

Thus if $(s-2) < (\rho_0 - b - 1)/(\rho_0 + \sigma)$ we see by choosing δ small enough that $\omega_s = \sup_{u_0 \in E} \omega_s(f_t'(u_0)) < 1$ for sufficiently large t. It follows from Theorem 12.2 that

$$\dim_H E \leq 2 + (\rho_0 - b - 1)/(\rho_0 + \sigma) \quad \text{where} \quad \rho_0 = b(r+\sigma)/4(b-1)^{1/2}.$$

$$(12.26)$$

With the usual values (12.21) for the Lorenz attractor E, we get $\dim_H E \leq 2.539$. These calculations may be refined to give better estimates; numerical evidence suggests that the dimension is actually about 2.05.

A similar method may be used to examine spatial attractors of certain other autonomous systems of differential equations.

The basic method may be extended to estimate dimensions of *functional attractors* of certain spatio-temporal partial differential equations. Thus we regard a solution $u(x, t)$ with $x \in D \subset \mathbb{R}^n$ and $t > 0$ as a point $u(\cdot, t)$ in a space of functions on $\mathbb{R}^n$, and look at the dimension of the set of functions that are limit points as $t \to \infty$ of such solutions. We indicate this procedure in the case of a simple non-linear partial differential equation.

A reaction-diffusion equation

Let D be a bounded open domain in $\mathbb{R}^n$ with smooth boundary ∂D, let p be a polynomial of odd degree with a positive leading coefficient, and fix $\epsilon > 0$. We consider the solutions $u(x, t)$ of the reaction-diffusion equation

$$\frac{\partial u}{\partial t} - \epsilon \nabla^2 u + p(u) = 0 \quad \text{for} \quad x \in D, t \geq 0 \tag{12.27}$$

with boundary condition

$$u(x, t) = 0 \quad \text{for} \quad x \in \partial D, t \geq 0 \tag{12.28}$$

and initial condition

$$u(x, 0) = u_0(x) \quad \text{for} \quad x \in D. \tag{12.29}$$

Equation (12.27) is known as the Allen–Cahn equation and has been used to model phase transitions. To apply the method to this situation, we replace $\mathbb{R}^n$ by the Hilbert space H of square integrable functions on D endowed with the usual inner product $\langle v_1, v_2 \rangle = \int_D v_1(x)v_2(x)dx$. We think of the time development of the solution (12.27)–(12.29) as a flow in H. Thus we write $f_t(u_0)$ for the solution $u(\cdot, t)$ at time t, regarded as a point in H, corresponding to the initial condition $u(\cdot, 0) = u_0$. Then (12.27) may be written in the form of (12.11) as

$$\frac{du}{dt}(t) = g(u(\cdot, t)) \quad \text{with} \quad g(u) = \epsilon \nabla^2 u - p(u), \tag{12.30}$$

where du/dt is now the Fréchet derivative of a Hilbert space valued function of t.

It may be shown that for each $u_0 \in H$, the problem (12.27)–(12.29) has a unique solution $u(x, t)$ that is continuous in x and t, with $f_t(u_0) = u(\cdot, t) \in H$ for all $t \geq 0$. Moreover, $u(\cdot, t) \in H^1$ where H^1 is the space of Fréchet differentiable functions with derivative in H.

The system (12.27)–(12.29) may be shown to have a maximal attractor E that is a compact subset of H such that $f_t(E) = E$ for all $t > 0$, with the Hilbert space distance $\text{dist}(f_t(u_0), E) \to 0$ as $t \to \infty$ for all $u_0 \in H$.

The dimension of E may be estimated in a way parallel to that followed for the Lorenz attractor. We are now working in an infinite dimensional space H, but nevertheless the singular values and singular value function $\omega_s(L)$ of a linear operator L on H may be defined for $s > 0$ in just the same way as in a finite dimensional space. Moreover, Theorem 12.2 remains valid with minimal modification to the proof. As before, we can examine the development of the infinitesimal vectors to get the first variation equation (12.15) for $\xi(t) = f_t'(u_0)\xi \in H$ which in this case is

$$\frac{d\xi}{dt} = g'(f_t(u_0))\xi(t) = \epsilon \nabla^2 \xi - p'(f_t(u_0))\xi \quad \text{for} \quad t > 0, \quad \xi \in H \tag{12.31}$$

with

$$\xi(t) = 0 \text{ on } \partial D \quad \text{and} \quad \xi(0) = \xi \text{ on } D. \tag{12.32}$$

(This formal assertion needs justification in the appropriate function spaces.)

Using the first variation equation we can examine the development of infinitesimal parallelepipeds spanned by $\xi_1(t), \ldots, \xi_m(t)$, where $\xi_i(t) \in H$ is the infinitesimal vector evolving from $\xi_i(0) = \xi \in H$. There is no problem in defining the m-dimensional volume of an m-dimensional parallelepiped in the Hilbert space H, and Lemma 12.3 and Proposition 12.4 are valid just as in the finite dimensional case.

To apply Proposition 12.4 in the Hilbert space situation we estimate $\text{Tr}(g'(f_t(u_0)) \circ Q_m(t))$ where $Q_m(t)$ denotes orthogonal projection onto $\text{span}\{\xi_1(t), \ldots, \xi_m(t)\}$. Fixing t for the time being, let $v_1, v_2 \ldots$ be an orthonormal basis of H, with $\text{span}\{v_1, \ldots, v_m\} = \text{span}\{\xi_1(t), \ldots, \xi_m(t)\} = \text{span } Q_m(t)H$. Since $\xi_j(t) \in H^1$ for all j and $t > 0$ we have $v_j \in H^1$ for all j. Then $Q_m(t)v_j = v_j$ (if $j \leq m$) $= 0$ (otherwise), so for $j \leq m$

$$\langle g'(f_t(u_0)) \circ Q_m(t)v_j, v_j \rangle = \langle g'(f_t(u_0))v_j, v_j \rangle$$
$$= \epsilon \langle \nabla^2 v_j, v_j \rangle - \langle p'(f_t(u_0))v_j, v_j \rangle$$
$$= \epsilon \langle \nabla^2 v_j, v_j \rangle - \int_D p'(f_t(u_0))v_j^2 dx$$

using (12.31). Since p is an odd degree polynomial with positive leading coefficient, there is a number $\kappa \geq 0$, depending only on p, such that $p'(s) \geq -\kappa$ for all $s \in \mathbb{R}$. Thus

$$\langle g'(f_t(u_0)) \circ Q_m(t) v_j, v_j \rangle \leq \epsilon \langle \nabla^2 v_j, v_j \rangle + \kappa$$

since $\int_D v_j^2 dx = 1$ by orthonormality of the v_i. Hence

$$\mathrm{Tr}(g'(f_t(u_0)) \circ Q_m(t)) = \sum_{j=1}^{m} \langle g'(f_t(u_0)) \circ Q_m(t) v_j, v_j \rangle$$

$$\leq \epsilon \sum_{j=1}^{m} \langle \nabla^2 v_j, v_j \rangle + \kappa m. \tag{12.33}$$

Let $0 < \lambda_1 < \lambda_2 < \ldots$ be the eigenvalues of $-\nabla^2$ for the Dirichlet problem in D (that is $\nabla^2 u + \lambda u = 0$ with $u = 0$ on ∂D). Then for any orthonormal sequence $v_1, v_2, \ldots \in H$ we have

$$-\sum_{j=1}^{m} \langle \nabla^2 v_j, v_j \rangle \geq \lambda_1 + \cdots + \lambda_m. \tag{12.34}$$

(This may be established by noting that

$$v_1 \wedge \cdots \wedge v_m \mapsto (-\nabla^2 v_1) \wedge v_2 \wedge \cdots \wedge v_m + \cdots + v_1 \wedge v_2 \wedge \cdots \wedge (-\nabla^2 v_m)$$

is an (unbounded) self-adjoint transformation on the m-fold exterior product of H which has $\lambda_1 + \cdots + \lambda_m$ as its least eigenvalue.) By Weyl's theorem (see (12.40)) the asymptotic distribution of the eigenvalues is given by $\lambda_j \sim c_0 \mathcal{L}^n(D)^{-2/n} j^{2/n}$, so summing, $\lambda_1 + \cdots + \lambda_m \geq c_1 \mathcal{L}^n(D)^{-2/n} m^{1+2/n}$, where c_0 and c_1 depend only on n, and $\mathcal{L}^n(D)$ is the n-dimensional Lebesgue measure of the domain D. Thus from (12.33) and (12.34)

$$\mathrm{Tr}(g'(f_t(u_0)) \circ Q_m(t)) \leq -c_1 \epsilon \mathcal{L}^n(D)^{-2/n} m^{1+2/n} + \kappa m. \tag{12.35}$$

If the integer m is chosen to make the right-hand side of this expression negative, less than $-\delta$, say, then from (12.19)

$$\omega_m(f_t'(u_0)) \leq e^{-\delta t}.$$

Then by Theorem 12.2, which remains valid in the Hilbert space setting, $\dim_H E \leq m$ where E is the functional attractor of the system. By equating the right-hand side of (12.35) to 0,

$$\dim_H E \leq 1 + c\epsilon^{-n/2} k^{n/2} \mathcal{L}^n(D), \tag{12.36}$$

where $c = c_1^{-n/2}$ depends only on n and the shape of D. (In fact c can be found quite accurately by careful consideration of the distribution of the eigenvalues of ∇^2.) Inequality (12.36) shows clearly the dependence of the dimension of the functional attractor on the volume of D, the constant ϵ in the partial differential equation (12.27), and κ which depends explicitly on the polynomial p.

The dimension of functional attractors of many other differential equations may be studied using extensions of the methods sketched in this section, for example the Navier–Stokes equation, pattern formation equations and non-linear Schrödinger equations. The functional attractor represents the permanent regime that can be observed when the system starts at any point in function space. Its dimension indicates the degree of complexity of the flow, and may be thought of as the number of degrees of freedom of the phenomenon that it represents.

12.2 Eigenvalues of the Laplacian on regions with fractal boundary

Let $D \subset \mathbb{R}^n (n \geq 1)$ be a bounded open region with boundary ∂D (we do not insist that D is connected). We consider the eigenvalue problem

$$\nabla^2 u = -\lambda u \quad \text{in} \quad D \tag{12.37}$$

with Dirichlet boundary condition

$$u(x) = 0 \text{ for } x \in \partial D \tag{12.38}$$

so that the (real) eigenvalues $0 < \lambda_1 < \lambda_2 < \ldots$ are those λ for which there is a non-trivial solution. The eigenvalues may be thought of as the principal frequencies of vibration of an (n-dimensional) membrane stretched across the region D.

We define the *eigenvalue counting function*

$$N(\lambda) = \#(k : \lambda_k \leq \lambda). \tag{12.39}$$

We are interested in the behaviour of $N(\lambda)$ for large λ, and in particular how this reflects the nature of the boundary of D. A classical result of Weyl states that if ∂D is sufficiently smooth then

$$N(\lambda) \sim c_n \mathcal{L}^n(D)\lambda^{n/2} \tag{12.40}$$

as $n \to \infty$, where $c_n = (2\pi)^{-n} \mathcal{L}^n(B)$, and B is the unit ball in $\mathbb{R}^n$ and $\mathcal{L}^n$ is n-dimensional volume. Moreover, if the boundary ∂D is sufficiently smooth,

$$N(\lambda) = c_n \mathcal{L}^n(D)\lambda^{n/2} + b_n \mathcal{L}^{n-1}(\partial D)\lambda^{(n-1)/2} + o(\lambda^{(n-1)/2}), \tag{12.41}$$

for a constant b_n depending only on n. Thus the 'surface area' of ∂D determines the second term in the asymptotic expansion of $N(\lambda)$; notice that the exponent $(n-1)/2$ is half the dimension of the (smooth) boundary.

We seek analogues of (12.41) for regions with *fractal* boundaries. In particular 'can one hear the dimension of a fractal?', that is, can the dimension of ∂D be recovered from a knowledge of the eigenvalues? Here we can do little more than indicate that there is often a connection between the dimension of the boundary and the second term of the expansion for $N(\lambda)$.

We first consider a one-dimensional version of the problem (12.37)–(12.38), which is rather artificial since the region is not connected, but which is

Figure 12.7 Two of the eigenfunctions of the Laplacian on the complement of the Cantor set. These may be thought of as resonances of a string stretched across the Cantor set.

reasonably tractable since it is possible to express $N(\lambda)$ in a closed form. We use the notation of Section 3.2. Let $A \subset \mathbb{R}$ be a bounded closed interval, and let $A_1, A_2 \ldots$ be a sequence of disjoint open subintervals in order of decreasing length with

$$|A| = \sum_{i=1}^{\infty} |A_i|. \tag{12.42}$$

We work on the region $D = \cup_{i=1}^{\infty} A_i$ which is a 'cut-out set' with boundary

$$\partial D = A \setminus \bigcup_{i=1}^{\infty} A_i; \tag{12.43}$$

recall that the box dimensions of ∂D were studied in Proposition 3.6. The eigenvalue problem on D may be thought of as finding the resonant frequencies of a string stretched across A and fixed at the points of ∂D so that it can vibrate independently on each interval A_i. Thus we seek solutions of

$$\frac{\partial^2 u}{\partial x^2} = -\lambda u \quad \text{in } D \quad \text{with} \quad u(x) = 0 \quad \text{on} \quad \partial D. \tag{12.44}$$

Non-trivial solutions occur at the resonant frequencies of each interval A_i: for $\lambda = (\pi k/|A_i|)^2$ $(k = 1, 2, \ldots)$, (12.44) has a sinusoidal solution of frequency $\pi k/|A_i|$ which vanishes off A_i (Figure 12.7). Counting these eigenvalues for all intervals A_i gives

$$\begin{aligned}
N(\lambda) &= \sum_{i=1}^{\infty} \#(k : (\pi k/|A_i|)^2 \le \lambda) \\
&= \sum_{i=1}^{\infty} \lfloor \pi^{-1} \lambda^{1/2} |A_i| \rfloor \\
&= \sum_{i=1}^{\infty} \pi^{-1} \lambda^{1/2} |A_i| - \sum_{i=1}^{\infty} \{\pi^{-1} \lambda^{1/2} |A_i|\} \\
&= \pi^{-1} \mathcal{L}^1(D) \lambda^{1/2} - \psi(\lambda) \tag{12.45}
\end{aligned}$$

using (12.42), where

$$\psi(\lambda) = \sum_{i=1}^{\infty} \{\pi^{-1}\lambda^{1/2}|A_i|\}. \tag{12.46}$$

(Here we use $\lfloor \ \rfloor$ to mean 'the greatest integer less than' and $\{ \ \}$ to mean 'the fractional part of'.) The first term in (12.45) is just the Weyl expression (12.40) when $n = 1$; we show that under certain conditions the remainder term $\psi(\lambda)$ is of order $\lambda^{s/2}$.

Proposition 12.5

Let $D = \cup_{i=1}^{\infty} A_i \subset \mathbb{R}$ be constructed as above. Let

$$N(\lambda) = \pi^{-1}\mathcal{L}^1(D)\lambda^{1/2} - \psi(\lambda). \tag{12.47}$$

Then

(a) if $|A_i| \asymp i^{-1/s}$ for some $0 < s < 1$ then $\psi(\lambda) \asymp \lambda^{s/2}$ as $\lambda \to \infty$,
(b) if $\underline{\dim}_B \partial D = \overline{\dim}_B \partial D = s$ where $0 < s < 1$ then $\lim_{\lambda \to \infty} \log\psi(\lambda)/\log\lambda = \frac{1}{2}s$.

Proof

(a) Given $\lambda > \pi^2|A_1|^{-2}$, let k be the greatest integer such that $\pi^{-1}\lambda^{1/2}|A_k| \geq 1$. From the hypothesis of (a)

$$c^{-1}\lambda^{s/2} \leq k \leq c\lambda^{s/2} \tag{12.48}$$

when $c > 0$ is independent of λ. Using (12.46)

$$\sum_{i=k+1}^{\infty} \pi^{-1}\lambda^{1/2}|A_i| \leq \psi(\lambda) \leq k + \sum_{i=k+1}^{\infty} \pi^{-1}\lambda^{1/2}|A_i|$$

Thus there exists $c_1 > 0$ such that

$$c_1^{-1}\lambda^{1/2}\sum_{i=k+1}^{\infty} i^{-1/s} \leq \psi(\lambda) \leq k + c_1\lambda^{1/2}\sum_{i=k+1}^{\infty} i^{-1/s}$$

and hence $c_2 > 0$ such that

$$c_2^{-1}\lambda^{1/2}k^{1-1/s} \leq \psi(\lambda) \leq k + c_2\lambda^{1/2}k^{1-1/s},$$

using the 'integral test estimate' $\sum_{i=k+1}^{\infty} i^{-1/s} \asymp k^{1-1/s}$. Incorporating (12.48) immediately gives $\psi(\lambda) \asymp \lambda^{s/2}$.
(b) By Corollary 3.8 the condition $\dim_B \partial D = s$ implies $\lim_{i \to \infty} \log|A_i|/\log i = -1/s$, so that for all $\epsilon > 0$ we have

$$i^{-\epsilon-1/s} \leq |A_i| \leq i^{\epsilon-1/s}$$

for sufficiently large i. Proceeding just as in (a) we get

$$\lambda^{s(1-\epsilon)/2} \leq \psi(\lambda) \leq \lambda^{s(1+\epsilon)/2}$$

if λ is large enough. $\square$

Thus, for highly disconnected regions in $\mathbb{R}$, the second term in the expansion of $N(\lambda)$ does indeed depend on the box dimension of the boundary.

These arguments may be taken further. For example if the condition (a) of Proposition 12.5 is strengthened to $|A_i| \sim c i^{-1/s}$ then

$$\psi(\lambda) = \pi^{-s} c^s \zeta(s) \lambda^{s/2} + o(\lambda^{s/2}) \qquad (12.49)$$

where ζ is the Riemann zeta function; this may be established by number-theoretic type estimates on the sum (12.46). For self-similar sets, this may be combined with the renewal theorem estimates of Chapter 7. Let ∂D be a self-similar set of dimension s constructed from $[0,1]$ so that the first step of the construction consists of intervals of lengths $r_1, \ldots, r_m$ with gaps of lengths $b_1, \ldots, b_{m-1}$ in between. Provided $\{\log r_1^{-1}, \ldots, \log r_m^{-1}\}$ is a non-arithmetic set it may be shown using Corollary 7.3 (see Exercise 12.5) that

$$|A_i| \sim i^{-1/s} s^{-1/s} \left(\sum_{i=1}^{m-1} b_i^s \Big/ \sum_{i=1}^{m} r_i^s \log r_i^{-1} \right)^{1/s}, \qquad (12.50)$$

where A_i is the i-th longest gap in ∂D, so

$$\psi(\lambda) = \pi^{-s} s^{-1} \zeta(s) \left(\sum_{i=1}^{m-1} b_i^s \Big/ \sum_{i=1}^{m} r_i^s \log r_i^{-1} \right) \lambda^{s/2} + o(\lambda^{s/2}). \qquad (12.51)$$

In the arithmetic situation, $\psi(\lambda) = \lambda^{s/2} p(\log \lambda) + o(\lambda^{s/2})$ for a periodic function p.

For plane regions D, we have the Weyl estimate $N(\lambda) \sim \frac{1}{4} \pi^{-1} \mathcal{L}^2(D) \lambda$ for large λ, and one might hope that the second term of the expansion would again reflect the fractality of ∂D. However, the situation is more complicated in the plane, since in general regions cannot be decomposed into parts which may be regarded as independent as was possible in one dimension. Nevertheless, some progress has been made, and the asymptotic behaviour of the eigenvalues of (12.37)–(12.38) on certain bounded regions D in $\mathbb{R}^n$ has been related to the *interior Minkowski dimension* of ∂D, given by

$$\dim_{\mathrm{I}} \partial D = n - \lim_{r \to 0} \log \mathcal{L}^n(E_r) / \log r \qquad (12.52)$$

assuming that this limit exists, where here E_r is the *interior r-neighbourhood* of ∂D defined as

$$E_r = \{x \in D : \mathrm{dist}(x, \partial D) \leq r\}, \qquad (12.53)$$

see Figure 12.8. (The interior Minkowski dimension may be thought of as a 'one-sided box dimension', compare (2.4)–(2.6).)

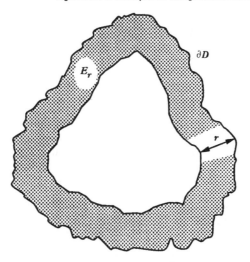

Figure 12.8   The interior r-neighbourhood E_r of ∂D

Most estimates of eigenvalue counting functions for plane regions are based on explicit calculations that are possible in the case of a square domain. For D a square of side a, the problem

$$\frac{\partial^2 u}{\partial x^2} + \frac{\partial^2 u}{\partial y^2} = -\lambda u \quad \text{in } D \quad \text{with} \quad u(x,y) = 0 \quad \text{on } \partial D \tag{12.54}$$

has eigenvalues $\lambda = \pi^2(k^2 + m^2)/a^2$ and corresponding eigenfunctions of the form $u(x,y) = \sin(k\pi x/a)\sin(m\pi y/a)$ for $k,m \in \mathbb{Z}^+$. Thus for the square of side a,

$$N(\lambda) = \#\{(k,m) \in \mathbb{Z}^+ \times \mathbb{Z}^+ : k^2 + m^2 \leq a^2 \pi^{-2}\lambda\},$$

that is the number of integer coordinate lattice points in $Q(0, a\pi^{-1}\lambda^{1/2})$ where $Q(0,r)$ denotes the strictly positive quadrant of radius r and centre the origin. It is easy to see that

$$\#\{\text{lattice points in } Q(0,r)\} = 0 \quad \text{if} \quad r < 1 \tag{12.55}$$

and

$$0 \leq \text{area } Q(0,r) - \#\{\text{lattice points in } Q(0,r)\} \leq 2r \tag{12.56}$$

for all $r > 0$, see Exercise 12.3. Thus for a square of side a

$$N(\lambda) = \tfrac{1}{4}a^2\pi^{-1}\lambda - \psi(\lambda) \tag{12.57}$$

where

$$0 \leq \psi(\lambda) \leq 2a\pi^{-1}\lambda^{1/2} \tag{12.58}$$

and
$$\psi(\lambda) = \tfrac{1}{4}a^2\pi^{-1}\lambda \quad \text{for} \quad \lambda < \pi^2 a^{-2} \tag{12.59}$$

(so that $N(\lambda) = 0$ for $\lambda < \pi^2 a^{-2}$).

Estimates of eigenvalue counting functions for (rather artificial) fractal regions consisting of an infinite sequence of disjoint squares may now be obtained in the same way as in Proposition 12.5. For example, let A_i be an open square of side a_i, where $a_1 \geq a_2 \geq \dots$, and suppose $D = \cup_{i=1}^{\infty} A_i$ is a disjoint union with D bounded. If $a_i \asymp i^{-1/s}$ where $1 < s < 2$ then $\dim_1 \partial D = s$, and by proceeding just as in Proposition 12.5 but using (12.57)–(12.59) we get

$$N(\lambda) = \tfrac{1}{4}\pi^{-1}\mathcal{L}^2(D)\lambda - \psi(\lambda) \tag{12.60}$$

where $\psi(\lambda) \asymp \lambda^{s/2}$. By using more delicate estimates for the number of lattice points in $Q(0,r)$ rather closer estimates for $\psi(\lambda)$ may be found. Moreover, if the squares A_i are arranged to abut and small gaps of rapidly decreasing lengths are made between neighbouring squares, these estimates may be extended to certain connected domains D.

An extension of this idea leads to a lower bound for $N(\lambda)$ for more general regions. We define the *upper interior Minkowski dimension* of the boundary ∂D of plane region by

$$\overline{\dim}_1 \partial D = 2 - \liminf_{r \to 0} \log \mathcal{L}^2(E_r)/\log r, \tag{12.61}$$

where E_r is the interior r-neighbourhood of ∂D.

Proposition 12.6

Let $D \subset \mathbb{R}^2$ be a bounded open region and let $s = \overline{\dim}_1 \partial D$. Then given $\epsilon > 0$,
$$N(\lambda) - \tfrac{1}{4}\pi^{-1}\mathcal{L}^2(D)\lambda \geq -\lambda^{s/2+\epsilon}, \tag{12.62}$$

for all sufficiently large λ.

Proof We term a square $(m_1 2^{-k}, (m_1 + 1)2^{-k}) \times (m_2 2^{-k}, (m_2 + 1)2^{-k})$ where $m_1, m_2 \in \mathbb{Z}$ an (*open*) *binary square* of side 2^{-k}. We take a *Whitney decomposition* of D into binary squares as follows. Let S_1 be the union of the binary squares of side 2^{-1} contained in D. Let S_2 be the union of the binary squares of side 2^{-2} contained in $D \backslash \bar{S}_1$. Continuing in this way, S_k is the union of the binary squares of side 2^{-k} contained in $D \backslash \cup_{i=1}^{k-1} \bar{S}_i$. Then the unions of squares $\cup_{i=1}^{k} \bar{S}_i$ give a sequence of increasingly good approximations to D from the inside, with $\bar{D} = \overline{\cup_{i=1}^{\infty} S_i}$, see Figure 12.9.

For $k \geq 2$, every square of S_k is contained in the interior neighbourhood $E_{2^{-k}(1+\sqrt{2})}$, since any square of S_k not in $E_{2^{-k}(1+\sqrt{2})}$ would be contained in a square of S_i for some $i < k$. Thus if n_k is the number of squares comprising S_k, we have by considering areas

$$n_k 2^{-2k} \leq \mathcal{L}^2(E_{2^{-k}(1+\sqrt{2})}) \leq 2^{-k(2-s-\epsilon)}$$

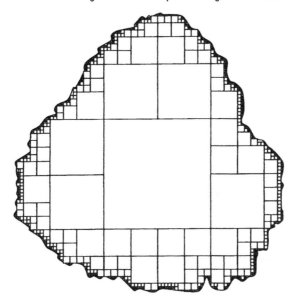

Figure 12.9 The Whitney decomposition of a domain into binary squares, used in estimating the asymptotic distribution of eigenvalues of the Laplacian

for all sufficiently large k, using (12.61), so

$$n_k \leq 2^{k(s+\epsilon)}. \tag{12.63}$$

A classical result on the eigenvalues of (12.37) satisfying the Dirichlet condition (12.38) is 'the larger the region the smaller the eigenvalues', that is if $D' \subset D$ then the k-th eigenvalue of D is no more than that of D' for $k = 1, 2, \ldots$. Thus, writing $D_k = \cup_{i=1}^{k} S_k$ and $N(\lambda)$ and $N_k(\lambda)$ for the eigenvalue counting functions of D and D_k, we have $N_k(\lambda) \leq N(\lambda)$ for all k. As D_k is a union of disjoint squares, we can find $N_k(\lambda)$ by counting the eigenvalues of each component square, as before. Using (12.56)–(12.58) we get that for k sufficiently large

$$N(\lambda) \geq N_k(\lambda)$$

$$\geq \sum_{i=1}^{k} \tfrac{1}{4}\pi^{-1}\lambda 2^{-2i} n_i - \sum_{i=1}^{k} 2\pi^{-1}\lambda^{1/2} 2^{-i} n_i$$

$$= \tfrac{1}{4}\pi^{-1}\lambda \mathcal{L}^2(D) - \tfrac{1}{4}\pi^{-1}\lambda \sum_{i=k+1}^{\infty} 2^{-2i} n_i - 2\pi^{-1}\lambda^{1/2} \sum_{i=1}^{k} 2^{-i} n_i$$

$$\geq \tfrac{1}{4}\pi^{-1}\lambda \mathcal{L}^2(D) - \tfrac{1}{4}\pi^{-1}\lambda \sum_{i=k+1}^{\infty} 2^{i(s-2+\epsilon)} - 2\pi^{-1}\lambda^{1/2} \sum_{i=1}^{k} 2^{i(s-1+\epsilon)}$$

$$\geq \tfrac{1}{4}\pi^{-1}\lambda \mathcal{L}^2(D) - c_1\lambda 2^{k(s-2+\epsilon)} - c_1\lambda^{1/2} 2^{k(s-1+\epsilon)}$$

with c_1 independent of k. Thus given $\lambda > 1$, taking k as the integer such that $2^{k-1} \leq \lambda^{1/2} < 2^k$ gives

$$N(\lambda) \geq \tfrac{1}{4}\pi^{-1}\lambda\mathcal{L}^2(D) - c_2\lambda^{(s+\epsilon)/2}.$$

Inequality (12.62) follows. $\quad\square$

These methods have been developed to give considerably more information about the 'discrepancy' $N(\lambda) - \tfrac{1}{4}\pi^{-1}\mathcal{L}^2(D)\lambda$. For example, the method of Dirichlet–Neumann bracketing may be used to show that the discrepancy also has $\lambda^{s/2}$ as an asymptotic upper bound. In special cases the coefficient of $\lambda^{s/2}$ has been calculated in terms of the geometry of ∂D to give an expression for $N(\lambda)$ with error $o(\lambda^{s/2})$. The work extends to regions $D \subset \mathbb{R}^n$ for all $n \geq 1$, as well as to eigenvalue problems for other elliptic partial differential equations.

12.3 The heat equation on regions with fractal boundary

The flow of heat into or out of a region may depend on the fractality of its boundary. We consider a heat conduction problem on a plane domain to illustrate how the heat flow across the boundary is related to its dimension.

Let D be a (bounded open) region in $\mathbb{R}^2$ with boundary ∂D. We assume that initially D has zero temperature throughout, and the boundary ∂D is maintained at unit temperature for all time. The region warms up as heat enters through the boundary and diffuses through D. (Imagine that an object at temperature zero is suddenly placed in a constant temperature oven.) We wish to estimate how rapidly D gains heat.

Formally, for a region D we write $u_D(x, t)$ for the temperature at $x \in D$ at time $t \geq 0$ and ∇^2 for the Laplacian operator. Then u_D satisfies the heat equation

$$\nabla^2 u_D(x, t) = \frac{\partial u_D}{\partial t}(x, t) \quad (x \in D, t > 0) \tag{12.64}$$

$$u_D(x, 0) = 0 \quad (x \in D) \tag{12.65}$$

$$u_D(x, t) = 1 \quad (x \in \partial D, t > 0), \tag{12.66}$$

where the initial condition (12.65) represents the initial zero temperature of D, and the boundary condition (12.66) represents the boundary being kept at unit temperature. The total heat content of D at time t is

$$h_D(t) = \int_D u_D(x, t)\mathrm{d}x. \tag{12.67}$$

We are interested in the behaviour of h_D for small t.

For plane domains D with smooth (say C^3) boundary, it has long been known that

$$h_D(t) = 2\pi^{-1/2}t^{1/2}\mathcal{L}^1(\partial D) + O(t) \tag{12.68}$$

as $t \to 0$, where $\mathcal{L}^1$ is 'length'. This estimate also holds if D has a polygonal boundary. (In fact the 'order t' term has been found explicitly in terms of the geometry of D in both smooth and polygonal cases.) The exponent $\frac{1}{2}$ in the leading term of (12.68) is characteristic of the boundary of D being one-dimensional. It turns out that for domains with fractal boundary the exponent is in general smaller, corresponding to a faster rate of heat flow through a 'larger' boundary. (This effect is relevant in meteorology in relation to the heat gain or loss by clouds, which might be regarded as having fractal boundaries).

Here we shall be content to show that the boundary dimension provides an upper bound to the heat gain, and then obtain more precise behaviour of $h_D(t)$ in the very special case of the von Koch snowflake domain. We require a bound for the solution of the heat equation in the case of a disc, which leads to a bound for the solution in a general region.

Lemma 12.7

The solution of (12.64)–(12.66) in the case of $D = B(z, r)$, the disc of centre z and radius r, satisfies

$$u_{B(z,r)}(z, t) \le 2e^{-r^2/4t}. \tag{12.69}$$

Proof The function defined by

$$u(x, t) \equiv (2\pi t)^{-1} \int_{\mathbb{R}^2 \setminus B(z,r)} \exp(-|x - y|^2/4t)dy \tag{12.70}$$

is continuous and easily seen to satisfy $\nabla^2 u = \partial u/\partial t$ for $x \in \mathrm{int}\, B(z, r)$ and $t > 0$, with $u(x, t) \to 0$ as $t \to 0$ for $x \in \mathrm{int}\, B(z, r)$, and $u(x, t) \ge 1$ for $x \in \partial B(z, r)$ and $t > 0$. (For this last property, integrating the Gaussian kernel gives $(2\pi t)^{-1} \int_{\mathbb{R}^2} \exp(-|x - y|^2/4t)dy = 2$; on replacing the domain of integration $\mathbb{R}^2$ by $\mathbb{R}^2 \setminus B(z, r)$, which contains a half-plane with boundary through each $x \in \partial B(z, r)$, at least half of this value of retained.) Since the temperature at a point of $B(x, r)$ given by the heat equation solution does not decrease if the boundary temperature is increased, $u_{B(z,r)}(x, t) \le u(x, t)$ for $x \in B(x, r)$, so

$$u_{B(z,r)}(z, t) \le (2\pi t)^{-1} \int_{\mathbb{R}^2 \setminus B(z,r)} \exp(-|z - y|^2/4t)dy$$

$$= t^{-1} \int_{\rho=r}^{\infty} \exp(-\rho^2/4t)\rho d\rho = 2e^{-r^2/4t}. \tag{12.71}$$

$\square$

Corollary 12.8

The solution of (12.64)–(12.66) *satisfies*

$$u_D(z, t) \leq 2e^{-\text{dist}(z, \partial D)^2/4t} \tag{12.72}$$

for $z \in \text{int } D$ *and* $t > 0$.

Proof Writing $r = \text{dist}(z, \partial D)$, we have $B(z, r) \subset D$. The solutions of (12.64)–(12.66) for the two regions $B(z, r)$ and D are related by $u_D(x, t) \leq u_{B(z,r)}(x, t)$ for $x \in B(z, r)$ (since the interior of $B(z, r)$ will heat up faster if its boundary rather than the more distant boundary of D is held at unit temperature). By (12.69)

$$u_D(z, t) \leq u_{B(z,r)}(z, t) \leq 2e^{-r^2/4t}. \quad \square$$

We bound the heat content of D in terms of the upper interior Minkowski dimension of ∂D, which we recall from (12.61) is given by

$$\overline{\dim}_I \partial D = 2 - \liminf_{r \to 0} \log \mathcal{L}^2(E_r)/\log r, \tag{12.73}$$

where E_r is the interior r-neighbourhood of ∂D.

Proposition 12.9

Let D be a plane region with $\overline{\dim}_I \partial D < s$. *There is a number c such that*

$$h_D(t) \leq ct^{1-s/2} \tag{12.74}$$

for all $t > 0$.

Proof From the definition of $\overline{\dim}_I \partial D$ there is a number c_1 such that $\mathcal{L}^2(E_r) \leq c_1 r^{2-s}$ for all $r > 0$. Using (12.72), integrating by parts and then substituting $u = r^2/t$,

$$h_D(t) = \int_D u_D(z, t)dz$$

$$\leq 2\int_D e^{-\text{dist}(z, \partial D)^2/4t}dz$$

$$= 2\int_{r=0}^{\infty} e^{-r^2/4t}d\mathcal{L}^2(E_r)$$

$$= \left[2e^{-r^2/4t}\mathcal{L}^2(E_r)\right]_0^{\infty} + t^{-1}\int_{r=0}^{\infty} re^{-r^2/4t}\mathcal{L}^2(E_r)dr$$

$$\leq c_1 t^{-1}\int_{r=0}^{\infty} re^{-r^2/4t}r^{2-s}dr$$

$$= \frac{1}{2}c_1 t^{1-s/2}\int_{r=0}^{\infty} e^{-u/4}u^{1-s/2}du$$

which is of the form (12.74). $\square$

Thus the (interior Minkowski) dimension of ∂D gives an upper bound for the heat content of D for small t. For many domains there is a lower bound of a similar order of magnitude, so that

$$h_D(t) \simeq t^{1-s/2} \tag{12.75}$$

for small t, where s is the interior Minkowski dimension (assumed to exist) of ∂D.

We shall prove this (and rather more) in one special case, namely for a domain bounded by the von Koch snowflake curve (consisting of three congruent snowflake curves joined end to end). We shall use the self-similarity of the boundary to write down a recursive relationship for the heat content. To do this we need two properties of heat equation solutions: the first concerns scaling a region and the second concerns dividing a region into subregions.

Lemma 12.10

Let D be a region and let D' be a similar copy of D scaled by factor λ. Then

$$h_{D'}(t) = \lambda^2 h_D(\lambda^{-2}t).$$

Proof We may assume that D' is obtained from D by scaling by a factor λ about the origin. Let u_D and $u_{D'}$ be the solutions to the problem (12.64)–(12.66) for the regions D and D' respectively. By differentiation $u_D(\lambda^{-1}x, \lambda^{-2}t)$ satisfies (12.64) for $x \in D'$ and $t > 0$, equals 1 if $x \in \partial D'$ and equals 0 at $t = 0$. Thus $u_{D'}(x, t) = u_D(\lambda^{-1}x, \lambda^{-2}t)$. Hence

$$h_{D'}(t) = \int_{D'} u_D(\lambda^{-1}x, \lambda^{-2}t)dx$$

$$= \lambda^2 \int_D u_D(y, \lambda^{-2}t)dy = \lambda^2 h_D(\lambda^{-2}t)$$

on substituting $y = \lambda^{-1}x$.   □

Lemma 12.11

Suppose a region D is divided into subregions $D_1, \ldots, D_m$ by a number of polygonal cuts, for example as in Figure 12.10(a),(b). Then

$$h_D(t) = \sum_{i=1}^m h_{D_i}(t) + O(t^{1/2}). \tag{12.76}$$

Proof This is a version of what is known as the 'principle of not feeling the boundary'. It expresses the plausible fact that the error in comparing the heat gained by D with the sum of the heat gains of the D_i considered individually is

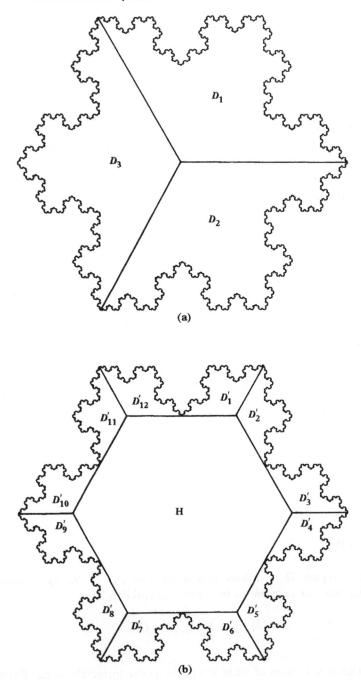

(a)

(b)

Figure 12.10 Decomposition of the von Koch snowflake domain into (a) three subdomains, and (b) 13 subdomains

no more than the heat that would flow across the polygonal cuts, which by (12.68) is $O(t^{1/2})$. We omit the technical details. $\square$

We use these two lemmas to estimate the heat content of the domain D bounded by the von Koch snowflake curve.   Recall that $\dim_H \partial D = \dim_B \partial D = \log 4/\log 3$.

Theorem 12.12

Let D be the von Koch snowflake domain. Then the heat content (12.67) of D in the problem (12.64)–(12.66) satisfies

$$h_D(t) = t^\alpha p(-\log t) + O(t^{1/2}) \tag{12.77}$$

as $t \searrow 0$, where $\alpha = 1 - \frac{1}{2}\log 4/\log 3 = 0.369$ and p is a positive continuous function of period $\log 9$.

Proof We cut up the domain D in two ways, as indicated in Figure 12.10. In Figure 12.10(a) D is divided into three congruent parts D_1, D_2, D_3 so by Lemma 12.11

$$h_D(t) = \sum_{i=1}^{3} h_{D_i}(t) + O(t^{1/2})$$

$$= 3h_{D_1}(t) + O(t^{1/2}). \tag{12.78}$$

In Figure 12.10(b) D is divided into twelve congruent parts $D'_1, \ldots, D'_{12}$ together with a hexagon H, so again by Lemma 12.11

$$h_D(t) = \sum_{i=1}^{12} h_{D'_i}(t) + h_H(t) + O(t^{1/2})$$

$$= 12h_{D'_1}(t) + O(t^{1/2})$$

since (12.68) is valid for a hexagon. Combining with (12.78)

$$h_{D_1}(t) = 4h_{D'_1}(t) + O(t^{1/2})$$

$$= \tfrac{4}{9}h_{D_1}(9t) + O(t^{1/2})$$

by Lemma 12.10, since D'_1 is similar to D_1 at scale $\frac{1}{3}$. Thus by (12.78)

$$h_D(t) = \tfrac{4}{9}h_D(9t) + q(t) \tag{12.79}$$

for $t > 0$, where $q(t) = O(t^{1/2})$ for small t.

Just as in the renewal theory examples of Section 7.2, on writing

$$t = e^{-\tau}, \quad f(\tau) = e^{\alpha\tau}h_D(e^{-\tau}), \quad g(\tau) = e^{\alpha\tau}q(e^{-\tau}), \tag{12.80}$$

where $\alpha = 1 - \frac{1}{2}\log 4/\log 3$, (12.79) becomes

$$f(\tau) = f(\tau - \log 9) + g(\tau). \tag{12.81}$$

Since h_D is bounded and continuous, so are q, f and g, with

$$|g(\tau)| \leq c e^{\tau(\alpha - 1/2)} \qquad (12.82)$$

for $\tau \geq 0$, for some constant c.

Define $p : \mathbb{R} \to \mathbb{R}$ by

$$p(\tau) = \sum_{i=1}^{k} g(\tau + i \log 9) + f(\tau).$$

By virtue of (12.82) this series is uniformly absolutely convergent, so p is continuous. Using (12.81)

$$p(\tau) = p(\tau + \log 9),$$

so p has period $\log 9$. Moreover

$$|f(\tau) - p(\tau)| \leq \sum_{i=1}^{\infty} |g(\tau + i \log 9)|$$

$$\leq c \sum_{i=1}^{\infty} e^{(\alpha - 1/2)\,(\tau + i \log 9)}$$

$$\leq c_1 e^{(\alpha - 1/2)\tau}$$

for some c_1, since $\alpha < \frac{1}{2}$. Transforming back using (12.80) gives (12.77). $\square$

Notice that (12.81) is a very special case of the renewal equation in the form (7.19) with just one 'time'. Applying Corollary 7.3 directly to (12.81) gives $h_D(t) \sim t^\alpha p(-\log t)$, but the above analysis gives the error estimate $O(t^{1/2})$.

With considerably more effort (12.77) may be improved to

$$h_D(t) = t^\alpha p(\log t) - t q(\log t) + O(e^{-1/(1152t)})$$

where p and q have period $\log 9$, an estimate that is remarkable for the exponential error bound.

12.4 Differential equations on fractal domains

In the previous two sections we considered differential equations on regions in $\mathbb{R}^n$ with fractal boundary. However, there are certain circumstances when it is appropriate seek solutions of a 'differential equation' defined on a set that is itself a fractal, for example when modelling conduction of heat or electricity through a highly porous material.

There are many technical difficulties in working with differential equations on fractals, not least in defining differential operators such as the Laplacian on fractal regions. By way of introduction to this complex subject we give a brief account of the heat equation and the distribution of eigenvalues of the

Laplacian on the Sierpinski triangle which has good enough regularity and connectivity properties to allow reasonable progress.

Most approaches to differential equations on a fractal E depend on discrete difference equations on graphs that approximate E. The aim is to take a limit of the difference equation solutions normalised in such a way to give a non-degenerate solution of the limiting 'differential equation' on E.

First we consider how to define Brownian motion on the Sierpinski triangle E, leading to solutions of the heat equation on E. Recall that one way of defining standard Brownian motion on $\mathbb{R}$ is as a limit of suitably scaled random walks, see FG, Section 16.1. Let $X_k(t)$ be the random walk on the set $\{j2^{-k} : j \in \mathbb{Z}\}$ starting with $X_k(0) = 0$ and taking steps at time intervals of 4^{-k}, so that given the position $X_k(m4^{-k})$ of the walker at time $m4^{-k}$, his position $X_k((m+1)4^{-k})$ at time $(m+1)4^{-k}$ is equally likely to be $X_k(m4^{-k}) - 2^{-k}$ or $X_k(m4^{-k}) + 2^{-k}$. It may be shown that as $k \to \infty$ the sequence of random walks $X_k(t)$ converges to a process $X(t)$ on $\mathbb{R}$, namely one-dimensional *Brownian motion*. (Scaling the time variable as the square of the space variable is essential for convergence to a non-degenerate process.) Thus if k is large $X_k(t)$ and $X(t)$ look very similar on all but the finest scales. The increments of Brownian motion, $X(t+h) - X(t)$, are normally distributed with mean 0 and variance h for all t and $h \geq 0$ so 'typically' the motion travels distance $h^{1/2}$ in a time interval of duration h. Moreover, Brownian motion has independent increments, that is, there is no historical memory of the path. More generally for Brownian motion on $\mathbb{R}^n$ (which may be constructed as the limit of scaled random walks on n-dimensional cubic lattices) $X(t+h) - X(t)$ has mean 0 and $|X(t+h) - X(t)|$ has variance h.

We attempt to mimic this construction of Brownian motion on the Sierpinski triangle. For the time being, it is convenient to take E to be the *extended* Sierpinski triangle, thus E extends outwards to infinity using self-similarity, see Figure 12.11. (This avoids the need to regard the three corner vertices of the usual 'bounded' Sierpinski triangle as exceptional.) There is a natural sequence of geometrical graphs $E_0, E_1, \ldots$ that approximate the extended Sierpinski triangle, see Figure 12.11. As geometrical sets, $E_0 \subset E_1 \subset E_2 \subset \ldots$ and $E = \overline{(\cup_{k=1}^{\infty} E_k)}$. We write V_k for the set of vertices of E_k. Thus the graph E_k has edges of length 2^{-k} and each vertex in V_k is adjacent to four others. For $k = 0, 1, 2, \ldots$ the vertices of V_{k+1} are obtained by augmenting V_k by additional vertices at the midpoints of the edges of E_k, with appropriate additional edges added to form E_{k+1}.

For $k = 0, 1, 2 \ldots$ we define random walks $X_k(t)$ on V_k by travelling along the edges of E_k, taking steps at time intervals of α_k (where α_k is to be specified) and starting at $X_k(0) = 0$. Thus if $X_k(m\alpha_k)$ is the vertex of V_k occupied by the random walker at time $m\alpha_k$, then $X_k((m+1)\alpha_k)$ is one of the 4 vertices of V_k adjacent to $X_k(m\alpha_k)$ in E_k chosen with equal probability $\frac{1}{4}$ independently of all previous steps.

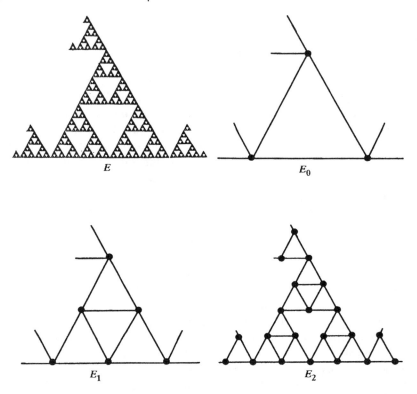

Figure 12.11   The extended Sierpinski triangle and approximating graphs

For $k \geq 1$ this random walk $X_k(t)$ on E_k induces a random walk on E_{k-1}: simply note the sequence of vertices of V_{k-1} visited by $X_k(t)$ (ignoring consecutive occurrences of the same vertex) and regard these as the sequence of vertices visited by a random walk on E_{k-1}. By the symmetry of E_k there is equal probability of moving to each of the 4 adjacent vertices of V_{k-1} in this random walk on E_{k-1}, so this induced random walk is just $X_{k-1}(t)$ undertaken with steps of varying time interval. For the random walks X_k to have a chance of converging to a reasonable limiting process we should ideally choose the time intervals $\alpha_1, \alpha_2, \ldots$ so that, for each k, the time for $X_k(t)$ to move from a vertex of V_{k-1} to a neighbouring vertex of V_{k-1} is α_{k-1}. We can at least achieve this *on average* by ensuring that the expected time of such a step is α_{k-1}. It follows from the following lemma that this requires that $\alpha_{k-1} = 5\alpha_k$. (Whilst a coefficient 4 might be expected since each vertex of E_k is adjacent to four others, the geometry of the Sierpinski triangle makes 5 the right number.)

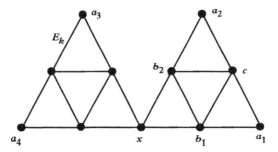

Figure 12.12   The random walk on E_k from x

Lemma 12.13

Let $X_k(t)$ be the random walk on E_k as above, let x be a vertex of V_{k-1} and let A be the set of four vertices in V_{k-1} adjacent to x. Then, conditional on the random walker being at x, the expected number of steps to reach a point of A is 5

Proof   By the symmetry of E_k, the portion of E_k near x is always equivalent to that shown in Figure 12.12, where $A = \{a_1, a_2, a_3, a_4\}$ and b_1, b_2 and c are as indicated. Write $\mathsf{E}(p, A)$ for the expected number of steps in a random walk on E_k to get from a point $p \in V_k$ to a point of A. Given that a random walker starts at x, symmetry allows us to assume that the first step is to b_1 when determining the expected number of steps $\mathsf{E}(x, A)$ to reach a point of A from x. Thus

$$\mathsf{E}(x, A) = 1 + \mathsf{E}(b_1, A)$$
$$\mathsf{E}(b_2, A) = \mathsf{E}(b_1, A) = 1 + \tfrac{1}{4}\mathsf{E}(x, A) + \tfrac{1}{4}\mathsf{E}(b_2, A) + \tfrac{1}{4}\mathsf{E}(c, A) + \tfrac{1}{4} \times 0$$
$$\mathsf{E}(c, A) = 1 + \tfrac{1}{2} \times 0 + \tfrac{1}{4}\mathsf{E}(b_1, A) + \tfrac{1}{4}\mathsf{E}(b_2, A)$$

on examining the probabilities of the possible steps from b_1, b_2 and c. Solving these equations gives that $\mathsf{E}(x, A) = 5$. $\square$

Thus if the random walk X_k on E_k is undertaken with time interval α_k between each step, this induces a random walk on E_{k-1} with mean time interval $5\alpha_k$ between each step. To take a non-degenerate limit of X_k as $k \to \infty$, the random walk on E_{k-1} induced by X_k must, at large scales, be close to the random walk X_{k-1}, so to achieve this we need $\alpha_{k-1} = 5\alpha_k$ for each k.

Hence to ensure scaling compatability, we set $\alpha_k = 5^{-k}$ for $k = 0, 1, 2, \ldots$. Then it may be shown, analogously to standard Brownian motion on $\mathbb{R}$ or $\mathbb{R}^n$, that the sequence of random walks $X_k(t)$ converges to a random process $X(t)$ which we call *Brownian motion* on the extended Sierpinski triangle E.

The basic properties of standard Brownian motion on $\mathbb{R}$ are mirrored in Brownian motion on E but the exponents are different. One might expect that the dimension $\dim_H E = \log 3/\log 2 = 1.585$ would be crucial for determining the progress of $X(t)$ through E. In fact, what is more relevant is what is known as the walk dimension d_w which reflects the rate at which $X(t)$ moves at different scales. From Lemma 12.13, $X_k(t)$, and thus in the limit $X(t)$, takes on average time $5\alpha_k = 5^{-k+1}$ to move between adjacent vertices of V_{k-1} which are distance 2^{-k+1} apart, so $X(t)$ 'typically' moves distance $h^{\log 2/\log 5}$ in time interval h. This leads to defining the *walk dimension* $d_w = \log 5/\log 2 = 2.322$; it is at least plausible that the mean square of the increments satisfes

$$\mathsf{E}(|X(t+h) - X(t)|^2) \asymp h^{2/d_w} \qquad (12.83)$$

for $h > 0$. (This should be compared with $\mathsf{E}(|X(t+h) - X(t)|^2) = h$ for standard Brownian motion — the loss of equality is an inevitable consequence of the fractality of the domain.)

This and much more can be proved rigorously, in particular there are key estimates for the probability that $X(t)$ lies in a set A given that $X(0) = x$. Let μ be the restriction of s-dimensional Hausdorff measure to E, where $s = \dim_H E = \log 3/\log 2$ (so μ is the natural locally finite measure on E). There exists a *transition density* $p_t(x, y)$ that determines the probability density for the position y reached after time t by Brownian motion starting at x:

$$P(X(t+h) \in A | X(t) = x) = \int_A p_h(x, y) \mathrm{d}\mu(y) \qquad (12.84)$$

if $x \in E$ and A is a measurable set. A knowledge of $p_t(x, y)$ allows the statistics of the motion to be studied. By careful analysis of the underlying random walks X_k it may be shown that there are constants $c_1, c_2, c_3, c_4 > 0$ such that for all $x, y \in E$ and $t > 0$

$$c_1 t^{-s/d_w} \exp(-c_2(|x - y|t^{-1/d_w})^{d_w/(d_w-1)}) \leq p_t(x, y) \qquad (12.85)$$
$$\leq c_3 t^{-s/d_w} \exp(-c_4(|x - y|t^{-1/d_w})^{d_w/(d_w-1)}).$$

(This should be compared with standard Brownian motion on $\mathbb{R}^n$, where $d_w = 2$, $\mu = \mathcal{L}^n$ and the transition density is given by

$$p_t(x, y) = (2\pi t)^{-n/2} \exp(-|x - y|^2/2t); \qquad (12.86)$$

(12.85) is the analogue of the familiar Gaussian kernel in this setting.)

By parallel methods to those for standard Brownian motion, compare FG, Section 16.1, it may be shown that with probability 1 the sample path $X(t)$ satisfies a Hölder condition

$$|X(t_1) - X(t_2)| \leq c|t_1 - t_2|^\gamma \qquad (12.87)$$

for all $\gamma < 1/d_w$, and $0 \leq t_1, t_2 \leq T$ where c depends on γ and T. Moreover, the Hausdorff dimension of the sample path is $\log 3/\log 2$, so the path 'fills' the set E.

Brownian motion on $\mathbb{R}^n$ is intimately connected with solutions of the heat equation. Diffusion of heat on $\mathbb{R}^n$ may be thought of as the aggregate effect of a large number of 'heat particles' following independent Brownian paths. If ν is the heat distribution on $\mathbb{R}^n$ at $t = 0$, then the temperature at point x at time t is

$$u(x, t) = \int p_t(x, y)\mathrm{d}\nu(y) \tag{12.88}$$

where p_t is the standard transition density (12.86). It may be checked by differentiation that

$$\frac{\partial p_t}{\partial t} = \frac{1}{2}\nabla_x^2 p_t$$

and thus (12.88) satisfies the heat equation

$$\frac{\partial u}{\partial t} = \frac{1}{2}\nabla^2 u$$

on $\mathbb{R}^n$, with $\int_A u(x, t)\mathrm{d}x \to \nu(A)$ as $t \to 0$.

In a similar way, Brownian motion on the extended Sierpinski triangle E may be regarded as modelling heat diffusion on E. Thus an initial heat distribution ν on E would yield the temperature distribution (12.88) at time t, but with p_t now the transition density on E given by (12.84).

To obtain a meaningful analogue of the heat equation on E which has (12.88) as solution, we must say what is meant by the Laplacian ∇^2 on E. Again we use discrete approximation. Recall that the Laplacian on $\mathbb{R}$, $\mathrm{d}^2/\mathrm{d}x^2$, is the limit of differences

$$\frac{\mathrm{d}^2 f}{\mathrm{d}x}(x) = \lim_{h \to 0} h^{-2}[(f(x+h) - f(x)) + (f(x-h) - f(x))] \tag{12.89}$$

$$= \lim_{h \to 0} h^{-2} \sum_{y=x\pm h} (f(y) - f(x)).$$

For a continuous $f : E \to \mathbb{R}$ we use discrete approximation via the geometric graphs E_k, see Figure 12.11. Writing $C(E)$ for the continuous functions on E, we define $\nabla^2 f \in C(E)$ by the requirement that, for every bounded set A

$$\lim_{k \to \infty} \sup_{x \in A \cap V_k} \left| 5^k \sum_{w \in V_k(x)} (f(w) - f(x)) - \nabla^2 f(x) \right| = 0, \tag{12.90}$$

where $V_k(x)$ comprises the four vertices of V_k adjacent to x. (Note that ∇^2 is defined only on a subspace of $C(E)$.) The number '5' is exactly what is needed for (12.90) to be meaningful; as indicated below, this is a consequence of the walk dimension of E being $\log 5/\log 2$.

For given k, let $x, y \in V_k$. On the assumption that Brownian motion on E starting at x at time 0 reaches one of the four adjacent vertices of V_k at time

$\delta t = 5^{-k}$ (we know this is true *on average*), consideration of transition densities to y gives

$$p_{t+\delta t}(x,y) \simeq \sum_{w \in V_k(x)} \tfrac{1}{4} p_t(w,y)$$

where $x_1, \ldots, x_4 \in V_k(x)$ are adjacent to x. Thus

$$(p_{t+\delta t}(x,y) - p_t(x,y))/\delta t \simeq \tfrac{1}{4} 5^k \sum_{w \in V_k(x)} (p_t(w,y) - p_t(x,y)),$$

so letting $\delta t \to 0$ and invoking (12.90) gives

$$\frac{\partial p_t}{\partial t}(x,y) = \tfrac{1}{4} \nabla^2 p_t(x,y).$$

We infer that u given by (12.88) satisfies the heat equation on E

$$\frac{\partial u}{\partial t} = \tfrac{1}{4} \nabla^2 u, \tag{12.91}$$

where $p_t(x,y)$ is the transition density for Brownian motion on E. Clearly, $p_t(x,\cdot)$ is concentrated near x when t is small (since $X(t)$ will not have moved far), so $\int_A u(x,t) d\mu(x) \to \nu(A)$ as $t \to 0$. Thus the Brownian motion integral (12.88) solves (12.91) on E for a heat distribution ν at $t = 0$. With considerable effort, these arguments can be made rigorous.

We move to the related problem of finding eigenvalues of the Laplacian on a fractal domain. Here we need the domain to be bounded, so from now on we take E to be the (usual non-extended) Sierpinski triangle and adapt our previous notation in an obvious way to the bounded setting. Thus E_k is the graph with a finite vertex set V_k and edges of lengths 2^{-k} that approximates E. The definition of the Laplacian (12.90) is modified slightly to require $\nabla^2 f \in C(E)$ to satisfy

$$\lim_{k \to \infty} \sup_{x \in V_k \backslash V_0} \left| 5^k \sum_{w \in V_k(x)} (f(w) - f(x)) - \nabla^2 f(x) \right| = 0. \tag{12.92}$$

where $V_k(x)$ is the set of vertices in V_k adjacent to x, other than the vertices of V_0. We are interested in eigenfunctions of the Dirichlet problem in this context, that is, for functions vanishing on V_0, the three corners of E. The eigenvalues of

$$\nabla^2 u = -\lambda u \quad \text{with} \quad u \in C(E) \quad \text{and} \quad u(x) = 0 \quad \text{for} \quad x \in V_0 \tag{12.93}$$

may be shown to be real and non-negative, and in the spirit of Section 12.2 we seek estimates for the size of the k-th eigenvalue.

We define the *spectral dimension* of E as $d_s = 2 \log 3 / \log 5 = 1.365$. It may be shown that the eigenvalue counting function $N(\lambda)$ satisfies

$$N(\lambda) \equiv \#\{\text{eigenvalues of (12.93) at most } \lambda\} \asymp \lambda^{d_s/2}. \tag{12.94}$$

(This should be compared with Weyl's theorem for open domains in $\mathbb{R}^n$ where d_s is replaced by n, see (12.40).) Roughly speaking, this is because the Sierpinski triangle splits into sub-triangles which are joined only at the corners and which therefore are essentially independent, similarly to the eigenvalue problem on the cut-out sets in $\mathbb{R}$, see Proposition 12.5.

To get some feel for this, let F_1, F_2, F_3 be the three similarity transformations of ratio $\frac{1}{2}$ that map E onto its three principal component triangles. We note that if u is an eigenfunction of (12.93) with eigenvalue λ and $u(x) \to 0$ sufficiently rapidly as x approaches a point of V_0, then for all $i = (i_1, \ldots, i_p)$, where $i_j \in \{1, 2, 3\}$, the functions

$$u_i(x) = u(F_i^{-1}(x)) \quad (\text{if } x \in F_i(E)) \tag{12.95}$$
$$= 0 \quad (\text{if } x \in E \backslash F_i(E))$$

are eigenfunctions with eigenvalues $5^p \lambda$ where $F_i = F_{i_1} \circ \cdots \circ F_{i_p}$. To see this, note that for x a point of $F_i(E) \cap E_k$ that is not a vertex of V_p and for $k \gg p$,

$$\nabla^2 u_i(x) \simeq 5^k \sum_{w \in V_k(x)} (u(F_i^{-1}(w)) - u(F_i^{-1}(x))$$
$$= 5^p \times 5^{k-p} \sum_{w' \in V_{k-p}(F_i^{-1}(x))} (u(w') - u(F_i^{-1}(x))$$
$$\simeq 5^p \nabla^2 u(F_i^{-1}(x))$$
$$= 5^p \lambda u(F_i^{-1}(x)) = 5^p \lambda u_i(x).$$

It may be shown that there exists an eigenfunction u that tends to 0 sufficiently rapidly near V_0 with eigenvalue $\lambda_0 \geq 0$, so there are at least 3^p independent eigenfunctions with eigenvalue 5^p for $p = 1, 2, \ldots$. Thus $N(5^p \lambda_0) \geq 3^p$ so $N(\lambda) \geq \text{const} \lambda^{\log 3/\log 5}$, which is half of (12.94). The opposite inequality depends on showing that a complete family of eigenfunctions may be obtained from a basic non-negative eigenfunction u using (12.95) for all $i \in I$.

Much more precise information may be obtained on the asymptotics of $N(\lambda)$. The renewal theorem method (see Section 7.2) gives a positive function p with period $\log 5$ such that $N(\lambda) \sim p(\log \lambda) \lambda^{d_s/2}$.

The approach described here for defining Brownian motion and the Laplacian on the Sierpinski triangle extends to any *post-critically finite* self-similar set, that is, roughly speaking, any self-similar set E defined by an IFS $\{F_1, \ldots, F_m\}$ for which $F_i(E) \cap F_j(E)$ is finite whenever $i \neq j$. For example, for the 'hexakun' of Figure 12.13 we have the Hausdorff dimension $\dim_H E = \log 6/\log 3 = 1.631$, walk dimension $d_w = (\log 360 - \log 19)/\log 3 = 2.678$ and spectral dimension $d_s = 2 \log 6/(\log 360 - \log 19) = 1.218$. In these examples the walk dimension indicates the scaling behaviour of the rate of diffusion through E and the spectral dimension indicates the coefficient needed in the definition of the Laplacian (12.92), which is twice the exponent in the power law for the eigenvalue counting function. In general $d_w = 2 \dim_H E/d_s$,

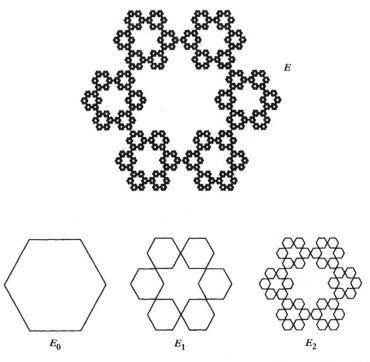

Figure 12.13 The 'hexakun' E is a post-critically finite self-similar set. Heat diffusion and the Laplacian on E may be set up using the approximating graphs E_k in a similar way to the Sierpinski triangle

just as in the case of standard Brownian motion and the Laplacian on $\mathbb{R}^n$ where $d_s = \dim_H E = n$ and $d_w = 2$.

In this section we have just touched on the involved topic of partial differential equations on fractal domains. There are further difficulties in constructing partial differential operators on many domains, even as 'simple' as the Sierpinski carpet (which is not post-critically finite). With the variety of fractal domains and physical processes that might be modelled by appropriately defined differential equations, this is an area with exciting potential for future development.

12.5 Notes and references

The methods of Section 12.1 have been applied to a wide variety of systems. The basic estimate for the dimension of an invariant set, Theorem 12.2, was given by Douady and Oesterlé (1980). This theory is developed in much greater detail in the books by Ladyzhenskaya (1991) and Temam (1988). These books contain many further applications to dynamical systems and differential equations, as well as historical information and many further references.

The suggestion that the second term in the asymptotic expansion of the eigenvalue counting function $N(\lambda)$ might reflect the fractal dimension of the domain boundary originated with Berry (1979), and this led to the question 'Can one hear the dimension of a fractal?' discussed by Brossard and Carmona (1986). The problem for one-dimensional domains, including the relationship with the Riemann zeta function, was studied by Lapidus and Pomerance (1993); see also Lapidus and Maier (1995) for a connection with the Riemann hypothesis. Bounds for the discrepancy in the eigenvalue counting function in higher-dimensional domains (also valid for more general elliptic equations) were obtained by Lapidus (1991). More exact estimates have been obtained for certain domains, see Fleckinger-Pelle and Vassilev (1993) and Chen and Sleeman (1995). Surveys of this subject are given by Lapidus (1993) and Sleeman (1995).

For the heat loss estimate (12.68) for the heat equation on polygonal domains, see van den Berg and Srisatkunarajah (1990). The asymptotic form for the snowflake domain was derived by Fleckinger, et al. (1995), and the estimate (12.75) for more general domains by van den Berg (1994).

Diffusion processes on the Sierpinski triangle were constructed by Goldstein (1987) and Kusuoka (1987), and a detailed analysis of the transition densities was given by Barlow and Perkins (1988). Barlow and Bass (1992) extended the theory to the (non-post-critically finite) Sierpinski carpet. Kigami (1989) defined the Laplacian on the Sierpinski triangle and Fukushima and Shima (1992) determined its eigenvalues and eigenfunctions. Kigami and Lapidus (1993) found the asymptotic distribution of eigenvalues for general post-critically finite self-similar sets. For a survey of this material see Kigami (1995).

Exercises

12.1 Show that the singular value function ω_s given by (12.2) is submultiplicative. (Hint: show this first in the case of s an integer.)

12.2 Improve the estimate for the dimension of the Hénon attractor by working with the second iterate $f^2(x, y)$ rather than $f(x, y)$. (Some numerical calculations are needed here.)

12.3 Verify (12.56). (Hint: use that the average number of lattice points in the quadrant $Q(x, r)$ as x ranges over the unit square equals the area of $Q(0, r)$, and consider the extreme values of $Q(x, r)$.)

12.4 Show that the eigenvalue counting function for a domain D that is a disjoint union of squares of sides $a_i \asymp i^{-1/s}$ is given by (12.60) with $\psi(\lambda) \asymp \lambda^{s/2}$.

12.5 Prove (12.50) by using self-similarity to obtain a recurrence for $\#\{i : |A_i| \geq r\}$ and applying Corollary 7.3.

12.6 Let D be the domain formed as the union of the bounded components of the complement of the Sierpinski carpet E of unit side-length. Show that the eigenvalue counting function $N(\lambda) \geq \frac{1}{4}\pi^{-1}\lambda - c\lambda^{-s/2}$ for some constant c, where $s = \dim_B E = \log 8/\log 3$.

12.7 Heat is distributed on a plane domain according to a finite measure ν at time $t = 0$, and diffuses according to the two-dimensional heat equation $\nabla^2 u = \partial u/\partial t$. Verify that

$$u(x, t) = (4\pi t)^{-1} \int \exp(-|x - y|^2/4t) d\nu(y)$$

gives the temperature at x at time t. Now take ν to be the restriction of s-dimensional Hausdorff measure to a self-similar set $E \subset \mathbb{R}^n$ satisfying the strong separation condition, where $s = \dim_H E$. Show that $u(x, t) \asymp t^{s/2-1}$ for small t for ν-almost all x. (Hint: Use Proposition 6.5.)

12.8 Let D be the union of the bounded components of the complement of the Sierpinski triangle, so that ∂D is the Sierpinski triangle. Show that in this case the heat content $h_D(t)$ at time t in the problem (12.64)–(12.66) satisfies $h_D(t) \asymp t^{1-s/2}$ for small t where $s = \dim_B \partial D = \log 3/\log 2$.

12.9 Extend the proof of Lemma 12.13 to show that $E(t^N) = t^2(4 - 3t)^{-1}$ is the probability generating function of the random variable N, the number of steps of a random walk on E_k taken to travel between adjacent vertices of V_{k-1}. Check that this gives $E(N) = 5$.

12.10 Use (12.85) to establish the almost sure Hölder condition (12.87) for Brownian motion on the Sierpinski triangle.

12.11 Let $E \subset \mathbb{R}^3$ be the Sierpinski tetrahedron (the three-dimensional analogue of the Sierpinski triangle, obtained from a tetrahedron by repeatedly removing inverted tetrahedra). Mimic the Sierpinski triangle theory to infer that $\dim_H E = \dim_B E = 2$, $d_w = \log 6/\log 2$ and $d_s = 2\log 4/\log 6$.

12.12 Let E be the hexakun of Figure 12.13. Verify the values stated for $\dim_H E, d_w$ and d_s in this case.

References

Arbeiter, M. and Patzschke, M. (1996) Random self-similar multifractals. *Math. Nachr*, **181**, 5–42.

Bandt, C., (1989) Self-similar sets 3. Constructions with sofic systems. *Monatshefte für Mathematik* **108**, 89–102.

Bandt, C., Flachsmeyer, J. and Haase, H. (eds.) (1992) *Topology, Measures and Fractals*. Akademie Verlag.

Bandt, C., Graf, S. and Zähle, M. (1995) *Fractal Geometry and Stochastics*. Birkhäuser.

Barlow, M.T. and Bass, R.F. (1992) Transition densities for Brownian motion on the Sierpinski carpet. *Probab. Th. Rel. Fields* **91**, 307–330.

Barlow, M.T. and Perkins, E.A. (1988) Brownian motion on the Sierpinski gasket. *Probab. Th. Rel. Fields* **79**, 543–624.

Barnsley, M.F. (1988) *Fractals Everywhere*. Academic Press.

Barnsley, M.F. and Demko, S.G. (1985) Iterated function systems and the global construction of fractals. *Proc. Roy. Soc.* **A399**, 243–275.

Barnsley, M.F. and Hurd, L.P. (1993) *Fractal Image Compression*. A.K. Peters.

Bedford, T. (1986) Dimension and dynamics of fractal recurrent sets. *J. London Math. Soc.* (2) **33**, 89–100.

Bedford, T. (1991) Applications of dynamical systems to fractals - a study of cookie-cutter Cantor sets. In *Fractal Geometry and Analysis* (eds. J. Bélair and S. Dubuc), pp 1–44, Kluwer.

Bedford, T. and Fisher, A. (1992) Analogues of the Lebesgue density theorem for fractal sets of reals and integers. *Proc. London Math. Soc.* (3) **64**, 95–124.

Bedford, T., Keane, M. and Series, C. (1991) *Ergodic Theory - Symbolic Dynamics and Hyperbolic Spaces*. Oxford University Press.

Bedford, T. and Urbanski, M. (1990) The box and Hausdorff dimension of self-affine sets. *Ergodic Th. Dynam. Sys.* **10**, 627–644.

Bélair, J. and Dubuc, S. (eds.) (1991) *Fractal Geometry and Analysis*. Kluwer.

Bellman, R. (1960) *Introduction to Matrix Analysis*. McGraw-Hill.

Berg, van den M. (1994) Heat content and Brownian motion for some regions with a fractal boundary. *Probab. Th. Related Fields* **100**, 439–456.

Berg, van den M. and Srisatkunarajah, S. (1990) Heat flow and Brownian motion for a region in $\mathbb{R}^2$ with a polygonal boundary. *Probab. Th. Related Fields* **86**, 41–52.

Berry, M.V. (1979) Distribution of modes in fractal resonators. In *Structural Stability in Physics* (eds. W. Güttinger and H. Eikemeier), pp 51–53, Springer-Verlag.

Besicovitch, A.S. and Taylor, S.J. (1954) On the complementary intervals of linear closed sets of zero Lebesgue measure. *J. London Math. Soc.* **29**, 449–459.

Billingsley, P. (1965) *Ergodic Theory and Information*. John Wiley.

Birkhoff, G.D. (1931) Proof of the ergodic theorem. *Proc. Nat. Acad. Sci. U.S.A.* **17**, 656–660.

Bowen, R. (1975) Equilibrium States and the Ergodic Theory of Anosov Differmorphisms. *Lecture Notes in Mathematics* **470**, Springer-Verlag.

Brossard, J. and Carmona, R. (1986) Can one hear the dimension of a fractal? *Commun. Math. Phys.* **104**, 103–122.

Brown, G., Michon, G. and Peyrière, J. (1992) On the multifractal analysis of measures. *J. Stat. Phys.* **66**, 775–790.

Cawley, R. and Mauldin, R.D. (1992) Multifractal decomposition of Moran fractals. *Adv. Math.* **92**, 196–236.

Chen Hua and Sleeman, B.D. (1995) Fractal drums and the n-dimensional modified Weyl-Berry conjecture. *Commun. Math. Phys.* **168**, 581–607.

Cherbit, G. (ed) (1991) *Fractals: Non-integral dimensions and applications*. John Wiley.

Cooper, D. and Pignataro, T. (1988) On the shape of Cantor sets. *J. Differential Geom.* **28**, 203–221.

Cutler, C.D. (1986) The Hausdorff dimension distribution of finite measures in Euclidean space. *Can. J. Math.* **38**, 1459–1484.

Cutler, C.D. (1990) Connecting ergodicity and dimension in dynamical systems. *Ergodic Th. Dynam. Sys.* **10**, 451–462.

Cutler, C.D. (1992) Measure disintegrations with respect to σ-stable monotone indices and pointwise representation of packing dimension. *Supp. Ai. Rend. Cir. Math. Palermo* (II) **28**, 319–340.

Cutler, C.D. (1995) Strong and weak duality principles for fractal dimension in Euclidean space. *Math. Proc. Cambridge Philos. Soc.* **118**, 393–410.

Devaney, R.L. and Keen, L. (eds.) (1989) Chaos and Fractals - The Mathematics Behind the Computer Graphics. *Proc. Symp. Applied Math.* **39**, American Mathematical Society.

Doob, J.L. (1994) *Measure Theory*. Springer-Verlag.

Douady, A. and Oesterlé, D. (1980) Dimension de Hausdorff des attracteurs. *C.R. Acad. Sci. Paris, Sér* A**290**, 1135–1138.

Dunford, N. and Schwartz, J.T. (1958) *Linear Operators*, Pt I. Interscience.

Edgar, G.A. (1990) *Measure, Topology, and Fractal Geometry*. Springer-Verlag.

Edgar, G.A. (1993) *Classics on Fractals*. Addison-Wesley.

Edgar, G.A. and Mauldin, R.D. (1992) Multifractal decompositions of digraph recursive fractals. *Proc. London Math. Soc.* (3) **65**, 604–628.

Evertsz, C.J.G. and Mandelbrot, B.B. (1992) Multifractal measures. Appendix B in *Chaos and Fractals* (H.-O. Peitgen, H. Jürgens, and D. Saupe), Springer-Verlag.

Falconer, K.J. (1985) *The Geometry of Fractal Sets*. Cambridge University Press.

Falconer, K.J. (1988) The Hausdorff dimension of self-affine fractals. *Math. Proc. Cambridge Philos. Soc.* **103**, 339–350.

Falconer, K.J. (1989) Dimensions and measures of quasi self-similar sets. *Proc. Amer. Math. Soc.* **106**, 543–554.

Falconer, K.J. (1990) *Fractal Geometry — Mathematical Foundations and Applications*. John Wiley.

Falconer, K.J. (1992a) The dimension of self-affine fractals II, *Math. Proc. Cambridge Philos. Soc.* **111**, 169–179.

Falconer, K.J. (1992b) Wavelet transforms and order-two densities of fractals. *J. Statistical Physics* **67**, 781–793.

Falconer, K.J. (1994) Bounded distortion and dimension for non-conformal repellers. *Math. Proc. Cambridge Philos. Soc.* **115**, 315–334.

Falconer, K.J. (1995a) Sub-self-similar sets. *Trans. Amer. Math. Soc.* **347**, 3121–3129.

Falconer, K.J. (1995b) On the Minkowski measurability of fractals. *Proc. Amer. Math. Soc.* **123**, 1115–1124.

Falconer, K.J. and Marsh, D.T. (1992) On the Lipschitz equivalence of Cantor sets. *Mathematika* **39**, 222–233.

Falconer, K.J. and O'Neil, T.C. (1996) Vector-valued multifractal measures. *Proc. Royal Soc. London* A **452**, 1–26.

Falconer, K.J. and Springer, O.B. (1995) Order-two density and measures with non-integral dimension. *Mathematika* **42**, 1–14.

Falconer, K.J. and Xiao, Y.M. (1995) Average densities of the image and zero set of stable processes. *Stochastic Processes Appl.* **55**, 271–283.

Fan, A.H. (1994) On ergodicity and unidimensionality. *Kyushu J. Math.* **48**, 249–255.

Feder, J. (1988) *Fractals*. Plenum Press.

Federer, H. (1969) *Geometric Measure Theory*. Springer-Verlag.

Feller, W. (1966) *An Introduction to Probability Theory and its Applications*. Vol. 2, 2nd Edition, John Wiley.

Fisher, G. (1995) *Fractal Image Compression: Theory and Applications*. Springer-Verlag.

Fleckinger, J., Levitin, M. and Vassiliev, D. (1995) Heat equation on the triadic von Koch snowflake: asymptotic and numerical analysis. *Proc. London Math. Soc.* (3) **71**, 372–396.

Fleckinger-Pelle, J. and Vassilev, D. (1993) An example of two term asymptotics for the "counting function" of a fractal drum. *Trans. Amer. Math. Soc.* **337**, 99–116.

Fleischmann, M., Tildesley, D.J. and Ball, R.C. (eds.) (1990). *Fractals in the Natural Sciences*. Princeton University Press.

Frisch, U. and Parisi, G. (1985) Fully developed turbulence and intermittancy. In *Turbulence and Predictability of Geophysical Flows and Climate Dynamics* (eds M. Ghil, R. Benzi and G. Parissi), pp 84–88, North Holland.

Frostman, O. (1935) Potential d'équilibre et capacité des ensembles avec quelques applications à la théorie des fonctions. *Meddel. Lunds Univ. Mat. Sem.* **3**, 1–118.

Fukushima, M. and Shima, T. (1992) On a spectral analysis for the Sierpinski gasket. *Potential Analysis* **1**, 1–35.

Goldstein, S. (1987) Random walks and diffusions on fractals. In *Percolation theory and ergodic theory of infinite particle systems, IMA Math. Appl.* **8**, 121–128, Springer.

Grimmett, G.R. and Striziker, D.R. (1992) Probability and Random Processes. 2nd Edition, Clarendon Press.

Haase, H. (1992) A survey of the dimensions of measures. In *Topology, Measures and Fractals, Math. Research* **66**, 66–75, Academie Verlag.

Halsey, T.C., Jensen, M.H., Kadanoff, L.P., Procaccia, I. and Shraiman, B.J. (1986) Fractal measures and their singularities: the characterization of strange sets. *Phys. Rev.* **A33**, 1141–1151.

Hastings, H.M. (1993) *Fractals: A User's Guide for the Natural Sciences.* Oxford University Press.

Hausdorff, F. (1919) Dimension und äusseres Mass. *Math. Ann.* **79**, 157–179.

Hentschel, H. and Procaccia, I. (1983) The infinite number of generalized dimensions of fractals and strange attractors. *Physica* **8D**, 435–444.

Holschneider, M. (1995) *Wavelets—An Analysis Tool.* Clarendon Press.

Hu, X. and Taylor, S.J. (1994) Fractal properties of products and projections of measures in R^d. *Math. Proc. Cambridge Philos. Soc.* **115**, 527–544.

Hueter, I. and Lalley, S.P. (1995) Falconer's formula for the Hausdorff dimension of a self-affine set in R^2. *Ergodic Th. Dynam. Sys.* **15**, 77–97.

Hutchinson, J.E. (1981) Fractals and self-similarity. *Indiana Univ. Math. J.* **30**, 713–747.

Kahane, J.-P. (1985) *Some Random Series of Functions.* Cambridge University Press.

Kahane, J.-P. and Katznelson, Y. (1990) Décomposition des mesures selon la dimension. *Colloq. Math.* **58**, 269–279.

Kahane, J.-P. and Peyrière, J. (1976) Sur certaines martingales de Benoit Mandelbrot. *Adv. Math.* **22**, 131–145.

Katznelson, Y. and Weiss, B. (1982) A simple proof of some ergodic theorems. *Israel J. Math.* **42**, 291–296.

Kigami, J. (1989) A harmonic calculus on the Sierpinski spaces. *Japan J. Appl. Math.* **6**, 259–290.

Kigami, J. (1995) Laplacians on self-similar sets and their spectral distributions. In *Fractal Geometry and Stochastics, Progress in Probability* **37**, 221–238, Birkhauser.

Kigami, J. and Lapidus, M.L. (1993) Weyl's problem for the spectral distribution of Laplacians on p.c.f. self-similar fractals. *Commun. Math. Phys.* **158**, 93–125.

King, J. (1995) The singularity spectrum for general Sierpinski carpets. *Adv. Math.* **116**, 1–11.

Kingman, J.F.C. and Taylor, S.J. (1966) *Introduction to Measure and Probability.* Cambridge University Press.

Kusuoka, S. (1987) A diffusion process on a fractal. In *Probabilistic Methods in Mathematical Physics*, Proc. Taniguchi Symp., Katata 1985, pp 251–274, Kino Kuniya-North Holland.

Ladyzhenskaya, O. (1991) *Attractors for Semigroups and Evolution Equations.* Cambridge University Press.

Lalley, S.P. (1988) The packing and covering functions of some self-similar fractals. *Indiana Univ. Math. J.* **37**, 699–709.

Lalley, S.P. (1989) Renewal theorems in symbolic dynamics, with applications to geodesic flow, non-Euclidean tessellations and their fractal limits. *Acta Math.* **163**, 1–55.

Lalley, S.P. (1991) Probabilistic methods in certain counting problems of ergodic theory. In *Ergodic Theory, Symbolic Dynamics and Hyperbolic Spaces*, pp 223–258, Oxford University Press.

Lapidus, M.L. (1991) Fractal drum, inverse spectral problems for elliptic operators and a partial resolution of the Weyl-Berry conjecture. *Trans. Amer. Math. Soc.* **325**, 465–529.

Lapidus, M.L. (1993) Vibrations of fractal drums, the Riemann hypothesis, waves in fractal media and the Weyl-Berry conjecture. In *Ordinary and Partial Differential Equations IV*, Pitman Research Notes in Mathematics 289, pp 126–209, Longman Scientific.

Lapidus, M.L. and Maier, H. (1995) The Riemann hypothesis and inverse spectral problems for fractal strings. *J. London Math. Soc.* (2) **52**, 15–34.

Lapidus, M.L. and Pomerance, C. (1993) The Riemann zeta function and the one dimensional Weyl-Berry conjecture for fractal drums. *Proc. London Math. Soc.* (3) **66**, 41–49.

Leistritz, L. (1994) *Geometrische und analytische Eigenschaften singularer Strukturen in* R^d. Ph.D. Dissertation, University of Jena.

Levitin, M. and Vassiliev, D. (1996) Spectral asymptotic, renewal theorem, and the Berry conjective for a class of fractals. *Proc. London Math. Soc.* (3) **72**, 188–214.

Lindvall, T. (1977) A probabilistic proof of Blackwell's renewal theorem. *Ann. Probab.* **5**, 482–485.

Lopes, A.O. (1989) The dimension spectrum of the maximal measure. *SIAM J. Math. Anal.* **20**, 1243–1254.

Mandelbrot, B.B. (1972) Renewal sets and random cutouts. *Z. Warsch. Verw. Geb.* **22**, 145–157.

Mandelbrot, B.B. (1974) Intermittent turbulence in self-similar cascades: divergence of high moments and dimension of the carrier. *J. Fluid Mech.* **62**, 331–358.

Mandelbrot, B.B. (1975) *Les Objects Fractals: Forme, Hasard et Dimension.* Flammarion.

Mandelbrot, B.B. (1982) *The Fractal Geometry of Nature.* W.H. Freeman.

Mandelbrot, B.B. (1991) Random multifractals: negative dimensions and the resulting limitations of the thermodynamic formalism. *Proc. Roy. Soc.* **A434**, 79–88.

Mandelbrot, B.B. and Riedi, R. (1995) Multifractal formalism for infinte multinomial measures. *Adv. Appl. Math.* **16**, 132–150.

Marstrand, J.M. (1954) Some fundamental geometrical properties of plane sets of fractional dimensions. *Proc. London Math. Soc.* (3) **4**, 257–302.

Marstrand, J.M. (1996) Order-two density and the strong law of large numbers. *Mathematika* **43**, 1–22.

Massopust, P.R. (1994) *Fractal Functions, Fractal Surfaces and Wavelets.* Academic Press.

Mattila, P. (1995a) *Geometry of Sets and Measures in Euclidean Spaces.* Cambridge University Press.

Mattila, P. (1995b) Cauchy singular integrals and rectifiability of measures in the plane. *Adv. Math.* **115**, 1-34.

Mattila, P. and Preiss, D. (1995) Rectifiable measures in R^n and existence of principal values for singular integrals. *J. London Math. Soc.* (2) **52**, 482–496.

Mauldin, R.D. and Williams, S.C. (1988) Hausdorff dimension in graph directed constructions. *Trans. Amer. Math. Soc.* **309**, 811–829.

McCluskay, H. and Manning, A. (1983) Hausdorff dimension for horseshoes. *Ergodic Th. Dynam. Sys.* **3**, 251–260.

McLaughlin, J. (1987) A note on the Hausdorff measures of quasi-self-similar sets. *Proc. Amer. Math. Soc.* **100**, 183–186.

Mehaute, A. (ed.) (1991) *Fractal Geometries: Theory and Applications.* Penton Press.

Moran, P.A.P. (1946) Additive functions of intervals and Hausdorff measure. *Proc. Cambridge Philos. Soc.* **42**, 15–23.

Olsen, L. (1994) *Random Geometrically Graph Directed Self-similar Multifractals.* Longman Scientific and Technical.

Olsen, L. (1995) A multifractal formalism. *Adv. Math.* **116**, 82–196.

Olsen, L. (1996) Multifractal dimensions of product measures. *Math. Proc. Cambridge Philos. Soc.* **120**, 709–734.

Parry, W. (1981) *Topics in Ergodic Theory.* Cambridge University Press.

Parry, W. and Pollicott, M. (1990) Zeta Functions and the Periodic Orbit Structure of Hyperbolic Dynamics. *Astérisque* **187–188**, Société Mathématique de France.

Patzschke, N. and Zähle, M. (1993) Fractional differentiation in the self-affine case III. The density of the Cantor set. *Proc. Amer. Math. Soc.* **117**, 132–144.

Peruggia, M. (1993) *Discrete Iterated Function Schemes.* A.K. Peters.

Petersen, K. (1983) *Ergodic Theory.* Cambridge University Press.

Peitgen, H.-O., Jürgens, H. and Saupe, D. (1992) *Chaos and Fractals: New Frontiers of Science.* Springer-Verlag.

Peitgen, H.-O. and Saupe, D. (eds.) (1988) *The Science of Fractal Images.* Springer-Verlag.

Pietronero, L. and Tosatti, E. (eds.) (1986) *Fractals in Physics.* North Holland.

Pollicott, M. (1992) *Lectures on Pesin theory and ergodic theory on manifolds.* Cambridge University Press.

Preiss, D. (1987) Geometry of measures in $\mathbf{R}^n$: distribution, rectifiability and densities. *Ann. of Math.* **125**, 537–643.

Rand, D. (1989) The singularity spectrum $f(\alpha)$ for cookie-cutters. *Ergodic Th. Dynam. Sys.* **9**, 527–541.

Riedi, R. (1995) An improved multifractal formalism and self-similar measures. *J. Math. Anal. Appl.* **189**, 492–490.

Rogers, C.A. (1970) *Hausdorff Measures*. Cambridge University Press.

Rogers, C.A. and Taylor, S.J. (1959) Additive set functions in Euclidean space. *Acta Math.* **101**, 273–302.

Rogers, C.A. and Taylor, S.J. (1962) Functions continuous and singular with respect to a Hausdorff measure. *Mathematika* **8**, 1–31.

Ruelle, D. (1978) *Thermodynamic Formalism: the Mathematical Structures of Classical Equilibrium Statistical Mechanics*. Addison-Wesley.

Ruelle, D. (1982) Repellers for real analytic maps. *Ergodic Th. Dynam. Sys.* **2**, 99–107.

Rudin, W. (1973) *Functional Analysis*. McGraw-Hill.

Salli, A. (1985) Upper density properties of Hausdorff measures on fractals. *Ann. Acad. Sci. Fenn. Ser. A.I. Math. Dissertationes* **55**.

Sinai, Y.G. (1972) Gibbs measures in ergodic theory. *Russian Math. Surveys* **27**, 21–70.

Sleeman, B.D. (1996) Some contributions to the Weyl-Berry conjecture for fractal domains. *Int. J. Appl. Sci. Comp.* **2**, 344–361.

Smith, J.M. (1991) *Fundamentals of Fractals for Engineers and Scientists*. John Wiley.

Stoyan, D. and Stoyan, H. (1994) *Fractals, Random Shapes and Point Fields: Methods of Geometric Statistics*. John Wiley.

Strichartz, R.S. (1993) Self-similar measures and their Fourier transforms III. *Indiana Univ. Math. J.* **42**, 367–411.

Sullivan, D. (1983) Conformal dynamical systems. In *Geometric Dynamics*; Lecture Notes in Mathematics, **1007**, 725–752, Springer-Verlag.

Tél, T. (1988) Fractals, multifractals and thermodynamics. *Zeit. Naturforsch* **43A**, 1154–1174.

Temam, R. (1988) *Infinite-dimensional Dynamical Systems in Mechanics and Physics*. Springer-Verlag.

Tricot, C. (1982) Two definitions of fractional dimension. *Math. Proc. Cambridge Philos. Soc.* **91**, 54–74.

Tricot, C. (1995) *Curves and fractal dimensions*. Springer-Verlag.

Vicsek, T. (1992) *Fractal Growth Phenomena*, 2nd Edition, World Scientific.

Walters, P. (1982) *An Introduction to Ergodic Theory*. Springer-Verlag.

Wicks, K.R. (1991) Fractals and Hyperspaces. *Lecture Notes in Mathematics* **1492**, Springer-Verlag.

Williams, D. (1991) *Probability with Martingales*. Cambridge University Press.

Zähle, U. (1984) Random fractals generated by random cutouts. *Math. Nachr.* **116**, 27–52.

Index

Entries in **bold type** refer to definitions.

Printed and bound by CPI Group (UK) Ltd, Croydon, CR0 4YY

16/04/2025

14658543-0004